Contents

Living on the Shores of Hawai'i

Natural Hazards, the Environment, and Our Communities

CHARLES FLETCHER
ROBYNNE BOYD
WILLIAM J. NEAL
AND VIRGINIA TICE

A Latitude 20 Book

UNIVERSITY OF HAWAI'I PRESS
Honolulu

Publication of this book has been assisted by the
University of Hawai'i Sea Grant College

© 2010 University of Hawai'i Press
All rights reserved
Printed in the United States of America
15 14 13 12 11 10   6 5 4 3 2 1

Library of Congress Cataloging-in-Publication Data

Living on the shores of Hawai'i : natural hazards, the environment,
and our communities / Charles Fletcher...[et al.].
    p.   cm.
"A latitude 20 book."
Includes bibliographical references and index.
ISBN 978-0-8248-3433-3 (pbk. : alk. paper)
    1. Coastal zone management—Hawaii.   2. Land use—Environ-
mental aspects—Hawaii.   3. Natural disasters—Environmental
aspects—Hawaii.   I. Fletcher, Charles H.
    HT393.H3L58 2011
    333.91'709969—dc22
                                        2010023993

University of Hawai'i Press books are printed on acid-free paper
and meet the guidelines for permanence and durability of the
Council on Library Resources.

Designed by Santos Barbasa Jr.

Printed by Everbest Printing Co., Ltd.

For Ruth

Acknowledgments

This book is a contribution to the "Living with the Shores" series funded by the Federal Emergency Management Agency.

Our very special appreciation is extended to Mark and Joann Schindler, who made it possible to bring this project to completion. We are deeply grateful. CF also wishes to thank Abby Sallenger, John Wehmiller, Kim Luis, Orrin Pilkey, Sam Lemmo, Orson van de Plassche, Neil Frazer, Greg Moore, Brian Taylor, Jacob Moller, Frank Smith, Dorian Simunovich, Kaiwi Nui, Kevin Bodge, Bruce Richmond, Mike Field, Dennis Pierce, Rhodes Fairbridge, Mark Merrifield, Kem Lowrey, Ralph Goto, and all my students past, present, and future. Special thanks to Ruth, Kellen, Katie, and Chase for tolerating Dad's long weekends at the computer.

The art component of this text would not have been possible without the genius and generosity of Nancy, Brooks, and Matt. Several maps and perspective diagrams are the work of Matthew Barbee, Perspective Cartographic LLC. Thank you to the Hawai'i Sea Grant College. Many thanks also to those who provided photos and assistance, such as Keoki Stender, Yuko Stender, Mike Field, Matt Dyer, Steve Dollar, Zoe Norcross-Nu'u, Gordon Tribble, Kevin Kodama, Susie Cochran, Bob Richmond, Dennis Oda, F. L. Morris, Bruce Casler, Morphing Puppy, DeSoto Brown, Ms. Angel, Konaboy, Gerard Fryer, Cindy Knapman, Taiko1, Carl Berg, Don MacGowan, Delwyn Oki, crew of USS *Abraham Lincoln,* Andy Short, Dolan Eversole, Chris Conger, Matt Patrick, Christine Heliker, T. Orr, Craig Kojima, and Sherri Yoshioka.

We are indebted to several key agencies that have supported the research of the University of Hawai'i Coastal Geology Group: U.S. Geological Survey; Hawai'i Sea Grant College; National Science Foundation; Hawai'i Department of Land and Natural Resources; Kaua'i, Maui, and Honolulu counties; The Hawai'i Lifeguard Association; U.S. Army Corps of Engineers; Hawai'i Coastal Zone Management Program; Federal Emergency Management Agency; National Oceanic and Atmospheric Administration; and the H. K. L Castle Foundation.

Several people provided important reviews of early drafts: a special thanks for this is extended to Kim Luis, Neil Frazer, Steve Dollar, Gerard Fryer, Gordon Tribble, and Tom Schroeder.

Thank you to our editors, Keith Leber, Ann Ludeman, and Eileen D'Araujo, for their patient guidance through the publication process; to two anonymous reviewers for important suggestions; and to the University of Hawai'i Press for publishing the work.

CHAPTER 1

Introduction

'O ka makapō wale nō ka mea hāpapa i ka pō uli
"Only the blind grope in the dark night"

Hawai'i's beauty is not subtle. Towering peaks, deep gorges, and azure waters paint stunning scenes. These mirror a biological diversity that reflects the unique setting of the land, ocean, and climate. Island beauty extends from mountaintop to coast and plunges into an equally exquisite underwater world. The people of Hawai'i, also, are known for their open hearts, welcoming and generous attitude, and for sustaining an integrated

Fig. 1.1. Ironically, because we live too near our favorite beaches we often end up destroying them. (Photo by M. Dyer)

and unified multicultural society unrivaled on the planet. Yet Hawai'i is not vast, and almost half the land and nearly every community lies within 5 miles (8 km) of the shoreline. Indeed, no point on the Islands is farther than 28 miles (45 km) from the sea, which means that all locations in Hawai'i are influenced by the ocean; the entire state is coastal.[1]

Although Hawai'i is promoted as a tranquil paradise, it is also the home of cataclysmic events such as tsunamis, hurricanes, rockfalls, flash floods, and earthquakes. Communities in the Islands have always existed in a balance between life-sustaining resources and the powerful geologic forces that shape the land: flowing lava wipes out a subdivision, a hurricane demolishes thousands of homes, an earthquake brings down power across an entire island, a tsunami takes dozens of lives.[2] These are just a few of the events continually shaping these restless islands. Residents, visitors, and the elected guardians of paradise are concerned with these hazardous processes and aware of the fact that these recurring perils are part of the natural landscape of Hawai'i. We ignore the role of natural hazards at our own risk. However, at times, efforts to mitigate these threats have been at a heavy cost to the environment to which we are inextricably linked.

Concern for the natural equilibrium of the Islands has dissipated with the rise of Western culture. It is normal for humans to alter the environment. Resource husbandry has coexisted hand in hand with environmental damage since the first human foot was placed on Hawaiian soil. Researchers have documented dramatic and widespread impacts to both lowland and upland ecologies[3] due to the expansion of the Native Hawaiian population, especially in the centuries immediately preceding the first Western visit to the Islands by Captain James Cook in 1778. However, although the damaging imprint of the native hand was significant, the grip of Western land abuses has necessitated a near total reliance on imported goods and services because the land is no longer being used to support us.

A fundamental shift in land-use patterns accompanied the change from Hawaiian to Western culture. The former practice of resource stewardship for human use under the rules set by natural rhythms was replaced by a continental ethic of managing resources from afar with rules not attuned to the local character of the land. This is exemplified by the system of modern land management, which is implemented on a permit-by-permit basis following the dictates of a legal framework that attempts to match land use to the capacity of the environment. However, modern stewardship is instituted through a blanket of jurisdictions that covers entire islands simultaneously and too often without acknowledgment of local variations in climate, ecology, ocean processes, geology, or human community. Alternatively, the traditional *ahupua'a* system was place-based, with rules and customs tied to individual watersheds, shorelines, reefs, winds, and human needs. With the power of management embedded in the local populace, economies of scale were achieved and sustainability more likely to be attained.

Rarely a day goes by in Hawai'i without the media reporting on environmental and land-use issues stemming from social debate. Most stories revolve around conflicting interests in land development, shoreline use, public access, water resources, and other controversies. Contradictory claims, often springing from differing personal priorities, confuse the gravity of the issues. Will that housing project block my access to the sea? Is global warming going to flood my home? Are the reefs overfished? Is the water clean where I surf?

Despite similarities among resource management jurisdictions (federal, state, county), there is a deficit of coordinated, place-based planning among agencies. Bringing multiple agencies together to coordinate resource management has been historically infrequent and undervalued. Additionally, a lack of considering local factors can result in unintended and, in some cases, irreversible consequences. Consider the following story of the cascading effects of uncoordinated, nonplace-based management.

> As a shoreline erodes because of past sand mining by the local sugar company and removal of the coastal sand dune to accommodate development, waves encroach on a house that has been in one family for generations. The owner's first choice may be constructing a seawall to protect the family heritage, but state authorities disagree because they realize a wall may cause beach loss,[4] accelerate erosion on neighboring property, and interfere with public access.
>
> A consultant hired by the family recommends beach nourishment with sand dredged from offshore—an expensive and temporary fix. Environmentalists envision that this will damage the reef, surfers worry that it will change wave characteristics, and fishers say dredging sand will alter the food chain and impact the fishing. In the end, because the state owns the beach and the county manages the adjacent land, the owner decides not to go to the state but instead gets a seawall permit from the county simply by squeezing the wall a few feet closer to the family home.

But erosion continues on the seaward side of the wall and within 5 years the beach is gone and water laps against the stone face. Waves bouncing off the seawall carry away sand on the adjacent seafloor. What was naturally a soft sandy substrate for shallow marine organisms becomes rocky and covered with fleshy algae. Environmentalists think that the algae bloom is fueled by nutrients running off the paved watershed, seeping from in-ground septic systems among the beachfront homes, and diffusing from coastal rocks where treated community sewage is injected into the ground. Scientists think that the algae blooms because it is an alien species filling a "niche" in the ecosystem where there is no natural competition. They also cite the lack of hard data supporting the "nutrient-rich runoff" hypothesis of the environmentalists.

Impacts to the coastal ecosystem on state submerged lands are substantial despite the fact that state resource managers had little say in the awarding of a county permit for the wall. But that original wall was the first domino. All seawalls produce "flanking," a phenomenon where land next to a wall experiences accelerated erosion. Neighboring property now erodes and seawalls are built up and down the coastline—some are permitted, many are not. Surfers can't get to the ocean anymore because with no beach there is no ocean access, the water seems dirty, and the local surf break is rarely used.

In 20 years more than a half mile (0.8 km) of white sandy beach is gone, replaced by solid walls made of black volcanic rock. Did the seawalls cause the beach loss, or are they an innocent symptom of preexisting erosion? There is no public access along the coast. The ecosystem consists of fleshy algae, dirty turbulent water, and a desert of limestone that once had a sparse but healthy community of reef organisms growing on it. It has been years since any surfers or fishers have been seen in the area, and the neighborhood is mostly new residents or transient vacation renters unaware that there was once a beach.

Now after heavy rainstorms the Department of Health posts signs warning you to stay out of the polluted water. The family home has been sold, torn down, and a large compound has been built in its place surrounded by high walls blocking the neighborhood view of the ocean. More than half the beachfront

Fig. 1.2. At first glance this may seem like a healthy beach. But seawalls and homes have replaced the sand dunes that formerly lined the shore, and the resulting sand deficiency has led to beach erosion. At high tide waves reach to the wall and the beach is only in existence for part of a tidal cycle; an example of beach loss. Hawai'i has always had narrow beaches. But the past practice of excavating sand dunes and building seawalls, without regard for how these activities impact the beach, has led to widespread beach loss on every island. Beach loss is symbolic of other losses in Hawai'i that impact our quality of life, the environment, and our safety. These losses include wetlands, aquatic ecosystems, reef fish, open views, access to the ocean, food and energy security, cultural losses, and others. (Photo by A. Short)

homes are vacation rentals, and the profits go offshore to the mainland, where the new owners live.

SYMPTOMS OF A LARGER PROBLEM

The scenario just described is not fictitious. Many popular beaches and beautiful views are deep into this land-use sequence—on O'ahu: Lanikai, Kāhala, 'Ewa, and Mokulē'ia; on Maui: Kalama, Kahana, and Honokōwai; on Kaua'i: Wailua, Kapa'a, and Pākalā. Many other shores have started down variations of the same path: Sunset Beach, Kailua, Punalu'u, Lā'ie, and Waimānalo all on O'ahu. On Kaua'i the neighborhoods at Hā'ena, Anahola, and others have leveled the dunes, and off-island owners hire experts to secure permits for the biggest homes as close as possible to the water. Beaches silently disappear annually among our Islands while beachfront homes are sold fractionally to families that visit only once a year.[5] Coastal erosion has been met with mitigation that damages the environment, and a caring and watchful community has been replaced by commerce focused on profit and homeowners that are not residents.

But beach loss, seawalls, and fractured communities are symptoms of a larger problem:

- Streams now flow between high cement dikes that cut our communities in half because we have built homes on dangerous floodplains.
- Thirty percent of wetlands on O'ahu have been lost over the years because we did not understand and protect their function as cleansing and nutrient-rich environments.[6]
- Many reefs are overfished; it is rare to see an adult of some species because some people feel that they are entitled to unabated fishing.
- Coastal sand dunes have slipped into oblivion on O'ahu, Maui, and Kaua'i because we wanted to build high-priced storm- and tsunami-vulnerable homes in their place.

Is this the inevitability of human land use? No, it is the product of ignoring nature's signals when we change the land; signals that are largely known by those familiar with the environment. There is a better way, and with Hawai'i's heritage of island stewardship, we have the tools and knowledge to lead the world and set the example.

At the same time, however, there are deep and complex ironies at work:

- It is ironic that beach-sustaining dunes are allowed to be flattened and replaced by hotels and buildings for tourists, damaging the very sandy shores tourists have come to see.
- It is ironic that regulations enacted to protect our beaches, in practice, act as a guideline for damage rather than preservation.
- And it is ironic that the more federal, state, and county governmental agencies get involved to help protect the coast, the less protected the coast is, due to fractured enforcement, a lack

Fig. 1.3. Because we have built communities on former floodplains, many streams are now confined to channels such as this. These damage aquatic ecosystems and send pollutants directly into coastal waters, impacting water quality at beaches and probably harming reefs and other marine ecosystems. (Photo by R. Richmond)

of coordination and integration, and the insensitivity of jurisdictional rules to place-based needs.

A RUINED SHORELINE

Hawai'i is filled with beachless beach parks where seawalls now welcome visitors instead of sandy strands: Swanzy (O'ahu), Kalama (Maui), Honokōwai (Maui), and Kapa'a (Kaua'i) are just a few. Walls a football field long are built and maintained by the various counties to stave off erosion. Because the structure impounds the sand, it acts to accelerate erosion, both in front of it and around adjacent properties that formerly shared sand with the now-extinct beach. When property owners next to the park ask for permits to build walls to stop the erosion of their lands they are subject to embarrassing and expensive public hearings. Why? Because coordinated place-based planning is undervalued.

If some of these parks were given a beach by tearing down the wall and nourishing the sand-starved shore, benefits to the local community might be substantial. Neighboring properties might see beaches return in front of their walls. The park could become a beach destination rather than a cement fortification.

Consider our earlier case study.

> Unaware that a beach once graced this shore, the new owners see no problem. They make a tidy income from their rental property. Tourists purchasing the image of paradise are happy in their vacation home for a 2-week visit. Marketing images of white sands and blue seas in their heads obscure the fact that their seawall is a poor substitute—they are happy enough to hop in the car to find other beaches on the island. There is no obvious impact to the visitor industry: tourists to the Islands in 2007 exceeded 7 million, an all-time record.

Does the resort industry care about beach loss? Elaborate pools with waterfalls, grottos, caves, and even kiddie sand pools with artificial beaches filled with Texas gravel divert the visitor from noticing that there is no beach on the adjoining shore.

Only a small percentage of the voting population in Hawai'i cites the environment as their primary concern, pointing to the economy instead. Many fail to see that tourism profits live (and may one day die) on the beauty and health of our environment. But more important, beach loss is a bellwether for the decline in our quality of life.

THE MANY VOICES OF HAWAI'I

The paradox of environmental loss, under a management system viewed by many as one of the most stringent in the nation, is the result of a powerful discrepancy between various priorities. The chapters that follow discuss a wide range of environmental issues in Hawai'i with an eye toward more sustainable resolution of such conflicts in the future.

Present in any discussion of managing natural resources in Hawai'i is the fact of climate change. Hawai'i's climate is changing in ways that are consistent with global warming. In Hawai'i:

- Air temperature has risen
- Rainfall and stream flow have decreased
- Rain intensity has increased

- Sea levels and sea-surface temperatures are rising
- The ocean is acidifying

If these trends continue, scientists anticipate that Hawai'i's water resources and forests, coastal communities, and marine ecology will be increasingly affected.

- Water is key to agriculture, and efforts to advance local agriculture will be set back by declining rainfall, base flow, and stream discharge.
- Heavy rains will have nowhere to drain when high ocean waters flood the storm-drain system, and storm surge, tsunami, and high waves will penetrate further inland with each year.
- Acidification and rising sea-surface temperatures threaten our marine ecology, the economy, and tourism.

Solving these problems requires a strong partnership between scientists and planners and the communities they serve. There is a compelling need for sustained and enhanced climate monitoring and assessment activities, and for focused research to produce models of future climate changes and their impacts on Hawai'i and the Pacific. The roadmap for these efforts must come from an informed and involved community working to leave future generations a resource-rich Hawai'i.

This book comprises 12 chapters. Chapter 2 traces the history of the Hawaiian Islands and the increased tempo of destructive land-use practices with the arrival of Western immigrants. Chapter 3 discusses volcanoes and explores the risk of placing homes and families in locations vulnerable to natural hazards. Chapter 4 explains both earthquakes and tsunamis and the potential harm to a complacent public. Chapter 5, on hurricanes, assesses our risk and volunteers mitigation techniques. Chapters 6, 7, and 8 survey various water issues, including scarcity, flooding, and pollution. Chapter 9 goes over climate change and the possible outcomes of projected sea rise for Hawai'i. Chapter 10 explains beach erosion and loss. The geologic history and current status of coral reefs and the problems of overfishing and ocean acidification are discussed in chapter 11, and chapter 12 provides a summary.

This book also highlights trends and the connection between the environment, communities, and natural hazards. With concepts grounded in long-standing Hawaiian resource practices that acknowledge the variability of the environment from one place to another and the need to understand local environmental and community character, it may be possible to more effectively integrate sister agencies and embrace local communities in resource management. By combining tools, such as interagency management teams and councils of local ecosystem experts, and conservation funds designed to purchase key lands with sensitive environments, the tide of environmental loss might be turned.

Responsibility for the environmental decay may also lie with a regulatory system that views natural resources as a patchwork of parcels. The land, sea, and sky form a continuous chain of environments that share water, dissolved compounds, organic matter, mineral particles, and energy. These environments fall under various jurisdictions of government, federal, state, and county, that are not well integrated and that typically have differing mandates. Even within one jurisdiction, different agencies may have competing agendas. For instance, the Hawai'i Department of Transportation wants seawalls to protect coastal roads yet must get a permit from the Hawai'i Department of Land and Natural Resources whose mandate is to protect submerged lands (beaches and reefs).

To perpetuate our environment various layers of government can unite under a common vision that is focused, with laserlike attention, on

Fig. 1.4. Much of world-famous Waikīkī Beach has no beach at high tide. (Photo by D. Eversole)

place-based assets. The rules that result from that union need effective enforcement, ideally with participation of the community of users themselves. It is easy to point to developers as culprits in this process, but that is naive. Most developers operate legally within our regulatory environment, and many fully understand that undermining the environment that is their main attraction is not of long-term benefit. The failure to protect our Islands occurs where regulations do not intersect to promote a common vision (despite the existence of community plans where explicit visions are defined), where good laws are weakly enforced (usually due to lack of sufficient resources), and where rules and regulations do not fit the many and unique personalities of our environment.

Our land is filled with many voices that do not always speak in unison. As the line of tension tightens between conflicting standards, differences in understanding these Islands, and contradictions between ideals and actions, we put at risk our economy and the quality of our lives. Sound decision making starts with asking the right questions. This book attempts to describe the context of sustainable answers and, in so doing, directly influence the future that Hawai'i's voters choose for the shores of Hawai'i.

CHAPTER 2
History of the Land

He aliʻi nō ka ʻāina, ke kauwā wale ke kanaka
"The land is the chief, the people merely servants"

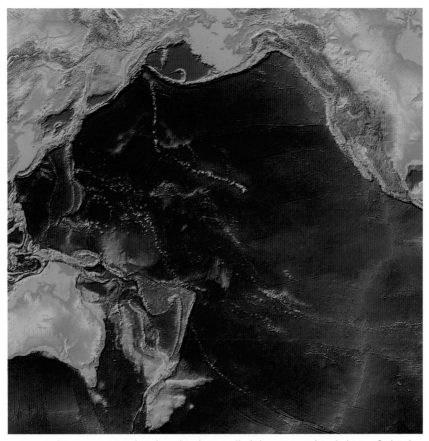

Fig. 2.1. The Hawaiian Archipelago has been called the most isolated chain of islands in the world. (National Atmospheric and Oceanic Administration, National Geophysical Data Center. Matthew Barbee, Perspective Cartographic LLC)

The early settlement history of Hawai'i is still not completely resolved. Some believe that the first Polynesians arrived in Hawai'i in the third century AD from the Marquesas and were followed by Tahitian settlers in AD 1300 who conquered the original inhabitants. Others believe that there was only a single, extended period of settlement.[1] Regardless, it was more than 1,500 years ago that voyagers in large double-hulled canoes set sail (probably from the Marquesas Islands), crossing more than 2,000 miles (3,220 km) of unforgiving open ocean. The reward for these explorers was an isolated homeland of fertile islands in the trackless waters of the North Pacific. Once here, the Hawaiian culture developed and flourished as the inhabitants were compelled to adapt to the environmental challenges of this now legendary and lonely archipelago.

One key to the early success of *kānaka maoli* (Native Hawaiians) was their creation of the *ahupua'a* system, a land management design based on geographic principles taken from the natural boundaries of their environment.[2] Paying close attention to the language of land and ocean, and by studying the effects of their actions, the Hawaiian population grew exponentially.

The navigational prowess of the Hawaiian ancestors led them to the discovery and settlement of an island paradise, but this Eden had another face. Daily life for the first Hawaiian settlers was carried out under the constant threat of volcanic eruptions, earthquakes, storms, unpredictable seas, and flash floods. Hawaiian deities, or *'aumākua,* embody this fragile yet intricate relationship of people with their environment. Reflecting the broad spectrum of natural elements, *'aumākua* take on the personalities of land (*'āina*), sea (*kai*), and sky (*lani*). Their actions are the forces that shaped and often threatened daily life.[3]

Hawaiian oral history tells of Haumea (Earth Mother) and Wakea (Sky Father) and their large family of five sons and eight daughters. Two of the daughters were Pele, fire goddess of volcanoes, and Na-maka-o-kaha'i, her jealous older sister and goddess of the sea. The waves of the sea goddess scour the lava, Pele's home, eroding it and crushing it to sand as she chases Pele from island to island. As these two clash along the shores of Hawai'i, mortals at times seem caught in between. Pele, known both as a creator and destroyer, now makes her home deep in Halema'uma'u at Kīlauea, the most active volcano on Earth.[4]

Today, the people of Hawai'i face the same challenges as our predecessors of perpetuating a healthy natural environment and managing the hazards of the land, sea, and sky. However, the problems are becoming increasingly complex as a growing population, now 1.25 million and projected to increase to 1.6 million by 2030, expands within the hazardous coastal zone, creating a greater demand on limited resources such as fresh water and open land, and increasing the need for careful husbandry of island ecosystems.[5]

SHAPED BY NATURAL FORCES

Situated almost in the center of the North Pacific, Hawai'i is one of the most isolated places on Earth, with Los Angeles 2,400 miles (3,862 km) to the east and Japan 3,800 miles (6,115 km) to the west. Hawai'i is also the southernmost state in the United States, at the same general latitude as Hong Kong and Mexico City. Totaling 6,425 square miles (16,640 km²) of land, the 47th smallest state consists of eight large islands including Hawai'i, Maui, Kaho'olawe, Lāna'i, Moloka'i, O'ahu, Kaua'i, and Ni'ihau, as well as 124 uninhabited small islands, emerged reefs, and shoals.

Vast contrasts in climate combined with dramatic physical forces such as weathering and erosion, flooding, volcanic eruptions, and massive landslides enabled the growth of tropical rain forests, freshwater streams, grasslands, deserts, and even tundra in the relatively compact area of the Hawaiian chain. Variations in temperature and landscape are striking: temperatures can drop below freezing or soar to 90°F (32°C), all within a few miles. On the island of Kaua'i, barren, dry deserts lie mere miles from the summit of Mt. Wai'ale'ale, one of the wettest places on Earth.[6]

Because no point on any of the Hawaiian Islands is far from the sea, however, the buffering effect of the Pacific Ocean's prodigious heat engine limits annual temperature variation in any one location. Absorbing heat during the day and releasing it against the chill of night, the ocean moderates daily, seasonal, and interannual temperature fluctuations. As a result, the climate in any one spot is more regulated and uniform than climates in similar continental latitudes such as Spain and Florida. In Honolulu, there is an approximately 11°F (6°C) temperature range between summer and winter.[7] Balmy conditions, found throughout the year at sea level, continue to entice new residents to Hawai'i every year.

THE APPEARANCE OF LIFE

The Hawaiian Islands' extreme isolation led to the magnificent community of plants and animals found at the time of the Polynesians' first landing. It is amazing that of all the island life that existed then, over 90% of the species were unique to these Islands.[8] The extraordinary, lush, volcanic pinnacles were unparalleled and alone—a secluded outback adrift in the Pacific. This authentic natural paradise allowed each growing thing the opportunity to develop adhering only to the dictates and limitations of the law of evolution.

Hawai'i's capacity to support this diversity, both on land and in the sea, was as great as anywhere on Earth. As the sunken seamounts of the aging islands moved inexorably northwest on the great Pacific Plate, the spores of marine plants and the larvae of corals, fishes, and other marine animals drifted back to colonize the waters of the younger islands. This process fostered the evolution of hundreds of marine species over millions of years. In the struggle to adapt or vanish from existence, more than 25% of the marine species in the Hawaiian chain evolved into endemics, organisms found nowhere else in the world.[9]

Along the shores, seeds of plants washed up to colonize the virgin soils. With time, a process of adaptation and natural selection produced new species. About 90% of the 1,000 or so native flowering plants and 70% of the 150 species of ferns are endemic success stories. About 25% of these are dune-loving shoreline plants.[10] Birds, also responsible for dispersing seeds of some species, surreptitiously made their way to the Islands as well. The Islands' amazing avifauna once consisted of more than 140 species of native birds.[11] A complex pattern of life unfolded as pollinators and dispersers refined their skills to adapt to the varying micro-climates and isolated topography of the high volcanic islands. Today, well over half of Hawai'i's original bird species are extinct, due in part to the introduction of insect-borne diseases and ground-nesting predators such as the rat, the mongoose, and the feral cat.

Much of the mountainous interior of the Islands is uninhabitable, which has led to settlement primarily on the coast of most islands and along the ridgelines of O'ahu around Honolulu. Trapped between the mountains and the sea, the Honolulu metropolitan area has one of the highest population densities in America, in company with Boston, Chicago, and Oakland.[12] Add the growing number of tourists every year (from close to 300,000 in 1960 to over 7 million in 2007[13]), and the pressure on our natural environment and resources becomes severe.

HAWAI'I'S INTRODUCTION TO WESTERN CULTURE

The arrival of European mariners to the remote Pacific in the late 1700s set in motion many of the dilemmas that Hawai'i faces today. Captain James Cook's sighting of O'ahu on January 18, 1778, opened the Oceanic culture to global influences and European diseases: missionaries, whalers, traders, and naturalists flocked to Hawai'i in search of adventure, riches, and religious converts. Muskets and cannonballs introduced by Europeans transformed island warfare and contributed to the violent unification of Hawai'i by King Kamehameha I in 1795.[14]

Following Kamehameha's conquest, agricultural and aquacultural production intensified, radically altering the coastal region and lower valleys from marginal hinterlands to a system of irrigated pond-fields and permanent residential sites. On April 14, 1820, the brig *Thaddeus* arrived from Boston with the first Christian missionaries. Accompanying the missionaries, and the flood of Westerners that followed, were numerous diseases never encountered before by native people, including mumps, cholera, typhoid, smallpox, various sexually transmitted diseases, measles, and others. As a result, the Native Hawaiian population was decimated from between 100,000 to over 1 million people (estimates differ)[15] in 1778 to only 50,000 people by the end of the nineteenth century.[16]

In the course of this near extinction, Western trade flourished. Hawai'i's international reputation grew as a stopover point between whaling voyages from the Arctic to Japan. Between 1829 and 1843, more than 1,700 whaling ships visited the waters around Hawai'i, and their crews fueled the contagion consuming the Hawaiian people.[17]

Whaling commerce in Hawai'i was rivaled by a vigorous lumber trade. Tracts of Hawaiian forest were burned to identify sandalwood by its scent. This coveted wood was cut and traded for silk and porcelain from Canton. The Chinese called Hawai'i "Than Heung Sahn," the Sandalwood Mountains. Aware of the costs of deforestation, Kamehameha I imposed a *kapu* (restriction) on extraction of this sought-after wood, but after his death frantic trade resulted in the eradication of forest tracts and left the *ali'i* (ruling chiefs) indebted to foreign traders as they borrowed on the promise of future lumber. In December 1826, the Kingdom of Hawai'i enacted its first written law, a tax that required each man to supply approximately 67 pounds (30.4 kg) of sandalwood annually to the *ali'i* to decrease the mounting debt.[18]

Two major famines occurred at that time as able-bodied men took to the forests to find sandalwood. Agricultural and fishing duties were abandoned as the need to pay the tax, and the desire to reap the profit of sandalwood sales, dictated daily life. The Hawaiian people suffered exposure, disease, malnutrition, and exhaustion as the last natural sandalwood forests were wiped out. As Island food production suffered from neglect, reliance on imported goods dominated, and Hawai'i lost the need for sustainable use of resources as it shifted to an economy based on trade.[19]

This trend has continued to the present day as islanders have largely lost the ability to feed themselves, relying instead on the daily lifeline of commercial shipping to provide basic goods.

But foreign abuses to the land were not the first among these Islands. Even ancient land use produced significant and long-lasting changes to the environment. Rats brought in Polynesian canoes flourished in the hills and plains and feasted on endemic plants, insects, snails, and birds and their eggs, substantially aiding in many extinctions. Wildfires were

AN ACT
Relating to the public Health.

WHEREAS, The Small-Pox is believed to exist in this Kingdom, and humanity and a just regard to life require that all who are affected with that disease should receive strict care and attention, and whereas it is desirable that the disease shall not extend through the Islands:

BE IT ENACTED, By the King, the Nobles and the Representatives of the Hawaiian Islands in Legislative Council assembled:

SECTION 1. That there shall be appointed by the King, with the assent of the Privy Council, a Commission consisting of three persons who shall act without pecuniary reward, "upon whom shall devolve all the powers and duties intended and expressed in the Act of 8th May, 1851, entitled a law establishing a Board of Health with power to extend the same to all parts of this Kingdom, in person or through their agents," and who are hereby authorized and empowered to provide for all persons, sick with the small pox, suitable medical attendance, food, lodgings and clothes, at the expense of the Hawaiian Government, and to make and publish such regulations for the public health as they may think wise and expedient, and enforce them by fines or otherwise through the Courts.

SECTION 2. For the purpose of carrying into effect the foregoing section a majority of the Commission thereby constituted are hereby authorized and empowered to draw from the Public Treasury such funds as may be necessary; and the Minister of Finance is hereby authorized and empowered to pay their drafts out of any monies belonging to the Government, provided their draft is ac-companied with an account current showing the objects for which the money has been used and satisfactory vouchers.

SECTION 3. In case any monies are expended to provide for the sick brought to this Kingdom in vessels from abroad, it shall be the duty of the Commission hereby constituted to demand the same of the Captain of the vessel bringing sick persons into the Kingdom, and unless the same is paid upon request, the Collector of Customs shall not grant a clearance to such vessel until the same is paid, and the Master shall be liable therefor, and may be sued for the same in the Courts of this Kingdom.

SECTION 4. This law shall take effect upon its passage and shall continue in force until the passage of a new law relating to the public Health, and all laws or parts of laws inconsistent herewith are hereby repealed.

Passed by the House of Representatives this sixteenth day of May A. D. 1853.

G. M. ROBERTSON,
Speaker of the House of Representatives.

Passed by the House of Nobles this 14th day of May A. D. 1853.

KEONI ANA,
President of the House of Nobles.

Approved May 16th, 1853.

KAMEHAMEHA.

KEONI ANA.

☞ In accordance with the provisions of the above act, it has pleased His Majesty and Privy Council to appoint G. P Judd, T. C. B. Rooke and W. C. Parke, as the Commissioners therein provided for.

Fig. 2.2. Disease accompanied the influx of foreigners to Hawaiʻi in the nineteenth century. (Image from Bishop Museum archives, Honolulu, Hawaiʻi)

purposefully set to clear lands; and alien "canoe plants" (brought with the discoverers) exploited available niches. Evidence suggests that some species of reef fishes and other marine edibles were depleted by the Native Hawaiian population during the last centuries before the arrival of Captain Cook. But by far the most extensive destruction was the near extinction of a major ecosystem, the lowland forests. Localized impacts among windward and leeward valleys pale beside the islandwide destruction of lowland native forests that took place in the middle period of Polynesian settlement of the Islands.[20] However, up until the Western conquest and

No. 5.

NOTICE.

OLELO   HOOLAHA.

WHEREAS, much difficulty is found in procuring aid to bury the dead, the Royal Commissioners of the Public Health hereby give notice that all able-bodied men, if recovered from the Small Pox, or already completely exposed thereto, are liable to be called on by them, by their Sub-Commissioners, by the Police, or by any of their Agents, to render assistance in burying the dead, without remuneration. Any person so called on, refusing to assist, shall be liable to a fine, not exceeding twenty-five dollars, or imprisonment not exceeding six months.

The Commissioners likewise give notice that they have authorized the destruction of Dogs in Honolulu and vicinity, wherever in the estimation of the Police they are liable to convey and communicate the Small Pox.

T. C. B. ROOKE,
July 18, 1853.   Chairman.

No ka mea, he nui ka pilikia, no ke kokua ole ia ke kanu ana o na kupapau, ke hoolaha aku nei na Komisina no ke ola o na kanaka, ma keia palapala; penei: E hiki no ia makou a i na hope Kamisina a i na luna o ia Oihana a i na makai no hoi ke kii aku i kela kanaka keia kanaka kino ikaika, ua ola ka Puupuu Liilii, a ua nui loa ka pili ana paha i keia mai, e kokua wale i ke kanu ana i na kapapau me ka ukuole. A o na mea a pau i kiiiaku pela, a hoole i ke kokua mai e hoopaiia oia i na dala aole e oi aku i iwakalua kumamalima, a i ole ia e paa i ka hale paahao aole e oi aku i na malama eono.

Ke hoakaka aku nei no hoi na Kamisina, na lakou i olelo e pepehiia, ma Honolulu a ma na wahi e kokoke ana, na ILIO a pau, ke manao na makai e hoolaha ana ka ilio i ka Puupuu Liilii iwaena o kanaka, no ka pili ana i keia mai.   T. C. B. ROOKE,
Luna Hoomalu.
Honolulu, 18 Julai, 1853.

Fig. 2.3. Every able-bodied man was forced into grave digging, and dogs were shot on sight, as a consequence of the epidemics sweeping through Hawai'i in the nineteenth century. (Image from Bishop Museum archives, Honolulu, Hawai'i)

Eastern arrivals, the consumptive impact on Hawaiian ecosystems was relegated largely to low elevations and at comparatively slow rates. This all changed with the arrival of foreigners whose appetite for expansion and subjugation, coupled with powerful engineering and agricultural tools, admitted of no limitation.

The relationship between the Hawaiian people and their natural environment changed with the influx of new settlers and traders, who viewed the Islands as an infinite resource ripe for the picking. Soon agriculture, particularly *kō* (sugar), linked the Hawaiian economy directly to the U.S. market. Pineapples emerged as Hawai'i's second major export crop, and the modest island of Lāna'i became the world's largest pineapple plantation.[21] In response to this land-use shift, thousands of imported workers came to the Islands: Chinese, Japanese, Portuguese, Russians, Puerto Ricans, Filipinos, and Americans. The new land-use and development methods accelerated the pattern of deforestation and soil erosion, permanently altering numerous tracts of land at increasingly higher

Fig. 2.4. Huge tracts of land, including entire watersheds, were altered by plantation agriculture. This photo shot in 1922 shows 1,500 acres (607 ha) planted under asphalt paper mulch by the Hawaiian Pineapple Company; view from the Pali, O'ahu. (Photo from Bishop Museum archives, Honolulu, Hawai'i)

elevations. At first barely noticeable but steadily growing throughout the twentieth century, sewage, pesticides, and other pollutants flowed into streams and through the porous volcanic soil, draining into coastal waters and aquifers. In many bays and estuaries with restricted circulation, reefs and wetlands were changed forever.

By the 1900s, the seeds that would drive future human development had been planted. Settlers adopted a philosophy of confronting, rather than bending with, nature. When Sanford Dole was appointed governor in 1900, just after Hawai'i had become a new territory of the United States in 1898, the young (and illegal)[22] government was ill prepared to make the tough decisions necessary for the long-term perpetuation of island resources.[23] Entrepreneurs from around the world were busy exploiting Hawai'i's natural commodities for economic gain, while large-scale engineering projects sought to divert fresh water to quench the thirst of agriculture. The spoil from dredging the Ala Wai Canal replaced the wetlands of Waikīkī as the new land-use ethic solidified.

STATEHOOD AND ITS CONSEQUENCES

Events of the early 1900s placed Hawai'i in a precarious situation. During the Spanish-American War and World War I, the Islands became strategic to the U.S. military. After the attack on Pearl Harbor in World War II, vast tracts of land were turned over to the U.S. armed forces and converted into military bases. Over 200,000 acres (80,937 ha) of the land on O'ahu was permanently converted into military warehouses, airstrips, housing, and marine facilities.[24]

The war effort brought a generation of service men and women through the Islands: a new wave of residents who would fuel postwar growth. Hawai'i joined the national trend of postwar affluence and increased leisure time, which led to America's rush to the shore on all of its coasts. Farmlands in middle America emptied as soldiers and sailors who had seen the world established new families along the shorelines of the nation. As with development in the low-lying barrier islands of the Atlantic and Gulf coasts and the urbanization of active fault zones in California, development patterns in Hawai'i ignored nature's warning signs as people settled in areas vulnerable to hazards and tested the carrying capacity of most environments.

Statehood arrived in 1959, generating more immigration and investment in the Islands. The consequences of postwar prosperity, immigration, and agricultural mechanization, without respect for the Islands' natural boundaries, are clearly visible today: 25% of the total length of beaches on Oʻahu have been lost;[25] over 30% of wetlands on Oʻahu have disappeared;[26] once meandering streambeds now flow between straight, concrete walls; and impermeable surfaces flush torrents of runoff into coastal waters with every season of rain.

Lured by the simplicity of sun, sea, and sand, mainland tourists and immigrants flocked to the Islands via jet plane. From 1960 to 2000, the population doubled as many visitors stayed and joined the Islands' cosmopolitan blend of Eastern and Western culture, dramatically changing the ethnic makeup of Hawaiʻi.[27] In 1853, nearly 96% of the population was Hawaiian, but by 1990 that had plunged to only 12%.[28] There is a sharp parallel between the shift in political control away from the native population and the Islands' environmental decline. Losses of wetlands, forests, streams, beaches, dunes, and the reef fishery are akin to cultural losses for the Hawaiian people, whose customs and ethnicity are intimately connected to the land.

Just as the early Hawaiians relied heavily on the sea for their sustenance and leisure, the people of Hawaiʻi today still depend on the beaches, reefs, and nearshore waters for their livelihood and pleasure. The marine tourism industry alone annually contributes over $800 million to state revenues, and in 2002 Hawaiʻi's coral reefs were valued at nearly $10 billion when combining recreational, amenity, fishery, and biodiversity values.[29] Managers must work nonstop to find the right mix of access and protection to ensure a healthy resource base along the coast.[30]

RELEARNING THE LESSONS OF THE *AHUPUAʻA*

Polynesians brought with them a "transported landscape."[31] That is, the plants and animals arriving with early Hawaiians resulted in large-scale changes to ecosystems on all the Islands. But, although the Polynesians certainly exerted pressure on the natural resources, a conservation ethic was transported with their Oceanic culture as well, and this helped to maintain the integrity of their relationship with the land. As a result, after more than a thousand years of impact, the Polynesians at the time of Western contact were still able to feast on the resources offered by their surroundings. Hawaiian traditions define a reciprocal relationship between people and the land[32] and the native culture evolved in the embrace of endemic ecosystems, both on the land and in the sea. As a result, Hawaiians developed an intimate relationship with their natural setting, marked by deep love, knowledge, and respect of these places. Exploring the Hawaiian connection to the land reveals a service bond: not land serving people, but people serving the land. A Hawaiian proverb

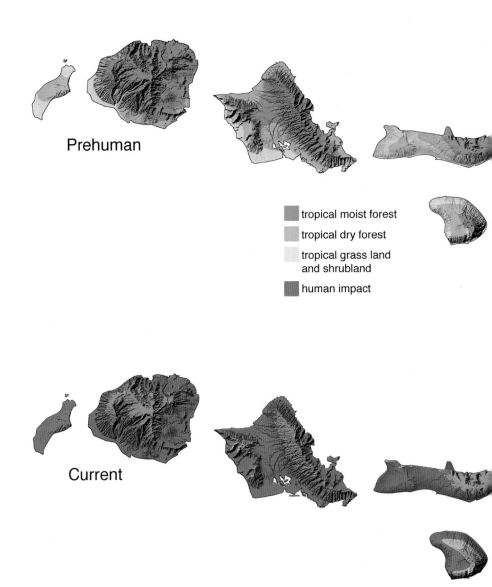

Prehuman

tropical moist forest
tropical dry forest
tropical grass land
and shrubland
human impact

Current

Fig. 2.5. Impacts to the natural ecosystem have been severe on every island. (SOEST Graphics, adapted from The Nature Conservancy, Hawaiian High Islands Ecoregion Project, "Major Habitat Types": http://www.hawaiiecoregionplan.info/MHT.html [last accessed October 28, 2009])

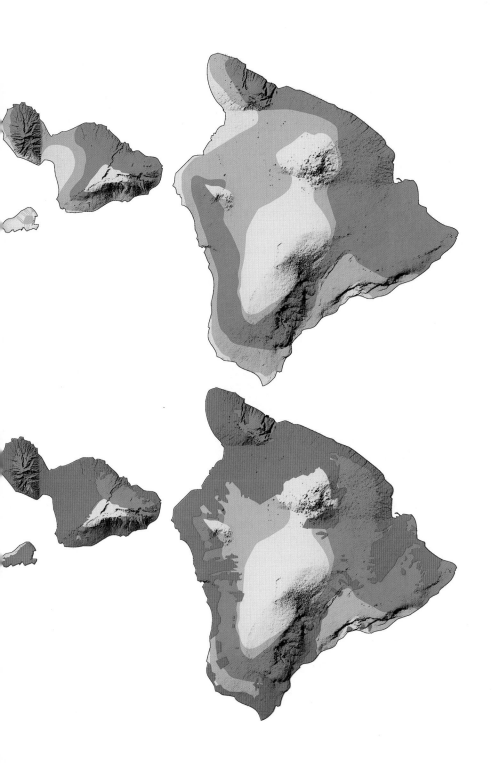

states *He ali'i nō ka 'āina, ke kauwā wale ke kanaka* (The land is the chief, the people merely servants).[33]

For instance, the traditional methods of capture and a system of *kapu* placed strict rules on the timing and the amount of fish captured, ensuring the availability of precious seafood. The *konohiki* (headman of an *ahupua'a* under a chief) and *ali'i* decided when to impose *kapu* by watching daily and seasonal marine patterns, and observing how the fish responded to harvest pressure. The moon was watched, and both fishing and planting times defined by its phases. This early form of adaptive management was based on a complex knowledge of the rhythms of the ocean and land, the physical and biological aspects of resources, and the right amount of resource use without damaging the ecosystem.

The island landscape was divided into a system of *moku,* or districts, each containing one or more *ahupua'a.*[34] These topographically designed commons ran across the watersheds to the sea and jointly provided each *moku* with access to a full range of island resources. This management system carries a powerful embedded message: the Islands' natural processes and resources are connected to the physical geography of the landscape. Knowledge of water quality and abundance, soil productivity, cultivation, fishing, winds, waves, and navigation were continually refined and were woven into the fabric of daily life.

In the time of Kamehameha III, the *ahupua'a* system was abandoned as tempting sums of money, disease, and death caused many Hawaiians to relinquish their land. Strict religious rules became increasingly violent as the commoners and ruling elite struggled for power. In the mid-1800s, landholdings of the elite were broken up and a new land tenure system, the Mahele, redistributed property and imposed new taxes.[35] Instead of securing land for Hawaiians, as Westerners had professed it would do, the Mahele alienated Hawaiians from their land, and the *ahupua'a* system was lost. Today, highway systems, water and utility lines, and government jurisdictional boundaries largely run parallel to the shore and contrary to the *mauka–makai* (ridgeline to reef) flow of natural energy and resources. Western attempts to tame the land have severed the bounds of the *ahupua'a* system.

Long after traditional stewardship was abandoned, modern residents are rediscovering the wisdom of the past and revitalizing the notion of the *ahupua'a* resource-management system. Hawaiian communities have revived the *ahupua'a* model as a means of addressing crucial resource-management problems. Community-driven *ahu moku* councils (*moku* stewardship groups) blend the best in conservation values and knowledge from Western and Hawaiian cultures.[36] On an island with strict boundaries imposed by Mother Nature, the *ahupuaʻa,* and *moku* within which they lie, requires residents and visitors to consider the consequences of their actions and to frame the future by becoming stewards of this archipelago. An updated and refined *ahupuaʻa* system within a twenty-first century setting can provide a template for living with the entire island system.

CHAPTER 3

Volcanism among the Islands

Kūkulu-o-ka-honua

"Pillars of Earth"

Fig. 3.1. Kīlauea Volcano on the south shore of the Big Island reveals the location of the Hawai'i hot spot. (Photo by M. Patrick, U.S. Geological Survey: http://hvo.wr.usgs. gov/kilauea/update/images.html [last accessed October 28, 2009])

In Hawai'i, humans live in a border area between the land, the sea, and the sky. We are vulnerable to hazards from all three of these environments, but it is the land we live on that is closest to the home we make, the job we hold, and the places our children go. Earthquakes, volcanic eruptions, tsunamis, and other geologic hazards are earthly events that may kill people and destroy livelihoods. When they occur, the world becomes a terrible place: the solid ground is suddenly no longer trustworthy, as if the very pillars of Earth were crumbling.

In European culture it was once believed that Vulcan, blacksmith of the Roman gods, had a subterranean forge and its chimney was the tiny island of Vulcano in the Mediterranean Sea north of Sicily. Vulcano's hot lava fragments and dark ash clouds were thought to come from Vulcan's furnace as he beat out thunderbolts for Jupiter, the king of the gods, and weapons for Mars, the god of war. The English word "volcano" comes from this island. Today we know that the lava on Vulcano actually comes from molten rock generated in Earth's interior below the Mediterranean seafloor.

The 132 outcroppings of land recognized as "Hawai'i" are merely the tips of a long ridge of volcanoes, known as the Hawai'i-Emperor Chain, extending 3,000 miles (4,800 km) from the Big Island of Hawai'i to the Aleutian Trench in the North Pacific. Pele's stepping-stones across the Pacific are also the highlights of a geologic story 80 million years long.

PLATE TECTONICS

To understand Hawai'i, we must understand the workings of the whole Earth. The solid, outer layer of Earth that consists of the crust, and the rigid upper part of the mantle, together form the lithosphere. The lithosphere, ranging from about 25 to 125 miles (40 to 200 km) thick on the continents and 30 to 60 miles (50 to 100 km) thick under the oceans, is broken into massive pieces called plates. Lithospheric plates move relative to one another and shift restlessly across the planet surface. Like a piece of toast afloat on a layer of peanut butter, these rigid slabs of rock float on the rest of the mantle, which reaches more than 1,800 miles (2,900 km) to the core at Earth's center.

Data suggest that the mantle is composed of solid rock that is very hot and able to slowly, plastically deform. Over millions of years it convects or mixes like a great lava lamp. Geologists hypothesize that massive plumes of hot rock rise through the mantle toward the lithosphere.

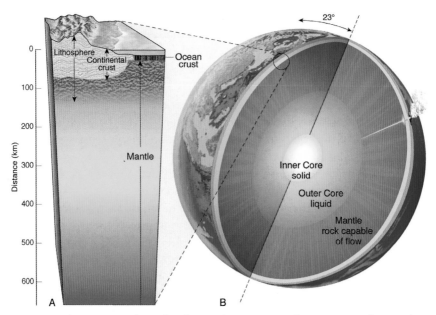

Fig. 3.2. Earth is organized into four layers: the inner core, the outer core, the mantle, and the crust. The uppermost mantle and the crust together form the lithosphere.(J. Wiley and Sons, Inc.; SOEST Graphics)

Melting occurs along the upper margin of a plume as it reaches the lower pressure at the base of a plate. The resulting magma (molten rock) can rise into the crust and generate active volcanism; such a point on Earth's surface is known as a hot spot.

Hawai'i is thought to be the site of such a plume, a quasi-stationary hot spot that brings magma to the surface.[1] It is this hot-spot volcanism that produced Mauna Kea, which stands more than 31,000 feet (9,450 m) from the seafloor to summit, taller than any other mountain on Earth from base to peak, and Mauna Loa, the most massive topographic feature on the planet.[2] Indeed, this hot spot has prodigiously given birth to every island in the Hawai'i-Emperor Chain as the Pacific Plate slowly passes across. Pele's home at Kīlauea, and the undersea volcano Lō'ihi,[3] are the focus of effort by the Hawai'i hot spot in recent centuries.

There are two other geologic settings (not found among the Hawaiian Islands) where volcanism is prolific. One type, subduction-zone volcanism, is produced when two plates collide and one of them dives below the other and is subducted (recycled) into the mantle. As the plate being recycled descends into the mantle it releases water into the hot rock at the base of the overlying plate, lowering its melting point and generating magma. This molten rock rises to the surface and forms a line of volcanoes above the zone where the two plates meet. The active volcanoes of the Cascade Range in Oregon and Washington, and the Indonesian Island Arc are examples of this type of volcanic action.

The other type is rift-zone volcanism, where two plates split apart and magma wells up from the underlying mantle. Rift-zone volcanism typically makes new seafloor (oceanic crust) and is recognized by a long valley on the ocean floor where heat and magma escape the upper mantle.[4] In the Pacific the rift zone is named the East Pacific Rise, and in the Atlantic it is called the Mid-Atlantic Ridge.

ISLAND EVOLUTION

In traditional teachings the demigod Māui, navigating beneath the sacred fishhook constellation Manaiakalani, raised (navigated to) the Hawaiian Islands one by one from the sea. He did this in sequence from the oldest to the youngest, demonstrating an ancient recognition of the Islands' age gradient foretelling the modern theory of plate tectonics.[5]

As molten rock pours out of Earth's crust at the Hawaiian hot spot, it piles upon itself on the deep seafloor and solidifies under the cooling influence of the surrounding ocean waters. Slowly over hundreds of thousands of years, volcanic rock accumulates under more than a mile (1.6 km) of seawater, eventually breaching the sea surface to become a high volcanic island composed of one or a number of shield volcanoes, named after its long low profile like a warrior's shield.

On the seafloor, the volcano does not grow as a neat layer cake of lava beds. Rather, submarine eruptions break into boulders of glass and ash that accumulate as a great pile of broken volcanic rock called talus. As the volcano erupts, it accretes steep aprons of glassy talus forming approximately 90% of the edifice, a fundamentally unstable foundation for such a monolith. Later, after the volcanic rock breaks above the ocean surface, layers of ʻaʻā (a type of lava recognized by the jagged boulders, known as "clinker," that form on its surface) and pāhoehoe (a lava type characterized by smooth, ropy folds) lava build upon one another constructing the massive shield. Most of a volcano's growth takes place in this shield-building stage as huge amounts of lava pour out over long spans of time. These young volcanoes are marked by the development of calderas, the immense summit depressions that form when lava drains out of a subterranean chamber and the surface collapses into the hollow cavity.

When a volcano enters the postcaldera stage, it is still active, but

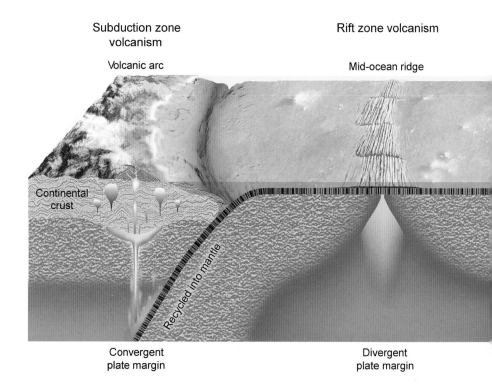

Subduction zone
volcanism

Rift zone volcanism

Volcanic arc

Mid-ocean ridge

Continental
crust

Recycled into mantle

Convergent
plate margin

Divergent
plate margin

no longer tapping deep magma chambers. Interacting with groundwater and charged with gas, post-caldera eruptions become more explosive in behavior with the formation of subsidiary cinder cones by high fire fountains of gas-charged magma. On Oʻahu, Diamond Head, Koko Crater, Punchbowl, and others are left as the legacy of these violent exhalations of the hot spot.

As the Pacific Plate rolls to the northwest, the volcanic island moves away from the heat source and thus ends major eruptive activity.[6] When this happens, a volcano enters the erosion stage. A combination of weathering, gravity, rainwater, and waves carves at the volcanic slopes, reducing its elevation and area. Away from the hot spot, the plate cools and subsides. Rock and vegetation decay, forming soil. Water, air, and animals add their chemistry, laying a blanket of organic detritus across the sterile volcanic landscape. Shaped by the land and in turn shaping the land, life, air, and water transport the island across an organic threshold as what was once an erupting volcano becomes a living community. Dense forests, gushing streams, deep watersheds, and broad fringing reefs stretch their living tendrils and fill every corner of the environment with the tissue of life.

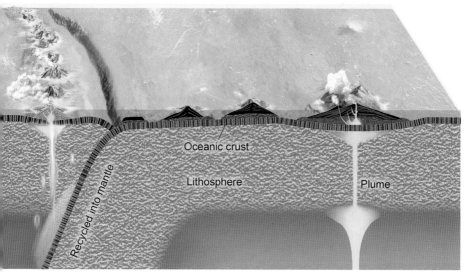

Subduction zone
volcanism

Island arc

Hot spot
volcanism

Shield volcano

Oceanic crust

Recycled into mantle

Lithosphere

Plume

Convergent
plate margin

Fig. 3.3. Earth's lithosphere is broken into plates that interact with one another along their edges. Two plates may diverge, converge, or slide past one another in opposite directions. In a general sense, there are three types of volcanic action: subduction-zone volcanism (formed at certain types of convergent margins), rift-zone volcanism (formed at divergent margins), and hot-spot volcanism (formed where a plate moves across a stable mantle plume). (J. Wiley and Sons, Inc.; SOEST Graphics)

In time, however, the volcanic peaks subside and the land wears down to the level of the sea. Although erosion will ultimately have its way, even in its burial beneath the sea a volcanic corpse sustains a thriving population. Coral reefs broaden their position at the sea surface, forming insular shelves and terraces, as the remnants of the old volcano form an underpinning for the coral and hard algae to construct an atoll upon its tomb. An atoll is an island of reef surrounding a lagoon that approximately marks the location of the former volcanic caldera.[7] Thus was born the Papahānaumokuākea Marine National Monument consisting of the Northwestern or Leeward Hawaiian Islands: Nīhoa, Necker, French Frigate Shoals, Gardner Pinnacles, Maro Reef, Laysan, Lisianski, Pearl and Hermes Atoll, Midway, and Kure.

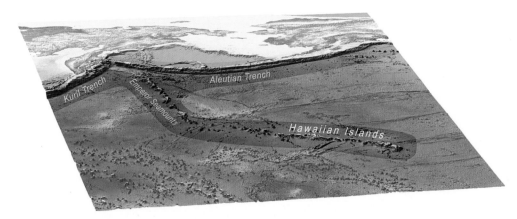

Fig. 3.4. The Hawai'i-Emperor Chain is considered the world's longest mountain range. (Matthew Barbee, Perspective Cartographic LLC)

The only visible evidence of a volcano's ancestry, these atolls ride continuously northwestward on the relentless Pacific Plate. Sometimes a lagoon will shallow and fill with calcium carbonate sediments, the skeletal fragments of reef organisms. But in the majority of cases the lagoon remains an open body of circulating water, protected by the surrounding reef from heavy wave barrage. The lush marine biomass of an atoll is remarkable; to call them rain forests of the sea is not overreaching. But the northern journey of the Pacific Plate cannot be denied, and ultimately the atoll enters the latitudes of cold water and reduced sunlight limits the growth of fragile coral polyps and marine algae. Atolls maintain their march to the northwest until they finally subside beneath the waves at the aptly named Darwin Point.[8] Currently located on the northern edge of Kure Atoll, the Darwin Point marks the limit of reef tolerance for northerly conditions. Beneath the waves, drowned atolls become seamounts. The oldest of these is the 80-million-year-old Meiji at the far end of the 3,000-mile (4,800-km)-long Hawai'i-Emperor Chain, considered the world's longest mountain range.[9] Eventually, the seamount will be subducted into the mantle and recycled beneath the Aleutian Islands from whence lava will erupt, a reminder that Earth is the ultimate recycler.

LIVING ON A VOLCANO

In Hawaiian culture, volcanic eruptions are attributed to the restless Pele, goddess of volcanoes. Pele arrived by canoe in Hawai'i from Kahiki, a

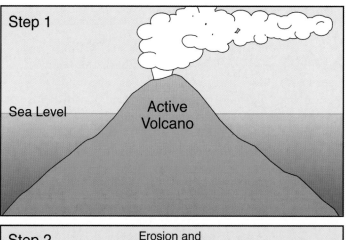

Step 1

Sea Level

Active Volcano

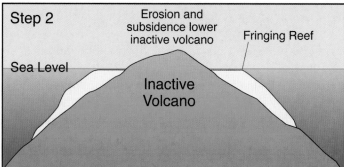

Step 2

Erosion and subsidence lower inactive volcano

Fringing Reef

Sea Level

Inactive Volcano

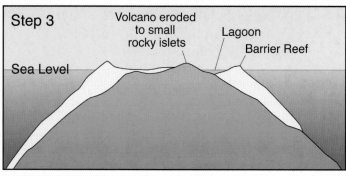

Step 3

Volcano eroded to small rocky islets

Lagoon

Barrier Reef

Sea Level

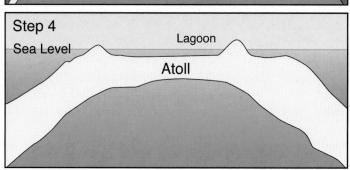

Step 4

Sea Level

Lagoon

Atoll

Fig. 3.5. Atolls are large, ring-shaped reefs with a central lagoon. Most atolls begin as fringing reefs on a subsiding, inactive shield volcano. The reef continues to grow vertically as the volcano subsides and erodes. Eventually, all surface evidence of the volcano disappears, and a central lagoon may mark the former position of the caldera before it submerged. (SOEST Graphics)

mythical land (interpreted by some as Tahiti) that was her birthplace. Originally landing on Ni'ihau, it was Pele's task to establish the family home and kindle volcanic fires. Oral history tells that Pele encountered groundwater that put out the volcanic flames she discovered with her powerful digging stick Pāoa.

Digging in the low coastal plains near the ocean, all her fire pits were so near the water that they burst out in great explosions of steam and sand, and quickly died. Moving from one island to the next, in frustration she at last found Kīlauea on the island of Hawai'i where she built a mighty enduring palace of fire at the summit crater of Halema'uma'u. This location, still considered the house of Pele, is a sacred spot among the Hawaiian Islands and should be visited with reverence because it is hallowed and dangerous. Kīlauea is the world's longest continuously erupting volcano. Since 1952, it has erupted 34 times, and since January 1983 the eruption has been continuous.[10]

The current eruption of Kīlauea has been relatively calm, without notable explosions or particularly violent behavior. But this has not always been the case. Even Kīlauea has its dangerous side. When hot internal magma encounters a large body of groundwater the union can be devastating. Sudden conversion of the water into steam can lead to violently explosive eruptions that are quite dangerous. These are known as phreatomagmatic eruptions, and Kīlauea has a history of such violence.[11] On May 10, 1924, a phase of violent eruptions began that sent columns of ash thousands of feet into the sky repeatedly throughout the next 18 days.[12] Explosions hurled rocks as large as 8 tons (7,257 kg) over half a mile (0.8 km) from the main vent. These boulders can be seen today littering the landscape around Halema'uma'u.

The words "volcanic eruption" conjure the classic image of a flaming geyser spouting out of a mountaintop with blazing, viscous rivers oozing down its slopes and a column of dark "smoke" rising into the sky. The image is instructive because volcanic rock fragments and lethal gases are common volcanic hazards in Hawai'i. The deadly pillar of gas and pyroclastic particles (rock fragments formed in a volcanic eruption) rushing from the central vent of a volcano is mostly made of water vapor and glass particles but can also include toxic gases. The gas plume rising from an active vent on Kīlauea consists of about 80% water vapor with lesser amounts of sulfur dioxide, carbon dioxide, and hydrogen. Small quantities of carbon monoxide, hydrogen sulfide, and hydrogen fluoride are also present. Extremely small amounts of mercury and other metals have been detected in gases emitted from vents along the east rift zone of Kīlauea.

TABLE 3.1   ACTIVE AND POTENTIALLY ACTIVE VOLCANOES IN HAWAI'I

Volcano	Eruption Type	Eruptions in Past 200 Years	Latest Activity	Remarks
Haleakalā	Lava, ash	1	1790	In last stage of volcanic cycle. Expected recurrence rate about 200–600 years.
Hualālai	Lava, ash	1	1800–1801	High hazard due to unusually fluid lava.
Kīlauea	Lava most common; ash rare	47	Ongoing from 1983	Explosive eruption at Kīlauea summit in 1790 killed approximately 80 Hawaiian warriors. Lava-flow hazard to four coastal areas in the nineteenth century, five in the twentieth century.
Kohala	Lava, ash	0	120,000 years ago	Volcanic cycle may not be completed, but eruption probability is low.
Lō'ihi	Lava	Not known	Not known	Submarine volcano; seismically active; youngest lava less than 1,000 years old.
Mauna Kea	Lava, ash	0	4,000 years ago	Frequency of activity before latest eruption is estimated to be about 300 years.
Mauna Loa	Lava	30	1984	Lava-flow hazard to eight coastal areas in the nineteenth century, eight in the twentieth century.

Note: See Pacific Disaster Center Web site: http://www.pdc.org/index.php (last accessed November 19, 2009).

Even in the nonexplosive phases of eruptions, these gases are released around the crater or fissure vents and are capable of causing fatalities.

The storied history of Kīlauea is full of warnings. In 1790, Keōua, chief of Puna, was leading his small army toward Ka'ū on a route near the Keanakāko'i Crater on the slopes of Kīlauea Volcano when an explosion occurred. Keōua split his army into three divisions to ensure that the whole army would not be harmed in case of catastrophe, and they proceeded after one another down the Ka'ū trail. Roughly 6 miles (10 km) from Halema'uma'u the third division found the entire second division lying motionless on the ground. None of the bodies showed signs of being burned or struck by rocks, yet everyone was dead. Apparently, Kīlauea's summit caldera had belched a pyroclastic surge cloud, an avalanche of poisonous gas and fine ash that raced down the slopes. These burning clouds are glassy, gassy avalanches moving on a thin cushion of air at high speeds. Eighty Hawaiian warriors were killed by the toxic gas, but their pigs were found wandering unharmed among the corpses because their low height kept them beneath the deadly cloud.[13]

When *pāhoehoe* lava enters the ocean for extended periods of time, new land is created in the form of a fan-shaped platform known as a lava bench. Lava pouring into the ocean from either surface flows or lava tubes cools rapidly, usually shattering in the cold water into sand- to block-size fragments. These accumulate along the shore to form a loose foundation that can support overlying lava flows, which build a bench above sea level. However, the bench is deceptive; although it looks like solid land, it will in fact eventually collapse into the sea, sweeping unwary visitors to their death. National Park Service personnel usually rope off these dangerous areas to protect the public. Nonetheless, foolhardy hikers are seen on these dangerous features and have been lost when they visited the wrong place at the wrong time.

Vog (volcanic smog) and laze (lava haze) are the most persistent hazards associated with gases. These hybrid words describe two common problems associated with Hawai'i's ongoing eruptions. Vog forms around craters and vents when sulfur dioxide (SO_2) gases mix with the atmosphere producing an aerosol of tiny, suspended droplets of sulfuric acid. As this drifts and accumulates downwind, a cloud of natural pollutants forms. In addition to the trace gases noted previously, mercury levels are frequently above normal in vog. The gas is not usually threatening over a short period of time except to persons with respiratory or heart problems. More typically, the acidic vog can produce acid rain that leaches heavy metals such as lead and copper from plumbing. These metals can find their way into our drinking water, a particular hazard for small children to regularly ingest as they grow.

Agriculture is exceedingly vulnerable to volcanic gas. Since March 12, 2008, sulfur dioxide emissions from a new gas vent in Halema'uma'u Crater have affected residents with respiratory problems, caused voluntary evacuations, and produced significant damage to agricultural farms and ranches.[14] The emissions have also led to concern for the long-term effects on health, water quality, and agriculture. The eruptions have produced massive volumes of sulfur dioxide gas, with daily emissions of roughly 1,000 tons (90,000 kg). This has devastated coffee, macadamia

Fig. 3.6. A lava bench at East Lae'apuki on the south shore of the Big Island collapsed during the night of August 13, 2007. (Photos by T. Orr, U.S. Geological Survey: http://hvo.wr.usgs.gov/archive/2007_09_08.html [last accessed October 29, 2009])

nut, protea, and other growing operations in the Ka'ū district, especially in the Ocean View region, located downwind of the eruption cloud. This problem has grown so severe that the governor of Hawai'i has convened a special task force of interagency personnel to develop strategies for managing the impacts related to volcanic emissions.[15]

Volcanic smog tends to accumulate during stagnation when trade winds fail, locally known as "kona weather," when air stalls against a mountainside from poor atmospheric circulation. Prevailing northeasterly winds on the Big Island usually carry the smog southwest, but it can also be carried across the southern end of the island and up along the Kona (western) coast. In contrast, Kona winds (blowing from the south and west) blow the noxious air into the populated areas on the east side of the island, including Hilo (and the rest of the state).

One telling story about the danger of volcanic gases involves a couple visiting the stark landscape where lava enters the sea on the south shore of the Big Island in 2002. Much of this terrain consists of young *pāhoehoe* that is precarious for hiking. Near the point where red lava flows into the sea, molten rock travels several tens of feet (10 m or so) underground through a network of lava tubes toward the shore. On that sad day, a sudden heavy rainsquall blew in from the ocean, and water running through the fractured rock beneath their feet turned to steam when it hit the subterranean lava. The ensuing cloud of scalding water vapor engulfed the hapless couple, killing both of them.[16]

If you are growing increasingly concerned for your safety, don't worry. It is not difficult to stay safe when you visit Kīlauea. Here are some suggestions:

- Stay clear of vents during any eruptive activity.
- Stay off coastal lava benches.
- Avoid topographic low spots in these same areas because heavier gases accumulate in low areas.
- Try to avoid visiting the region of active lava on windless days. If you have a health problem that is aggravated by smog, stay away until the air clears.
- Pay heed to information issued through local news media by the Hawaiian Vog Authority (HVA), established in 1990 to address associated problems.
- If you live on the Big Island and have a cistern water supply, have it checked for lead and mercury levels and other compounds as recommended by the Hawai'i Department of Health.

GROUND FRACTURES AND SUBSIDENCE

The body of a volcano is alive and breathing. As magma pours into its belly and later retreats back to Earth's interior, the volcano will expand and contract, tilt and vibrate, and portions will even shift laterally under the push of magma and the pull of gravity. Cracks and fractures in the ground commonly result from the inability of brittle rocks to stretch in compensation for the changing size of the volcano. These features are generally small, but fractures can occur in larger dimensions of several feet across and vertical offsets in the level ground are common. These cause serious damage to buildings, roads, and services such as water pipes and power lines.

As the flanks of a volcano inhale and exhale, ground subsidence, or sinking of the land, may occur. The scale of subsidence ranges from the entire island, to parts of a volcano's sides sinking, to the settling of

TABLE 3.2   VOLCANIC HAZARDS

Type	Description
Pyroclastic surges	Gravity-driven, rapidly moving, ground-hugging mixtures of hot gases, ash, and glass fragments.
Lahars	Avalanches of water, rock, and mud that rush down hills leading away from a volcano.
Landslides	Rapid downslope movements of rock, often with mud, soil, water, and other loose materials on hillsides.
Lava flows	Molten rock (magma) that pours or oozes onto Earth's surface.
Tephra	Glassy shards that are ejected from a volcano and form a sedimentary deposit.
Volcanic gas	Dissolved gases released to the atmosphere during an eruption: water vapor, carbon dioxide, carbon monoxide, sulfur oxides, hydrogen sulfide, chlorine, and fluorine.
Tsunamis	Ocean waves generated by a sudden displacement of water. May be caused by earthquake, landslide, or volcanic activity on the seafloor.

Note: See Pacific Disaster Center Web site: http://www.pdc.org/index.php (last accessed November 19, 2009).

a small area and local collapse of lava benches on the shore. Rapid subsidence is more directly related to volcanic or seismic activity and is a particular threat in coastal areas because of the possible inundation by a tsunami. Similarly, ground settling and fractures are also associated with earthquakes.

A particularly large type of ground fracture known as a rift zone is found on most Hawaiian volcanoes. Sharing the same name as the rift zone mentioned at the beginning of this chapter in connection with lithospheric plates, rift zones in Hawai'i are smaller than the aforementioned features, which mark the edges of massive, ocean-forming, continent-carrying plates. Rather, rifts in Hawai'i are places where the volcanic crust is pulling apart. Often two rifts radiate away from the summit of Hawaiian shield volcanoes. These do not point toward adjacent volcanoes but instead tend to run parallel to volcano boundaries. Because rifts mark openings in the crust they are natural conduits for subterranean magma migration away from the main magma chamber (usually located under the summit caldera). At their surface, rifts are characterized by a seismically active valley, numerous vents, fissures, cracks, cinder cones, pit craters, faulting, and various sources of lava. All of these are indications that magma prefers to intrude (underground) into rift zones and may be stored there for periods of up to a few years.

Because rifts are characterized by earthquakes and ground subsidence, they are an especially bad place to build a house. Yet it is amazing that an entire subdivision exists in Kīlauea's east rift zone where it enters the sea. Kapoho Vacationlands subdivision is built in the center of the east rift valley, and the subsiding ground has allowed flooding by the sea, forming Kapoho Bay. Add marine flooding to the heightened incidence of quakes, tsunamis, eruptions, and other hazards, and one begins to wonder about the long-term safety of residents in that area.[17]

On the scale of an entire island, subsidence is a natural chapter in the evolution of the landscape. Supported by heat from the hot spot and the fresh outpourings of lava, young volcanoes are able to build to great heights. But the mass of the enormous edifice eventually overwhelms the ability of the lithosphere to support its weight, and paralleling its growth is a process several hundred thousand years long of bending the lithosphere and slowly settling deeper into the upper mantle. As long as volcanism is active the mountain cheats this process and grows lava layer upon lava layer. But as the plate shifts farther from the hot spot, cooling, subsidence, and erosion take over as the main agents controlling the elevation and geography of the land surface.

VOLCANIC HAZARDS

At daybreak of March 25, 1984, the second day of the last Mauna Loa eruption, six large vents on the mountain's slopes were feeding numerous lava flows headed eastward down the mountain—straight for Hilo. For over a week, the town of Hilo watched in growing horror as hot rivers flowed toward their homes, one of them halting within a mere 1.2 miles (1.9 km) of the local prison. Residents were mesmerized by the smoke from burning vegetation, the lava's glow at night, and the explosions caused by methane gas along the advancing flow front. The river of flames crept closer to Hilo every day for a week. On April 4, the impending fire hazard changed course 4 miles (6.4 km) from the city's main road; everyone breathed a sigh of relief. On April 15, the eruption ceased.[18]

Thanks to many years of study by scientists at the U.S. Geological Survey Hawaiian Volcano Observatory, various state and county authorities, and the University of Hawai'i, volcanic eruptions are semipredictable. Accurate predictions rely in large part on data that stream in from state-of-the-art instruments monitoring the rumblings of Kīlauea and Mauna Loa. Precursor events such as earthquake swarms (hundreds of small earthquakes transpiring in a few days) and patterns of summit swelling and deflating can suggest the onset of a volcanic event. Even though the precise moment or type of eruption cannot be forecast, there is usually ample time to warn residents and take necessary precautions for reducing the impact of an eruption.

Once a lava flow slithers its fiery form down the mountain, evacuation is usually straightforward and methodical. At the same time, however, volcanologists warn people not to be complacent because if Pele is feeling particularly spirited that day, events like pyroclastic surges or random explosive events may occur without warning. It is also important to remember that volcanic activity cannot be prevented, so it is best to do some research before buying your dream home for a steal in Hilo, Puna, Ka'ū, or Kona. The U.S. Geological Survey provides a volcanic hazards map[19] and guide highlighting the island of Hawai'i's high-risk areas.

The Hawaiian Islands are beautiful, but we should never forget that they are a fleet of volcanoes that are incompletely understood and capable of surprising violence. As they pass from the youth of Kīlauea to the middle age of Mauna Kea and Haleakalā we should keep in mind that they will be the source of still more eruptions and the inevitable cause of earthquakes and tsunamis.

WHERE TO GET HELP

Here are some sources of information for you:

- U.S. Geological Survey Hawaiian Volcano Observatory:
  http://hvo.wr.usgs.gov/
  - What are volcano hazards?: http://pubs.usgs.gov/fs/fs002–97/
  - Vog: A volcanic hazard: http://hvo.wr.usgs.gov/volcanowatch/
    1996/96_05_29.html
- State of Hawaiʻi Governor's Office Vog information: http://hawaii.
  gov/gov/vog
- Hawaiʻi State Civil Defense: http://www.scd.state.hi.us/
- Pacific Disaster Center: http://www.pdc.org/iweb/pdchome.html
- U.S. Geological Survey Atlas of Natural Hazards in the Hawaiian
  Coastal Zone: http://pubs.usgs.gov/imap/i2761/
- *Purchasing Coastal Real Estate in Hawaiʻi:* http://www.soest.
  hawaii.edu/SEAGRANT/communication/pdf/Purchasing%20
  Coastal%20Real%20Estate.pdf
- *Hawaiʻi Coastal Hazard Mitigation Guidebook:* http://www.soest.
  hawaii.edu/SEAGRANT/communication/HCHMG/hchmg.htm
- *Homeowner's Handbook to Prepare for Natural Hazards:* http://
  www.soest.hawaii.edu/SEAGRANT/communication/Natural
  HazardsHandbook/naturalhazardprepbook.htm

CHAPTER 4

Earthquakes and Tsunamis

Nei ka honua, he ōlaʻi ia
"When Earth trembles it is an earthquake"

The largest locally generated Hawaiʻi tsunami in modern times was triggered by violent shaking on the south shore of the Big Island during a magnitude 7.2 earthquake (some geophysicists argue that the earthquake was larger, perhaps 7.7)[1] on November 29, 1975. Thirty-two campers

Fig. 4.1. Kalāhikiola Church on the Big Island, Hawaiʻi, was damaged during the October 15, 2006, earthquake. (Photo by Ms. Angel, Flickr: http://www.flickr.com/photos/18697834@N00/1998923404/ [last accessed October 29, 2009])

at Halapē experienced the earthquake and tsunami firsthand. The campers were able to stand during the initial violent shaking caused by the earthquake but soon lost their balance if they did not cling to trees or large rocks for support. A deafening roar rose from the steep cliffs above Halapē as rockfalls rumbled downhill. Many campers, frightened by the noise, moved closer to the beach to escape the falling rocks.

Fearing a tsunami, some campers ran toward the beach to check the water and found a slowly but noticeably rising sea. Within a minute or so the rising water accelerated, causing them to run back toward the rockfalls at the base of the cliffs. The ocean arose into a foamy, surging discharge of debris-filled seawater, knocking many of the campers off their feet and briefly submerging some as they fled for higher ground. They barely had time to catch their breath before a second tsunami struck, far more turbulent and higher than the first. A giant wave, reaching as high as 48 feet (14.6 m), carried every loose object in its path over 300 feet (91 m) inland. Trees, debris from the Halapē camping shelters, car-sized boulders, horses, and people were swept into a preexisting 20-foot (6-m) fracture in the ground and churned by the surging waters. One camper said he felt like he was "inside a washing machine."

In about 10 minutes, the major surge was over although several smaller waves repeatedly washed over the exhausted victims stranded in the fracture. One person was drowned or battered to death during the terrifying ordeal; another was swept out to sea and never found. Nineteen people were injured at Halapē; seven of them required hospitalization.[2]

The tsunami at Halapē Campground was generated by a local earthquake associated with the active volcanism at Kīlauea. This illustrates one of the greatest threats to coastal safety in Hawai'i: a locally generated tsunami hitting our crowded shores with little warning. Tsunamis from across the Pacific, although dangerous, can be spotted in time to call for a general evacuation of our coastal area using the system of sirens that exists throughout the Islands.[3] Evacuation of our coastal zone in such an event is by no means assured to be a smooth and problem-free process, and assessing the likely impact of a distantly generated tsunami remains a significant challenge to scientists, but at least the availability of some advance warning provides individuals time to consider appropriate action. An earthquake among the Hawaiian Islands that generates a tsunami will hit our shores in mere minutes, however, straining the capacity of our evacuation system by leaving each of us little time to realize the threat, evaluate a retreat, and evacuate the area.

HAWAIIAN EARTHQUAKES

Although the Islands' natural history is measured in protracted, mostly quiet, geologic time, the actual processes of landscape evolution are often abrupt and catastrophic. Earthquakes, volcanic explosions, tsunamis, great storms, floods, and landslides are near-instantaneous events. Repeated frequently through time, they become the artisans behind Hawai'i's spectacular scenery and Murphy's Law behind its catastrophes.

Do you know when the last major earthquake or explosive eruption occurred in Hawai'i? Without the urgency of a recent disaster, most of us bury past calamities in our subconscious and move on with our lives. We trust to leave our safety to the guiding hand of a vigilant government. Our trust is well placed with the committed and experienced scientists at the National Oceanic and Atmospheric Administration (NOAA) Pacific Tsunami Warning Center, where threats are monitored and the first alerts are sent out to a vulnerable public.[4] We are also well served by state and county Civil Defense offices, where evacuation and mitigation steps are planned and executed.[5] But is government protection adequate? Or are we building lives that are vulnerable to natural disasters? It is a sad fact

Fig. 4.2. A cliff collapsed and sent a cloud of silt across the water in Kealakekua Bay, Hawai'i, during the October 15, 2006, earthquake. (Photo by Konaboy, Flickr: http://www.flickr.com/photos/konaboy/270510964/ [last accessed October 29, 2009])

that disaster loss increases in the United States seem to be mostly tied to population growth and investment in high-risk areas.[6] As a reminder to us, and to our elected officials, of what we risk by poor planning it pays to recall catastrophic events that occurred in Hawai'i's history. This builds awareness in our consciousness of what may occur at any time and provides a guide to mitigating the damage that will be caused by future disasters. If you lead a life oblivious to the earthquake and tsunami threat, you are the most likely to succumb to their deadly reach.

The Great Ka'ū Earthquake of April 2, 1868, isn't exactly fresh in anyone's mind, but perhaps Pele feels that it was an impressive performance worth an encore. After 6 days of foreshocks, the earthquake, centered below the island of Hawai'i's south coast, unleashed its massive pent-up energy. The Ka'ū district was hardest hit, with damage to every stone wall and many stone houses. Wooden buildings shifted off their footings, and the majority of Big Island cisterns were either damaged or destroyed, triggering an islandwide drinking water crisis. Buildings from Hāmākua to Hilo collapsed as ground fissures split open in the streets. Across the island, hundreds of landslides tumbled down the steep slopes. Ground on the flanks of Kīlauea cracked, and the island's southeast coast abruptly dropped 6 feet (1.8 m). Crockery fell from kitchen shelves on Maui, and as far away as Kaua'i tall trees swayed violently.

The magnitude of that great earthquake—the largest earthquake to occur in the Hawaiian Islands in historic time—was estimated at 7.9 on the Richter scale. The fault movement that generated this shock wave was the result of molten rock rising into rift zones, swelling the volcano, and wedging its base seaward. For almost 90 seconds, the entire southeastern flank of Mauna Loa volcano bulldozed out across the seafloor, producing severe shaking and, by pushing water ahead of it, triggering a devastating tsunami with waters flooding as high as 50 feet (15 m). The tsunami wiped out all signs of life in a village called Keauhou. Half a dozen fishing communities along the south shore were abandoned due to heavy damage.

The death toll attributed to the tsunami was 77 people; but this may well be an underestimation. The earthquake triggered a mudflow in the Wood Valley of Ka'ū that killed an additional 31 people. Two more people perished from falling rocks near Hilo. Scientists at the Hawai'i Institute of Geophysics, who studied the likelihood of earthquakes of that intensity recurring in the Ka'ū-Puna area, have stated: "The potential for hazards is so obvious that the area should be considered off limits for normal permanent residency."[7]

SHAKY GROUND

Earthquakes in Hawai'i are closely related to volcanism. Like volcanoes, they too are an important component of the island-building processes that have shaped the Hawaiian archipelago. Like future volcanic action, earthquakes are more likely to occur among the younger islands than on the older ones. Yet, the threat of a large earthquake occurring in the area of Maui to O'ahu is far greater than the threat of a volcanic eruption. Given the large population in this region, and their general lack of preparation for seismic hazards, the net exposure to a chance of loss or damage (the "risk") is relatively high.

Hawai'i residents were reminded of this threat when we awoke to a pair of earthquakes on Sunday, October 15, 2006. The main earthquake occurred first at 7:07 a.m. This magnitude 6.7 shaker was followed 7 minutes later by another powerful shock registering magnitude 6.0. The first earthquake occurred 24 miles (39 km) beneath Kīholo Bay north of Kona on the Big Island, and the second was 12 miles (19 km) beneath the seafloor off Māhukona.

Geophysicists hypothesize that these two earthquakes were associated with stresses generated by the huge load of Mauna Loa on the lithosphere. As the weight of Mauna Loa pushes down, it flexes the underlying rock generating extension (stretching) in the deep lithosphere and compression (squeezing) in the shallower lithosphere. The main 6.7 magnitude earthquake was caused by extensional stress breaking rock in the upper mantle. The shallower 6.0 magnitude earthquake was compressional, an unexpected result because it was the product of a different mechanism, a different fault, and a different stress environment from the first one. These differing types of earthquakes carry some chilling news: they can occur at any time, on any side of the island, they do not give notice, and having a recent earthquake does not mean that underground stress is relieved. There can be decades between earthquakes—or minutes.

Deeper earthquakes such as these tend not to cause tsunamis. But they can nonetheless cause horrendous damage. The larger earthquake of October 15, 2006, knocked down buildings and damaged others. Dozens of homes were declared uninhabitable, all power on the island of O'ahu was knocked out for most of the day, emergency radio stations were inoperable, and bridges and roads were closed, thus isolating rural communities for days to weeks. In all, earthquake damage to the Big Island alone is estimated at $200 million.[8]

Most Hawai'i earthquakes occur in places of weak rock within or at

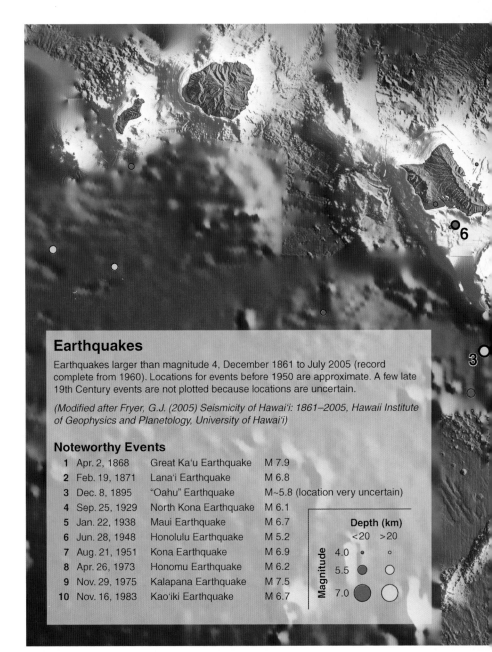

Earthquakes

Earthquakes larger than magnitude 4, December 1861 to July 2005 (record complete from 1960). Locations for events before 1950 are approximate. A few late 19th Century events are not plotted because locations are uncertain.

(Modified after Fryer, G.J. (2005) Seismicity of Hawai'i: 1861–2005, Hawaii Institute of Geophysics and Planetology, University of Hawai'i)

Noteworthy Events

1	Apr. 2, 1868	Great Ka'u Earthquake	M 7.9
2	Feb. 19, 1871	Lana'i Earthquake	M 6.8
3	Dec. 8, 1895	"Oahu" Earthquake	M~5.8 (location very uncertain)
4	Sep. 25, 1929	North Kona Earthquake	M 6.1
5	Jan. 22, 1938	Maui Earthquake	M 6.7
6	Jun. 28, 1948	Honolulu Earthquake	M 5.2
7	Aug. 21, 1951	Kona Earthquake	M 6.9
8	Apr. 26, 1973	Honomu Earthquake	M 6.2
9	Nov. 29, 1975	Kalapana Earthquake	M 7.5
10	Nov. 16, 1983	Kao'iki Earthquake	M 6.7

Depth (km)
<20   >20

Magnitude
4.0
5.5
7.0

Fig. 4.3. Hawaiian earthquakes larger than magnitude 4, December 1861 to July 2005; although the Big Island is the site of most seismic activity, Maui County and Oʻahu have also been significantly damaged by earthquakes. (Map by G. Fryer)

the base of the young shield volcanoes on the Big Island as they change shape to accommodate magma moving into and through the structure. However, earthquakes may also occur within Earth's crust and in the upper mantle beneath the Big Island and farther to the northwest at O'ahu, Moloka'i, Lāna'i, Kaho'olawe, and Maui. These islands are highly vulnerable to seismic shaking. In the past 150 years, several strong earthquakes have caused extensive damage to roads, buildings, and homes; triggered local tsunamis; and resulted in loss of life in Hawai'i.

Volcanism, in the form of magma breaking rock as it moves through the interior of a volcano, is the catalyst for about 95% of the earthquakes on the Big Island. However, earthquakes are not exclusive to the Big Island. In the central region encompassing Maui, Lāna'i, Moloka'i, Kaho'olawe, and O'ahu, seismicity is generally related to tectonic activity on the seafloor close to the Islands. An important exception to this is the high potential for volcanic-related seismicity on Maui's Haleakalā Volcano as magma moves within the towering slopes of the mountain. The northwestern, or Kaua'i-Ni'ihau, region has experienced damage from earthquakes originating farther south, but no significant seismic activity has historically originated among these northern islands. Consequently the earthquake risk for the Northwestern Hawaiian Islands and Kaua'i has been evaluated as minimal.

EARTHQUAKE DAMAGE

The amount of damage caused by an earthquake depends upon several factors, including the duration and strength of shaking, the type of soil in the affected area, and the strength and type of buildings within range. Generally speaking, Hawaiian earthquakes generate short episodes of shaking. However, on November 16, 1983, an earthquake on the Big Island caused the ground to shake for nearly an entire minute, and the 1972 Kalapana earthquake had a total duration of over a minute.[9] Predictably, the longer the ground shakes, the greater the damage done. The amount of shaking is amplified around thick, soft, wet soils, and the shallower an earthquake is, the more intense the shaking is likely to be near its epicenter (the point on Earth's surface directly above where faulting has occurred). If extensive shaking takes place soil may undergo liquefaction, a process by which water-saturated sediment temporarily loses strength and acts like a liquid instead of a solid. Buildings using the soil as a foundation will suffer cracked walls and buckled support beams, and may even collapse.

Buildings, roadways, bridges, the electrical grid, and buried pipes and sewer lines typically suffer enormous damage during an earthquake. The amount of damage to buildings varies depending on the type and construction. Wood-frame houses frequently suffer lateral displacement, collapsed supporting posts, warped windows and doors, and distorted frames. Most often this is because a building is poorly attached, or completely unattached, to its foundation. Some of these damaging effects can be decreased if a structure is anchored to bedrock instead of sitting on unconsolidated dirt (such as underlies most of the Waikīkī area).

Cinder-block buildings experience cracking and wall collapse, and building attachments such as drainpipes, chimneys, and ornamental facades separate from the walls. Water tanks and stone walls shift laterally and collapse. Although all structures remain at risk, there are ways to construct earthquake-resistant buildings by establishing a continuous load path from the roof to the foundation, by using flexible materials in the original design, and by making sure the structure is firmly attached to its foundation. A continuous load path is achieved by tying, or connecting,

Fig. 4.4. During an earthquake, buildings experience cracking and wall collapse, and building attachments such as drainpipes, chimneys, and ornamental facades separate from the walls. (Photo by Taiko1, Flickr: http://www.flickr.com/photos/mic eyelatt/1460471203/ [last accessed January 5, 2010])

every structural component of a building together and ultimately to the foundation. The result is that the building acts as a single unit to withstand the shaking and is therefore stronger than any single component.

The October 15, 2006, earthquakes spurred a statewide effort to adopt updated building codes to guide new construction and retrofit older buildings to a higher standard of engineering. All counties, even relatively earthquake-free Kaua'i, are considering or have adopted the updated international building code, employing the latest structural techniques to safeguard against damage due to seismic shaking.[10] These steps include constructing a continuous load path of connected building components from the roof to the foundation, doubly securing load-bearing walls and beams in the structure, and employing double-wall construction on all framing.[11]

You can mitigate the effects of earthquakes in advance by doing these, and other, simple but effective measures:

- anchor tall bookcases, computers, appliances, and propane tanks
- install latches in cupboards and cabinet doors
- secure gas and water lines
- brace weak walls
- mount pictures and mirrors to the wall

For more information review the Federal Emergency Management Agency brochure "Earthquake Safety Guide for Homeowners" at http://www.fema.gov/library/viewRecord.do?id=1449.[12]

EARTHQUAKE READINESS: ARE PRECAUTIONS BEING TAKEN?

Although seismologists analyze the frequency and patterns of earthquake occurrence, it is not possible to predict when an earthquake will occur nor how big it will be. It is also not possible to prevent an earthquake from happening. As with all geologic hazards, when facing the inevitable, mitigation becomes important. Considerable forethought was demonstrated on the Big Island by the implementation of earthquake-resistant building codes and disaster preparedness plans that are regularly updated as construction methods improve. This is an important type of mitigation that many communities employ.

An equally critical component of a successful mitigation strategy is convincing both the authorities and residents of islands other than the Big Island that they are at risk as well. After Hawai'i County, Maui County has the next greatest earthquake risk in the state—a sobering statistic that should not be taken lightly.[13] Oʻahu also faces seismic risk: it is currently being carried over an arch in the lithosphere caused by the weight of the Big Island, a motion that might result in small earthquakes.

The widespread destruction from the 1938 Maui and 1948 Oʻahu earthquakes, as well as the history of earthquakes on Lānaʻi and the Molokaʻi Fracture Zone, is a harbinger of events to come. The critical difference now is the increased density of population and property at risk. Much of the commercial district of Honolulu and Waikīkī rests on unconsolidated soil that will experience heightened ground motion when the next earthquake arrives. Anyone with a stake in the state's tourist economy should consider that it would take up to a decade to recover from a disaster in which widespread fatalities and infrastructure damage occur.

A STATE SCIENCE AGENCY

As the table of Hawaiʻi earthquake history clearly indicates, this state has a history of damaging seismic events. But who within state government is responsible for scientific analysis and monitoring of the seismic threat? The answer is that we largely look over the shoulders of generous researchers in federal agencies and at the University of Hawaiʻi. These scientists are not specifically tasked with the responsibility of assessing the hazard, yet they nevertheless volunteer their time and research results. Our official risk assessors are the folks at Hawaiʻi State Civil Defense who deal with this information gap by organizing experts from various agencies and departments into committees that meet and share ideas and data.

This is fine, but it is a Band-Aid analysis. Hawaiʻi needs a state science agency, a "geological survey," an office with responsibility to gauge and help prepare state citizens for seismic hazards as well as other natural threats (e.g., rockfalls, tsunamis, flash flooding, sea-level rise, coastal erosion, global warming, volcanism, and others). Such an agency would offer the advice that sister agencies need for decision making that is currently lacking. Civil Defense is meant for managing people before, during, and after a disaster, as well as mitigating hazards likely to visit their damage upon a region. It simply is not effective to ask them to serve as a scientific agency; they are not trained for it and it is not in their mission; they do so only out of necessity to fill an obvious gap in our system of public safety. Hawaiʻi is the only state lacking a geological survey office. The geological surveys in other states conduct the scientific research that underlies hazard mitigation as well as natural resource management. Natural resources include sand and mineral commodities; water supplies; soil; beach, reef, and dune environments; forest and estuarine environments; and others.

TABLE 4.1 HISTORY OF LARGE EARTHQUAKES IN HAWAI'I

Year	Date	Magnitude	Source
1868	March 25	6.5–7.0	Mauna Loa south flank
1868	April 2	7.5–8.1	Mauna Loa south flank
1871	February 20	6.7–6.9	South of Lāna'i
1918	November 2	6.2	Ka'ōiki, between Mauna Loa and Kīlauea
1919	September 14	6.1	Ka'ū District, Mauna Loa south flank
1926	March 19	>6.0	Northwest of Hawai'i Island
1927	March 20	6.0	Northeast of Hawai'i Island
1929	September 25	6.1	Hualālai
1929	October 5	6.5	Hualālai
1938	January 22	6.9	North of Maui
1940	June 16	6.0	North of Hawai'i Island
1941	September 25	6.0	Ka'ōiki
1950	May 29	6.4	Kona
1951	April 21	6.9	Base of Mauna Loa Volcano
	August 22	6.3	slipped seaward
1952	May 23	6.0	Kona
1954	March 30	6.5	Kīlauea south flank
1955	August 14	6.0	Pacific Plate
1962	June 27	6.1	Ka'ōiki
1973	April 26	6.3	Pacific Plate
1975	November 29	7.2 (probably 7.6 to 7.7)	Kīlauea south flank
1983	November 16	6.6	Ka'ōiki
1989	June 25	6.1	Kīlauea south flank
2006	October 15	6.7	Kīholo Bay, Hawai'i Island
2006	October 15	6.0	Māhukona, Hawai'i Island

Note: See Pacific Disaster Center: http://www.pdc.org/iweb/pdchome.html (last accessed November 19, 2009).

Managing our complex system of natural hazards and resources requires ongoing scientific research. Currently in Hawai'i, research is spread among numerous federal, state, and county agencies or not performed at all. A geological survey office would provide important coordination and advancement of this critical duty. A geological survey would be responsible for producing and disseminating scientific information, monitoring threatening natural hazards, and providing authoritative advice. To the detriment of public health and safety, we lack this organized science component.[14]

A SURGE OF WATER: TSUNAMIS

A tsunami is a series of great waves caused by violent movement of the seafloor—usually an earthquake, submarine landslide, underwater volcanic explosion, or faulting on the bottom of the ocean. Tsunamis differ from regular ocean waves that are generated by the wind because of their great speed (up to 590 miles [950 km] per hour), long wavelength (up to 120 miles [193 km] between wave crests), long period (varying from 5 minutes to a few hours, generally 10 to 60 minutes between wave crests), and low height in the open sea. The first wave may not be the largest, so the danger from a tsunami can last for several hours after the arrival of the first wave. Sometimes the trough of a tsunami arrives at a shoreline first, causing the water to recede, exposing the ocean floor. In other cases the tsunami crest may arrive first with no attendant retreat of the ocean. In coastal areas tsunami height can be as great as 100 feet (30 m) in extreme cases. They can move inland several hundred feet to over a mile (100 m to over 3 km), traveling up rivers and streams connected to the ocean and damaging bridges and buildings along the banks.

When we hear news reports of a tsunami, it is easy to be misled by accounts of the wave height. Let's say a tsunami is reported with a "6-foot wave height." Most people in Hawai'i would think to themselves, "That's not so high; 6-foot waves arrive on our shores a couple dozen times a year and don't even trigger a high surf warning." It is a mistake to compare a tsunami with a typical wind wave. A wind wave passes by in a few seconds; a tsunami wave takes many minutes to pass by. The equivalent of a 6-foot (1.8-m) tsunami wave would be a 6-foot raging river running onto the shoreline. Could you stand in a 6-foot-deep raging river? Of course not. The reported tsunami wave height is the predicted depth of the water as it flows across the shoreline. Anything over 1 foot (0.3 m) can potentially be a killer if you are caught in the swirling currents.

Everyone has viewed the terrible images from late December 2004, when lithospheric plates meeting on the seafloor of Sumatra unlocked and shifted the ocean bottom vertically over 50 feet (15 m). The resulting wave rocketed across the Indian Ocean and wiped out coastal communities thousands of miles (km) apart. The earthquake spawning that killer was the second largest ever recorded and ripped the seafloor along a fault that was nearly 1,000 miles (1,600 km) long. This earthquake had the longest duration of faulting ever observed, between 8 and 10 minutes, and the seafloor rupture traveled between approximately 4,700 and 6,300 miles (7,600 to 10,100 km) per hour from south to north. It caused the entire planet to vibrate as much as half an inch (1.3 cm) and triggered other earthquakes as far away as Alaska.[15]

We are all susceptible to the common misconception that tsunamis are monstrous waves the size of skyscrapers with the perfect shape of a barrel at Pipeline. However, these are inaccurate characterizations. A tsunami more closely resembles a rapidly approaching wall of foamy water

Fig. 4.5. The Indian Ocean tsunami of 2004 reached a height of over 100 feet (30 m) and killed more than 225,000 people in 11 countries. This town near the coast of Sumatra, Indonesia, lies in ruins on January 2, 2005. This picture was taken by a U.S. military helicopter crew from the USS Abraham Lincoln that was conducting humanitarian operations. (Photo by crew of USS Abraham Lincoln: http://commons.wikimedia.org/wiki/File:Sumatra_devastation1.jpg [last accessed October 29, 2009])

followed by a rising tide that floods the coast in a matter of minutes. It is not a huge, surfable wave. When a tsunami hits, it is as if a large river of ocean water were directed onto the shore lasting several minutes and potentially rising dozens of feet (tens of meters).

Tsunamis flood the land with a turbulent mass of white water known as a bore. The bore, looking like a large wave traveling through a surf zone, is not surfable because it is not a coherent body of water; it is largely a mass of aerated foam and debris that will not support the weight of a surfer. The bore, followed by a flood of seawater, pushes into vegetation, across roads, through shorefront houses—and all the while the water level rises. That is, at the peak, the beach may be under 5 to 10 or even 30 feet (1 to 10 m) of water. In the Indian Ocean tsunami, an extreme case, the water depth was 75 to 150 feet (22 to 46 m) in some locations. The water depths achieved onshore depend on the topography of the seafloor and the land.

The onward surge of water moves with great speed and results in property damage, injuries, and death when the tsunami is large. Reports from Indonesia show that when all this water starts flowing back out to sea, it floats houses, drags cars, erodes the soil, and carries helpless people in a massive filthy plume directed offshore hundreds of yards (m) to a mile (1.6 km) or more. Being caught in this swirling torrent is what causes many deaths by drowning. The true character of a tsunami is apparent when you consider the nature of the wave—the crest alone can take 10 minutes or more to pass an island, and the following trough may take half an hour or more.

Between 1929 and 1998, tsunamis caused 222 deaths in the Hawaiian Islands and over $50 million in reported damages.[16] The April 1, 1946, tsunami had the highest death toll of any modern natural hazard event in Hawai'i. The town of Hilo was hit by six to seven large waves with heights up to 25 feet (7.6 m) that surged in at 15-minute intervals. These waves pummeled the shoreline, punching several city blocks inland, demolishing Hilo's waterfront, ripping every house from its foundation and dragging them across the street, only to smash into other buildings. The death toll was 159 people.[17] That tsunami damaged all of the Hawaiian Islands but was particularly devastating to the island of Hawai'i, caught completely unaware early one morning.

The Hawaiian Islands have a long history of tsunamis because we lie in the middle of the Pacific Ring of Fire. Throughout the Pacific are active plate margins capable of suddenly moving the seafloor. In addition, we live on active volcanoes with unstable slopes, making us surrounded by dynamic geologic processes that cause tsunamis. It is not surprising,

TABLE 4.2   TSUNAMIS THAT HAVE HIT HAWAI'I WITH RUN-UP OF 6.6 FEET (2

Date	Place of Observation	Source
1812: December 21[a]	Hoʻokena, Hawaiʻi	California
1819: April 12	West Hawaiʻi	Chile
1837: November 7	Hilo, Hawaiʻi	Chile
1841: May 17	Hilo, Hawaiʻi	Kamchatka
1860: December 1	Māliko Bay, Maui	North Pacific[b]
1868: April 2	Keauhou Landing, Hawaiʻi	Kaʻū
1868: August 13	Hilo, Hawaiʻi	Chile
1869: August 24	Southeast Puna	South Pacific[b]
1877: May 10	Hilo, Hawaiʻi	Chile
1878: January 10	Māliko Bay, Maui	North Molokai[b]
1896: June 15	Keahuhou, Hawaiʻi	Japan
1903: November 29	Pelekunu, Molokaʻi	North Molokai
1906: August 17	Māʻalaea, Maui	Chile
1919: October 2	Hoʻōpūloa, Hawaiʻi	South Kona
1922: November 11	Hilo, Hawaiʻi	Chile
1923: February 3	Hilo, Hawaiʻi	Kamchatka
1933: March 2	Keauhou, Hawaiʻi	Japan
1946: April 1	Waikolu Valley, Molokaʻi	Aleutian Islands
1952: November 4	Kaʻena, Oʻahu	Kamchatka
1957: March 9	Hāʻena, Kauaʻi	Aleutian Islands
1960: May 22	Hilo, Hawaiʻi	Chile
1964: March 27	Waimea Bay, Oʻahu	Alaska
1975: November 29	Keauhou Landing, Hawaiʻi	South Puna

[a] Earliest tsunami for which definite information exists.

[b] Probable source

Source: George Pararas-Carayannis, 1969, *Catalog of Tsunamis in the Hawaiian Islands,* U.S. Coast and Geodetic Survey, May; Harold G. Loomis, *The Tsunami of November 29, 1975 in Hawaii,* 1975, Hawaiʻi Institute of Geophysics, December, pp. 1 and 10; D.C. Cox and J. Morgan, 1977, "Local

| Maximum Height in Hawai'i | | Deaths in Hawai'i | Damage in Hawai'i |
Meters	Feet		
2.5	8	—	Hut flooded
2.0	7	—	Houses destroyed
6.0	20	16	100 houses destroyed
4.6	15	—	Unknown
3.6	12	—	Houses, wharf destroyed
13.7	45	47	Severe in Puna and Ka'ū
4.6	15	—	Houses, bridges destroyed
8.2	27	—	Houses destroyed, roads washed out
4.8	16	5	Severe in Hilo
3.6	12	—	Scattered flooding, North Maui, North O'ahu
5.5	18	—	Houses, wharfs, stores destroyed
4.5	15	—	Houses destroyed on Maui, railroad washed out on O'ahu
3.6	12	—	Piers damaged
4.3	14	—	Wharf damaged, car swept away
2.1	7	—	Fishing boats swept away
6.1	20	1	$1,500,000
3.2	10	—	Boathouses, walls destroyed in Kona
16.4	54	159	$26,000,000
9.1	30	—	$1,000,000
16.1	53	—	$5,000,000
10.5	34	61	$23,000,000
4.9	16	—	$68,000
14.3	47	2	$1,500,000

Tsunamis and Possible Local Tsunamis in Hawaii," Hawai'i Institute of Geophysics, Report HIG 77-14, November; Doak C. Cox, 1987, *Tsunami Casualties and Mortality in Hawaii,* University of Hawai'i, Environment Center, June, p. 39; James F. Lander and Patricia A. Lockridge, 1989, *United States Tsunamis (Including United States Possessions) 1690–1988,* Publication 41–2, National Geophysical Data Center, August, pp. 17–77; U.S. Geological Survey, Hawaiian Volcano Observatory, records; Pacific Tsunami Warning Center, records. See Pacific Disaster Center: http://hawaii.gov/dbedt/info/economic/databook/dh2007/index_html (last accessed November 20, 2009).

then, that Hawai'i is one of the most tsunami-vulnerable locations on the planet. We average one tsunami every year and a damaging occurrence every 7 years.[18]

RING OF FIRE

Surrounding the Hawaiian Island chain is a vast circle of highly volatile subduction zones—places where moving lithospheric plates are recycled down into the mantle. When they produce melting, the magma feeds ranges of volcanoes (such as the Aleutian Islands and the Cascade Range), and the subduction process can generate large earthquakes. The Pacific Ring of Fire owes its name to the circle of volcanoes and earthquake zones along the Pacific Rim of Asia, New Zealand, and North and South America. Many of Earth's greatest earthquakes have occurred along the

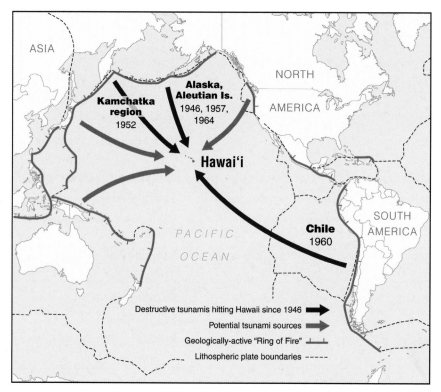

Fig. 4.6. The Pacific Plate is surrounded by subduction zones that produce earthquakes, volcanoes, landslides, and tsunamis; Hawai'i sits at the center of this ring. (SOEST Graphics)

borders of the Pacific Plate, and this trend will continue. Hawai'i sits at the center of this ring, a bull's-eye positioned as a potential target for the seismic ocean waves generated by seafloor movements in and around the entire Pacific Ocean.

Fortunately, the scientists at the Pacific Tsunami Warning Center[19] in 'Ewa Beach are constantly monitoring earthquakes and sea level around the Pacific (and the world) and will issue a tsunami warning if one is detected. On the other hand, if a locally generated submarine landslide, volcanic eruption, or seafloor fault occurs, a resulting tsunami could reach any of the seven major Hawaiian Islands within 5 to 40 minutes—little time for warning and evacuation. Nonetheless, the geophysicists at the warning center in 'Ewa Beach are tackling this problem by adding seismometers throughout the Islands and automating their procedures. Within seconds of a large earthquake in the Hawaiian Islands a computer program tells them the location. Knowing that the most likely earthquake to generate a locally damaging tsunami is a seaward sliding of the south flank of Kīlauea or of the south or west flanks of Mauna Loa, warning center scientists can rapidly evaluate the threat of a tsunami from an earthquake and sound the statewide sirens in less than 5 minutes. Considering that tsunami travel time from the Big Island south shore to Hilo is about 10 minutes, and to Waikīkī about 30 minutes, this should give time to get out of the water and into safe areas if people react immediately. But will they?[20]

COMPLACENCY KILLS

One of the greatest concerns of Civil Defense agencies everywhere is the threat of complacency. In Hawai'i, this problem has grown every year since 1964, the last time a damaging tsunami affected the whole state. Two generations of Hawaiian *kamaʻāina* (native-born residents) have come of age without any direct tsunami experience. Hundreds of thousands of *malihini* (new residents), lacking tsunami education or the cultural memory of their grandparents' experience, have immigrated to the Islands in the half century of quiescence. Millions of tourists come and go without any awareness of what the yellow or green sirens along the coast signify. In these decades, surfing and water sports have exploded in popularity, and at any given moment of the day many thousands of people are at play or work in coastal waters, confident that the safety of the shoreline is only a short swim or surf away.

Our Civil Defense officials fear that the next major tsunami will be a record catastrophe. This concern is absolutely justified. In a 1986 tsunami warning, Oʻahu motorists fled via routes such as the Kalanianaʻole

Highway through areas that would have been inundated had the tsunami been a major one.[21] In October 1994, an estimated 400 foolhardy surfers were on O'ahu's north shore and ignored a tsunami warning.[22] Others purposely paddled out in Waikīkī at the predicted time of arrival thinking that they would get the ride of their lives. They very well might have had their last ride, had not the tsunami been only a few inches high. Although it was minor, even a moderate event would have resulted in fatalities, and dozens of board riders would have died in a major event. Such complacency and willful ignorance will lead to unnecessary casualties.

Hawai'i residents are familiar with the siren test at 11:45 a.m. on the first working day of the month. This is your personal reminder, a moment to think: "What would I do right now if a tsunami were coming?" "Do my children know where to go?" All coastal residents should be familiar

Fig. 4.7. On April 1, 1946, the Aleutian Islands earthquake (7.8 magnitude) generated a tsunami that killed 159 people throughout Hawai'i. This photo shows the Hilo pier, looking north across Hilo Bay in the background. The courageous stevedore, Tony Correa, on the left side of the photo, was trying to free the dynamite-laden SS Brigham Victory from the pier when he was struck by the wave. The roof of the warehouse has collapsed and is surging toward him on the incoming wave. Beyond, the wave crest rushes past the pier (right to left) toward Hilo Bay shoreline. His body was found 2 days later floating in Reed's Bay, Hilo. This event provided the motivation for establishing the Pacific Tsunami Warning Center in 'Ewa Beach, Hawai'i. (Image from Bishop Museum archives, Honolulu, Hawai'i)

with the tsunami warning system (steady 3-minute siren tone alert signal for any emergency: tune in to local media for emergency information) and know what to do when a tsunami threatens. The tsunami hazard maps in the phone book provide excellent guidance on where our most vulnerable lands are located; when you hear the siren, head to higher ground. Remember, you may only have minutes to escape if the tsunami is locally generated.

ARE WE READY?

Hawai'i Civil Defense officials are well aware that a locally generated tsunami could cause damage along our crowded shoreline. But have we made adequate plans for evacuation considering that a wave could arrive in less than half an hour? The plan for residents of Waikīkī to evacuate vertically (head for the third floor and higher of buildings) during a tsunami was tested by a merchant in the International Marketplace—in the heart of Waikīkī.

Knowing that he would be expected to find his way to the third or fourth floor of the nearest building in the case of an emergency, he conducted a self-imposed drill. He walked from building to building around his business and discovered the following:

- a lack of street signs offering guidance
- no signs in building lobbies indicating evacuation routes and great difficulty finding stairwells
- some lobby stairwells locked
- many stairwells marked with signs warning that they were not to be used
- long waits for elevators
- hotel employees untrained in disaster preparedness

It is not hard to imagine the utter chaos that would result with thousands of tourists and residents in Waikīkī desperately trying to get out of the way of an approaching killer wave. A good step would be for the Waikīkī hotel owners themselves to think this problem through and then offer their suggestions to the appropriate authorities.

FACING THE INEVITABLE

So, the question remains: Have the people of the Hawaiian Islands learned the lessons that earthquakes and tsunamis have to teach? The events we consider to be "natural disasters" are simply natural processes that form and shape Earth over time. Given their inevitability, why do we fail to adequately incorporate them into community planning? If you live in a tsunami zone, you may not want the bearing walls in your home to face the ocean—they can be oriented perpendicular to the shore. Rather than build your home on a flat concrete slab—use post and pier construction so that the water can roll beneath your home. Why build next to the shore at all? Consider keeping some distance between our families and the next killer tsunami. Where we concentrate tourists in high hazard zones, let's not fail to provide adequate evacuation.

Natural processes only turn hazardous when people get in their path with inappropriate behavior and vulnerable roadways, subdivisions, schools, and entire communities. To reduce the risk associated with geologic events, it is important that we understand the natural hazards that threaten the Islands and why and how they occur in Hawai'i, so that we can prepare for those that are certainly yet to come. Here are some suggestions on what to do if a tsunami threatens.

Before an emergency threatens:

- Have a family disaster kit. Officials recommend that all items should fit in a 5-gallon (19-liter) bucket and include essentials for survival: food, water, and warmth. For a list of suggested items see http://www.scd.state.hi.us/documents/red_cross_kit_checklist.pdf.
- Discuss evacuation plans with your family so that everyone knows what to do, where to go, and how you will communicate and meet later.
- Check tsunami inundation and evacuation maps in the phone book to determine if your home, school, or business is in a flood-prone region.
- Know the best route to evacuate an area.

When a siren sounds:

- Listen to television and radio stations for instructions on how to proceed and whether to evacuate.
- Be prepared to walk out of the danger zone in case of traffic gridlock.
- Avoid telephone use so that circuits are not overloaded.

If you are instructed to evacuate:

- Wear protective clothing and sturdy shoes.
- Take your disaster supply kit.

- Use travel routes specified by local authorities and do not use shortcuts because certain areas may be impassable or dangerous.

After the tsunami:

- Do not return to low-lying areas until the all clear is sounded.
- Be prepared to identify yourself to emergency officers.
- Be alert to damaged buildings, downed electrical wires, debris, and fires.

Here are some sources of information for you:

- Pacific Tsunami Warning Center: http://www.prh.noaa.gov/ptwc/
- Hawai‘i State Civil Defense: http://www.scd.state.hi.us/
- Pacific Disaster Center: http://www.pdc.org/iweb/pdchome.html
- U.S. Geological Survey Atlas of Natural Hazards: http://pubs.usgs.gov/imap/i2761/
- University of Hawai‘i Sea Grant Publications:
  - *Purchasing Coastal Real Estate in Hawai‘i:* http://www.soest.hawaii.edu/SEAGRANT/communication/pdf/Purchasing%20Coastal%20Real%20Estate.pdf
  - *Hawai‘i Coastal Hazard Mitigation Guidebook:* http://www.soest.hawaii.edu/SEAGRANT/communication/HCHMG/hchmg.htm
  - *Homeowner's Handbook to Prepare for Natural Hazards:* http://www.soest.hawaii.edu/SEAGRANT/communication/NaturalHazardsHandbook/naturalhazardprepbook.htm

And always remember, if you feel an earthquake, immediately head for high ground—a locally generated tsunami may be headed your way.

CHAPTER 5

Hurricanes

'A'ohe wa'a ho'ohoa O ka la 'ino
"No canoe is defiant on a stormy day"

Few phenomena in nature compare to the destructive force of a hurricane. Called the greatest storm on Earth, a hurricane is capable of annihilating coastal areas with sustained winds over 100 miles (160 km) per hour, intense areas of rainfall, flooding ocean waters, and huge waves. In fact, during its life cycle a hurricane can expend as much energy as 10,000

Fig. 5.1. Hurricane 'Iniki was the costliest hurricane to strike the state of Hawai'i, causing $1.8 billion ($2.6 billion in 2004 U.S. dollars) in damage. Most damage was on the island of Kaua'i, where the storm destroyed thousands of homes and left most of the island without power, although O'ahu also suffered significant damage. 'Iniki was also responsible for six deaths.

nuclear bombs.[1] In Hawai'i, although hurricanes are not common, when they do approach and hit an island the results can be catastrophic. This is largely because we have not taken sufficient steps to protect ourselves, to factor in the potential for violent winds in the design of our buildings, to build up and away from potential storm surge, to bury power lines, and to plan for the isolation that communities experience immediately after a storm has passed. Because our power lines are strung on vulnerable poles instead of buried in the ground, Hawai'i communities suffer power outages after a hurricane. This means no water or ice, no communication, and, within a day or two, no food. Even downtown Honolulu would go dark, dry, and hungry with a direct hit by a category 4 hurricane; no area is immune. "We don't get hit often enough to justify the expense of buried power lines?" Global warming may change that. A hurricane is cause for us all to be humble in the face of nature's overpowering force. Don't forget that "No canoe is defiant on a stormy day."

Beyond the soft winds and tranquil north-shore waters of Hawai'i, late summer and fall usher in threatening tropical storms and hurricanes. The term "hurricane" is derived from Huracan, a god of evil recognized by the Mayan civilization in Central America and the Tainos of the Caribbean.[2] Hurricanes, tropical storms, and typhoons are collectively known as tropical cyclones. These disturbances are among the most devastating, naturally occurring hazards in coastal areas of the tropics and middle latitudes. Constant media reminders that it is "hurricane season" keep us all on guard and stimulate a somewhat closer watch among residents on the weather forecast portion of the evening news.

Tropical cyclones[3] are classified as follows:

- Hurricane: An intense tropical weather system with a well-defined circulation and maximum sustained winds of 74 miles (119 km) per hour or higher. In the western Pacific, hurricanes are called typhoons; in the Indian Ocean, they are called cyclones.
- Tropical storm: An organized system of strong thunderstorms with a defined circulation and maximum sustained winds of 39 to 73 miles (63 to 117 km) per hour.
- Tropical depression: An organized system of clouds and thunderstorms with defined circulation and maximum sustained winds of 38 miles (61 km) per hour or less.

Hurricanes tend to form around a preexisting atmospheric disturbance (an area of low pressure) in warm tropical oceans where there is

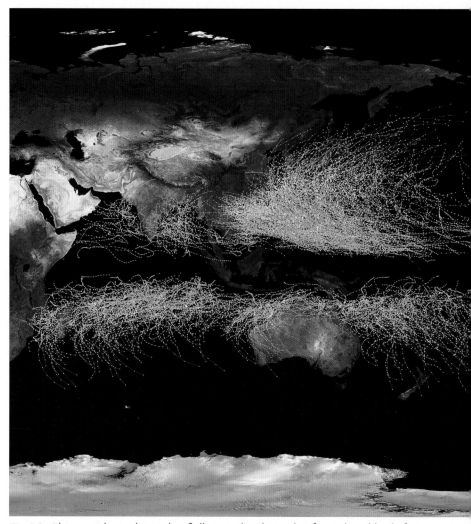

Fig. 5.2. This map shows the tracks of all tropical cyclones that formed worldwide from 1985 to 2005. Storms are plotted at 6-hour intervals and use a color scheme indicating the Saffir-Simpson Hurricane Scale. Hawai'i lies in the corridor between the East Pacific spawning grounds and the West Pacific. (Image from Wikipedia: http://en.wikipedia.org/wiki/Tropical_cyclone [last accessed November 4, 2009])

high humidity in the atmosphere, light winds above the storm, and a high rate of condensation in the atmosphere. If the right conditions last long enough, a hurricane can develop, producing violent winds, incredible waves, torrential rains, and floods.

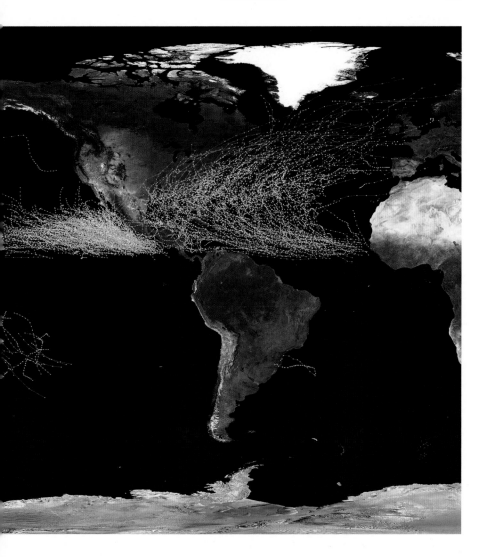

There are on average six Atlantic hurricanes each year; over a 3-year period, approximately five hurricanes strike the United States coastline from Texas to Maine. The U.S. Atlantic and Gulf of Mexico coasts are famous for their vulnerability to hurricane damage, but other areas are also at risk. In fact, there are three regions in the Northern Hemisphere known for their tendency to spawn hurricanes: the tropical Atlantic, the eastern tropical Pacific, and the western tropical Pacific. In the Southern Hemisphere there are two primary areas: the western tropical Pacific and the Indian Ocean.

HURRICANE INTENSITY

All hurricanes are ranked by the Saffir-Simpson Hurricane Scale according to the power of their winds.[4] Yet a storm's strength fluctuates. What appears to be a weak tropical depression may intensify into a full-blown hurricane. The vigilant National Weather Service's Central Pacific Hurricane Center (CPHC) monitors weather patterns and tropical low-pressure centers, and tracks any suspected tropical storms.[5] If need be, the CPHC is able to issue storm warnings at least several hours before a storm hits. When the shrill sirens blare out a state of emergency and call for evacuation, citizens are encouraged not to second-guess the CPHC's judgment, especially because high winds and flooding may occur before the storm's landfall.

Coastal flooding called storm surge almost always accompanies hurricanes. In Hawai'i, the threat is doubled because floodwaters arrive from both the sea and the land. The heavy rainfall that accompanies most hurricanes can overwhelm our tiny watersheds in a blink, and narrow streams turn into flooding, hazardous torrents. On the coastal plain at the base of the steep watersheds, standing water collects in poorly drained areas. This is especially the case where a low-lying coastal plain is flooded by high ocean waters accompanying a storm. Runoff simply cannot drain into the ocean when the storm-drain system is already flooded by high sea level. Compounding the situation, unlike continental coastal areas, the islands of Hawai'i do not have a continental shelf, which makes them prone to bigger waves and marine flooding that surges across coastal lands.

As if flooding and high winds were not enough, hurricanes may be accompanied by bursts of extremely intense winds. These concentrated

TABLE 5.1   SAFFIR-SIMPSON HURRICANE SCALE

Hurricane Category	Sustained Winds	Damage Level	Typical Storm Surge
1	74–95 miles (119–153 km) per hour	Minimal	5–7 feet (1.5–2.1 m)
2	96–110 miles (154–177 km) per hour	Moderate	7–12 feet (2.1–3.6 m)
3	111–130 miles (178–209 km) per hour	Extensive	12–15 feet (3.6–4.6 m)
4	135–155 miles (210–249 km) per hour	Extreme	15–20 feet (4.6–6.1 m)
5	>155 miles (>249 km) per hour	Catastrophic	20–24 feet (6.1–7.3 m)

winds, called microbursts, can reach over 200 miles (322 km) per hour and literally blow apart any structure in their path.[6] Another dangerous feature of hurricanes in our mountainous topography is the tendency for winds to accelerate as they climb steep slopes. As hurricane-force winds blow across the flat ocean and low-lying coastal plains, they encounter a sudden increase in elevation on the slopes of the high shield volcanoes. To maintain the volume of airflow within the storm system these leading winds must accelerate to climb the steep slopes. The wind gusts created by this process can approach and exceed 150 miles (241 km) per hour. Houses perched on ridgelines, the edge of steep slopes, and even typical valley hillsides can have their roofs torn off, leading to wall collapse and total structural failure.

These same winds can further accelerate as they cross the *pali* (the volcanic cliffs on every island) and hurtle downhill along steep contours. During Hurricane 'Iwa in 1982, the highest wind speeds were measured in Kāne'ohe even though the hurricane passed on the leeward side of the island.[7] In either case of upslope or downslope winds, homes among the hills get it coming and going. A good dictum for siting your house: "If you have a nice view, you are vulnerable to wind damage."

HURRICANES IN HAWAI'I

Hurricanes are relatively rare in Hawai'i, but should they run ashore on any of the islands their impact is so devastating that they top the list of local meteorological hazards. Not only is their immediate impact overwhelming, but the high winds, heavy rainfall, storm surge, and large surf they generate trigger a domino effect of other hazards, including flash flooding, mud slides, and other forms of mass wasting, coastal erosion, and coastal flooding. Usually it is our shoreline that receives a hurricane's most immediate and immense blow. Sitting in the crosshairs of an approaching storm, the shoreline endures the first and most severe pummeling from the wind, waves, and floods.

Hawai'i lies in the central Pacific, which, on average, has four to five tropical cyclones (depressions, storms, and hurricanes) every year. These numbers can range as high as 11 a year, such as in the 1992 and 1994 seasons, or as low as zero as in 1979. History shows that strong winds associated with hurricanes have struck all major Hawaiian islands, but true landfalls are not common. Hurricane Dot in 1959 and 'Iniki in 1992 are the only two that have actually run ashore in the past half century. Hurricane 'Iwa came close as it passed to the northwest of Kaua'i in 1982. However, a hurricane passing offshore does not have to make landfall to

**TABLE 5.2   TROPICAL CYCLONES IN THE CENTRAL PACIFIC,
1971–2008**

Parameter	Hurricanes	Tropical Storms	Tropical Depressions	Total
Total number	58	46	59	163
Average per year	1.6	1.2	1.6	4.4
Percentage of all systems	36%	28%	36%	

Source: National Weather Service, Central Pacific Hurricane Center, "Climatology of Tropical Cyclones in the Central Pacific Basin": http://www.prh.noaa.gov/cphc/pages/climatology. php (last accessed November 20, 2009).

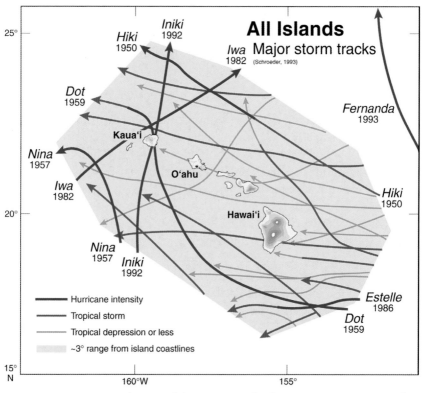

Fig. 5.3.   Major storm tracks around the Hawaiian Islands since 1949. (SOEST Graphics, from Fletcher et al., 2002, *Atlas of Natural Hazards in the Hawaiian Coastal Zone,* p. 17: http://geopubs.wr.usgs.gov/i-map/i2761/ [last accessed 5 November 2009])

cause catastrophic damage. The passing glance of a nearby system can generate large waves that erode beaches and sweep onto low-lying roads and properties, causing significant damage, and high winds that accelerate up and down Hawai'i's steep slopes.

Before the launch of the first weather satellite in April 1960, storm history in the trackless waters of the Pacific is sketchy and built mostly from aircraft and ship reports. The first officially recognized hurricane in Hawaiian waters was Hurricane Hiki, which passed to the northeast of O'ahu and Kaua'i in August 1950. Since 1950 four hurricanes have caused serious damage in Hawai'i. These include Hurricane Nina (1957), which produced record winds in Honolulu as it passed to the southwest of the state; Hurricane Dot (1959), which made landfall on Kaua'i; Hurricane 'Iwa (1982), which passed north of Kaua'i on November 23 and caused over $200 million in damage; and most recently Hurricane 'Iniki (1992), which made landfall on Kaua'i and caused $2.4 billion in damage.[8]

WHEN THEY HIT

When hurricanes move across the ocean surface and onto land, the heavy rain, strong winds, and large waves can damage buildings, trees, and cars. Accompanying the large waves is a high sea-level phenomenon called storm surge. Storm surge is a combined effect that includes low atmospheric pressure above the ocean surface causing a bulge of water that travels beneath a hurricane as well as the effect of wind pushing water in the direction of the hurricane's forward movement. Sea-level set up (a rise in water level on the shore due to wave momentum) also contributes to storm surge. The combined processes of wind, low pressure, and set up can raise the water level several feet (2 m or more) along a coastline. Add to this waves 30 to 45 feet (9 to 14 m) high (or more) formed by the winds of a hurricane and it is easy to see why these storms are so dangerous when they intersect the shore, especially a shoreline with low topography. Storm surge causes massive damage and flooding of low-lying coastlines. Damage is compounded if a hurricane should hit at high tide, as Hurricane 'Iniki did when it ran aground on Kaua'i.

Because the wind in a hurricane circulates counterclockwise in the Northern Hemisphere, as a storm moves across the ocean toward a shoreline it is the coast that gets hit by the forward right quarter of the storm that sustains the worst initial damage. Winds are greatest in that quarter because they are blowing in the same direction that the hurricane is traveling. Wind speed is enhanced because of the addition of the speed of the storm. Worst of all, the winds are blowing onshore in the forward right quarter, and therefore the storm surge is highest of anywhere under the storm. When Hurricane 'Iniki hit Kaua'i, this quarter of the storm made a direct hit to the Kukui'ula–Po'ipū–Makahū'ena Point region, and it was there that damage was greatest. Coastal homes were literally wiped off the map by a combination of high winds and storm surge that reached over 25 feet (7.6 m) in places.[9]

Hurricanes are not the only type of storm to cause damage in coastal areas, but they are the most severe. High winds and storm surge are a primary cause of hurricane-inflicted loss of life and property damage. When a hurricane makes landfall on low-lying coastal lands, ocean waters sweep across beaches, roads, and into adjacent communities. Houses float or are blown off their foundations. Debris torn loose is carried by the water and wind to smash into neighboring dwellings. Entire blocks of homes can be brought down in the hours that a hurricane batters the coast. In the forward right quarter of the storm, winds blow onshore at first, and when the eye of the storm passes they blow from the opposite direction, often completing the total damage to structures severely weakened by the first winds. Heavy rains compound the coastal flooding by saturating the ground and forcing nearby streams to overtop their banks. Coastal erosion compromises any buffering effect that beaches and wetlands may offer because high waves and storm surge can strip a beach of its sand and undermine homes, roadways, and businesses.

How can damage be mitigated? The simplest and most direct way is to avoid the hazard by not developing the shoreline. If we stopped putting communities on the edge of the ocean, we would greatly reduce the suffering, loss of life, and enormous damage caused by hurricanes and other types of coastal hazards. A second step is to design homes and buildings to better withstand high winds and high floodwaters. Generally speaking, the locations most at risk for these hazards are known, and building codes targeted for those spots could improve community resiliency.

STORM FORMATION

Tropical storms that arrive in the Hawaiian Islands usually form off western Mexico and Central America in the warm coastal waters of the eastern Pacific. The hurricane season officially starts on June 1 and extends through November 30. However, the stormiest period generally falls from July through September.

Hurricanes forming off the Americas are designated eastern Pacific hurricanes when they occur east of 140° W longitude. Storms occurring

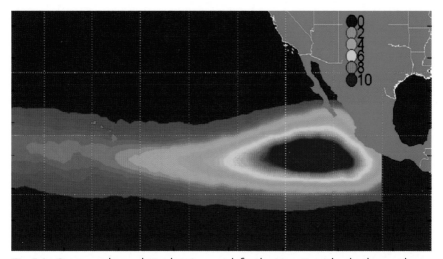

Fig. 5.4. Contours show relative hurricane risk for the Hawaiian Islands: the number of times a hurricane passes within 75 nautical miles (120 km) in an average 10-year period. (D. Hwang, 2005, *Coastal Hazard Mitigation Guidebook*, University of Hawaiʻi, Sea Grant College)

between 140° and the international date line are central Pacific hurricanes. This distinction determines which weather office tracks and forecasts the storms; eastern Pacific storms are tracked by the National Hurricane Center in Miami, and central Pacific storms are tracked by the Central Pacific Hurricane Center in Honolulu. West of the international date line, the hurricanes are called typhoons, and tracking is performed by the U.S. Joint Typhoon Warning Center and the Japan Meteorological Agency.

Tropical storms have been identified using both men's and women's names since 1978. A mix of English and Spanish names is used in the eastern Pacific Region, and Hawaiian names in the central Pacific Region. Storm names in the eastern Pacific are recycled every 6 years, and in the central Pacific four lists of names are used sequentially.

The conditions that lead to tropical cyclone formation in the eastern Pacific are the same as elsewhere in the world. With the onset of summer, a large area of warm water, high humidity, and light winds forms between 8° and 20° north of the equator with a temperature that exceeds 80°F (26.7°C). This heat fuels the energy of the storm.

The first sign of hurricane development is the appearance of a cluster of thunderstorms over the tropical oceans, called a tropical disturbance. Tropical disturbances most commonly form where surface winds (such as trade winds) converge near the equator, accelerate, and develop areas of turbulence. Given favorable conditions, a tropical disturbance can grow, indicated by falling air pressure on the ocean surface in the area around the storm and the development of a counterclockwise circulation of air. Surface pressure falls as water vapor condenses and releases heat into the air, causing it to rise. This lowers air pressure on the ocean surface, and surrounding air flows inward toward the area of low pressure.

As the warm air rises, it expands and cools, triggering more condensation, the release of more heat, and a further increase in upward airflow. This process essentially becomes a chain reaction as the rising column of air in the center of the storm causes surface pressure to lower even more, drawing air (wind) into the storm center. As the air flows in, it develops counterclockwise circulation in the Northern Hemisphere and clockwise circulation in the Southern Hemisphere: a consequence of the Coriolis effect caused by Earth's rotation. As the process of condensation, heat release, and air circulation continues, it generates more thunderstorms, more heat, lower surface pressure, stronger winds, and so on. Under the right set of circumstances, this process can build and gain strength.

At the top of the storm, atmospheric pressure begins to rise in response to the rising column of warm air. This causes air to flow outward (diverge) around the top of the disturbance. Like a chimney, this upper-level area "vents" the tropical system and prevents the air converging on the ocean surface from piling up around the center. If it did,

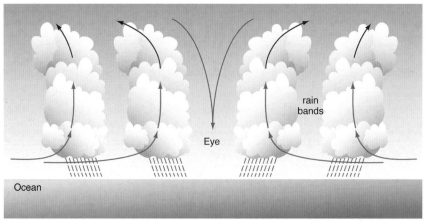

Fig. 5.5. Vertical cross section of air circulation, clouds, and precipitation in a hurricane. (SOEST Graphics, modified from Physical Geography.net, *Fundamentals* eBook: http://www.physicalgeography.net/fundamentals/7u.html [last accessed November 4, 2009])

surface pressure would rise inside the base of the storm and ultimately destroy it.

As a hurricane gathers strength, the center develops into a 5- to 40-mile (8- to 65-km)-diameter cloud-free area of sinking air and light winds, called the eye. The eye is the calmest part of the storm. This is because the strong cyclonic winds that accelerate toward the center never actually reach the exact center of the storm. Instead, the winds form a cylinder of intensely fast airflow, called the eye wall, surrounding a column of relatively calm, downward-flowing air. Like an ice skater whose body spins faster as his or her arms are drawn inward, air near the center speeds up as it spirals in toward (but never reaching) the eye. Hence, the winds form an intense high-velocity band ringing the eye characterized by tall thunderstorms, heavy rain, and destructive winds.

The most dangerous section of the storm is in the eye wall on the right side of the eye. This is where the wind blows in the same direction as the storm's forward motion, adding the forward speed of the storm to the speed of the wind. For example, in a hurricane moving due north at 20 miles (32 km) per hour with eye wall winds at 125 miles (200 km) per hour, the most intense airflow is found on the eastern side of the storm, where the winds blow north at 145 miles (232 km) per hour. On the west side of the eye the same winds would be southerly at 105 miles (168 km) per hour.

Even when conditions are ripe for storm formation on the ocean surface, high-altitude winds or descending columns of air can prevent a disturbance from developing. For example, in the region of 20° to 30° latitude in the North Pacific, the air aloft is often sinking because of the presence of a semipermanent high-pressure system called the North Pacific High.[10] As the air in the North Pacific High sinks, it warms. This prevents the formation of an upward-moving column of air necessary to produce a storm. In addition, strong upper-level winds in some areas tend to "decapitate" thunderstorms by dispersing rising air and preventing a storm from getting organized.

Three conditions usually spell a hurricane's demise:

- The storm moves over cooler water that can't supply warm, moist tropical air.
- The storm moves over land, also cutting off the source of warm, moist air.
- The storm migrates into an area where high-altitude airflow disrupts storm circulation.

TRACKING HURRICANES

Tropical storms formed in the eastern Pacific tend to drift westward or to the north into the open North Pacific. As the season progresses, more hurricanes steer northward paralleling the American Pacific coast and ultimately meeting their fate in the colder waters. It is the trade winds that steer most Pacific hurricanes from their place of birth westward into the open ocean. This track ensures that few hit land.

The storms that do make landfall usually strike along the Mexican coast. If a storm were to curl back toward the coast of California, its impact would be muted by the cold waters of the California Current that flow southward along the coastline with temperatures that rarely exceed 60°F (15.5°C). In fact, because it is protected by the current of cold water, hurricanes are unlikely to hit California.[11] The same cannot be said of tropical storms, however. In September 1939, winds of 50 miles (80 km) per hour struck the Los Angeles–Long Beach area as part of a tropical Pacific storm that produced 5.6 inches (14 cm) of rain in Los Angeles and over 11.6 inches (29 cm) at nearby Mount Wilson. The storm claimed 45 lives (mostly at sea) and caused $2 million of damage to crops and coastal development.[12] Has California ever been hit by a hurricane? The answer is "sort of." In 1858 the coastline from San Diego to Long Beach was raked with damaging winds, some reaching hurricane strength, generated by a low-pressure system that tracked offshore, probably without the eye actually making landfall.[13]

The majority of central Pacific hurricanes that capture our attention also originate in the warm, tropical waters off the Mexico coast. Every few years a storm will move along a path that is west by northwest toward Hawai‘i. However, most of these diminish before reaching the Islands because waters to our east and northeast are generally too cool to sustain a hurricane's intensity. They also come under the influence of the North Pacific High, where high-altitude airflow disrupts storm circulation.

Figuring out the history of hurricanes in the Pacific has been complicated by the lack of direct observations before the modern era of satellites. Most hurricanes spend their lives over waters of the East Pacific where few ships and even fewer islands are available to report their position and intensity. The majority of hurricanes went undetected before the 1960s. But as global weather satellites came on line, meteorologists were surprised to learn that eastern Pacific hurricanes are rather frequent phenomena with patterns of movement that follow semipredictable pathways.

Eastern Pacific hurricanes can make landfall along the Central American coast, potentially hitting the nations of Mexico, Guatemala, El Salvador, Nicaragua, Costa Rica, and Panama. When these storms do make landfall they bring with them strong winds, raging surf, storm surge, and heavy rains. These of course often lead to flash flooding, landslides, mud slides, and other associated hazards. In the downgraded form of tropical storms, East Pacific hurricanes can lead to heavy rainfall and gusty winds in the southwestern states of California, Arizona, and New Mexico and as far east as Oklahoma and Texas.

Hurricane John in 1994 is the longest-lived Pacific hurricane on record.[14] Over 31 days it tracked across 4,000 miles (6,400 km) of eastern and central Pacific waters. On two different occasions John's winds reached the alarming speed of 170 miles (273 km) per hour, it crossed the international date line twice (becoming Typhoon John in the process), and thankfully skirted the Hawaiian Islands to the south and west.

Lest you think there is no relationship between Atlantic and Pacific hurricanes, a few storms have tracked from the Gulf of Mexico into the Pacific, and (more rarely) from the Pacific into the gulf. Hurricanes that cross the Central American landmass typically lose energy quickly and degrade to tropical storms. But on occasion, a weakened storm reentering the ocean can strengthen and gather energy if it encounters a warm pool of water. Hurricane Cosme left the eastern Pacific in 1989 to become Atlantic Tropical Storm Allison.[15] There was even a "hurricane with three names" in 1961 when an Atlantic storm crossed into the Pacific, regained hurricane strength in the East Pacific warm pool, and doubled back into the Gulf of Mexico again.[16]

EL NIÑO AND HURRICANES

Weather and climate in the Pacific is noted for having a strong relationship to a phenomenon involving coupling of the atmosphere and the ocean known as the El Niño Southern Oscillation (ENSO). Along the equator, the western Pacific has some of the world's warmest ocean water, but in the eastern Pacific cool water wells up from the deep ocean, carrying nutrients that support large fish populations. Every 2 to 7 years, the strong westward-blowing trade winds subside, and warm water from the western Pacific migrates into the eastern Pacific, like water shifting in a giant bathtub. This condition is known as El Niño. When the southern trade winds are stronger than normal the condition is known as La Niña. The El Niño and La Niña disruptions of the ocean-atmosphere system in the tropical Pacific have important consequences for weather around the globe; they result in redistribution of rains, thus causing flooding and droughts. ENSO can influence climate and weather in Hawaiʻi; the relationship of ENSO to our water resources is discussed in chapter 6.

El Niño, in addition to influencing episodes of major drought, plays an important role in the location of hurricane genesis in the Pacific. During the onset year of an El Niño, changes in the wind pattern of the central Pacific create a zone of disturbances between equatorial west-flowing winds and subtropical east-flowing winds. Under the right conditions, these disturbances can grow into hurricanes.

El Niño warm phases in the Southern Oscillation have corresponded to some of the largest annual storm counts in the central Pacific.[17] However, the relationship is not unique because 1978, which was not an El Niño year, still had as many storms as the El Niño years of 1972, 1982, and 1992. Also, 1977 was an El Niño year, and the central Pacific storm count was zero that year.

Fig. 5.6. Hurricane Flossie on August 14, 2007; Flossie passed approximately 100 miles (161 km) south of the Big Island. In a typical year, four to five tropical cyclones form or cross into the central Pacific, two of which on average reach hurricane intensity. (National Oceanic and Atmospheric Administration, 2008: http://www.noaanews.noaa. gov/stories2008/20080519_pacific.html [last accessed November 4, 2009])

Pao-Shin Chu, a University of Hawai'i meteorology professor, reported in a 1997 study of tropical cyclones that El Niño years produce more Pacific tropical storms and that with some exceptions their tracks are more erratic than in non–El Niño years.[18] If this is the case, then terms like "100-year storm" or "50-year storm" are misleading and may lull people into thinking 'Iniki or 'Iwa, the last two hurricanes to strike Hawai'i, warrant our safety for many years to come.

Unfortunately, more often than we like or plan for, the path of a hurricane clips or intersects the Islands. Kaua'i's position farther west than the other Hawaiian Islands is therefore more likely to lie in the path of a hurricane hooking to the north after passing south of the Big Island. This effect was seen in Hurricanes Dot (1959), 'Iwa (1982), and 'Iniki (1992), all of which passed over the Kaua'i and Ni'ihau Island zone. More recently, Hurricane Flossie, a category 4 storm when it began life in the East Pacific, passed south of the Big Island in a downgraded state in August 2007. The storm produced strong swell that hit exposed shores of all the Islands and heavy rainfall on the Big Island that triggered a flash-flood watch. In August 2009 Hurricane Felicia approached the Islands from the southeast. Downgraded to a tropical depression, and finally to a remnant low before it hit, it nonetheless produced copious amounts of rainfall across several islands.

GLOBAL WARMING AND HURRICANES

The hypothesis that global warming will spawn more hurricanes and that more of them will be big storms is widely discussed by popular media. Proving it with direct observations of hurricane history, however, is turning out to be vexing. Warmer air should lead to warmer water, which should provide more heat to both build and supercharge storms. The problem is that the history of instrumented observations recording hurricanes is only as old as the global satellite network, and this is not long enough to confirm an unequivocal link to global warming, especially with the sometimes confusing linkage between storminess and the irregularly timed ENSO.

For example, one study in 2005 concluded that hurricanes and typhoons have become stronger and longer-lasting over the past 30 years and that these changes correlate with a rise in sea-surface temperatures over the same time period.[19] Kerry Emanuel, a professor of atmospheric science at the Massachusetts Institute of Technology, reported that the duration and strength of hurricanes increased by about 50% in the past three decades. This observation revised existing models for measuring storm strength, which suggested that the intensity of hurricanes and typhoons should only increase by 5% for every 1°C (1.8°F) rise in sea-surface temperature. Emanuel said "We've had half a degree [Celsius] of warming, so that should have led to a 2.5% increase [in intensity], which is probably not detectable. What we've seen is somewhat bigger than that, and we don't really know why."[20] Emanuel speculated that ocean temperatures may be increasing more quickly than atmospheric temperatures. "When that happens we've shown theoretically you get an increase in the intensity of hurricanes," he said.

However, by April 2008, Emanuel was changing his stand. He unveiled a novel technique for predicting future hurricane activity, suggesting that even in a dramatically warming world hurricane frequency and intensity may not rise substantially during the next two centuries.[21] Nonetheless, the new work convinced Emanuel that global warming may still play a role in raising the intensity of hurricanes. What that role is, however, remains far from certain.

Another study by scientists at the Georgia Institute of Technology and the National Center for Atmospheric Research in Boulder, Colorado, suggested that the number of category 4 and 5 hurricanes has almost doubled globally over the past three decades.[22] These results indicate that a substantial increase has occurred in all the major tropical storm basins of the world, with the greatest increases in the North Pacific, Indian, and Southwest Pacific oceans. However, a key question for all studies of the relationship between hurricanes and global warming is the quality and degree of agreement in the storm intensity and frequency data being used, a subject of vigorous debate in the hurricane research community. Indeed the Georgia study has been criticized for drawing conclusions from a historical database that may suffer from sampling errors.

Is global warming changing the intensity or frequency of hurricanes? According to the 2007 Fourth Assessment Report of the Intergovernmental Panel on Climate Change (IPCC-AR4), it is more likely than not (better than even odds) that there is a human contribution to the observed trend of hurricane intensification since the 1970s. In the future, "it is likely [better than 2 to 1 odds] that future tropical cyclones (typhoons and hurricanes) will become more intense, with larger peak wind speeds and more heavy precipitation associated with ongoing increases of tropical sea surface temperatures."[23] According to the IPCC-AR4, on a global scale, "[t]here is no clear trend in the annual numbers [i.e., frequency] of tropical cyclones."[24] However, the frequency of tropical storms has increased dramatically in the North Atlantic and this tends to drive public

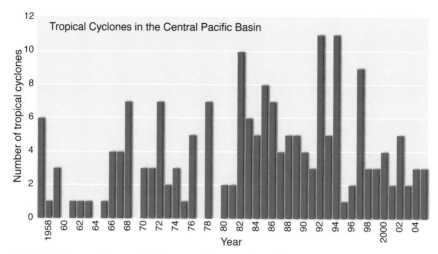

Fig. 5.7. History of tropical cyclones in the central Pacific. (SOEST Graphics, National Oceanographic and Atmospheric Administration, Central Pacific Hurricane Center, "Climatology of Tropical Cyclones in the Central Pacific Basin": http://www.prh.noaa. gov/cphc/pages/climatology.php [last accessed 5 November, 2009])

awareness of the potential role of global warming. Reasons for the North Atlantic increase are subject to intense debate among climate scientists. At least one peer-reviewed study indicates a significant statistical link between increasing intensity of the strongest storms and global warming,[25] but research to identify a mechanism explaining this link is ongoing. Notably, this study does not show a significant trend of growing storms in the Pacific, only in the Atlantic and the Indian oceans.

Relevant to this discussion is the actual number of tropical cyclones happening in recent years in the central North Pacific. In 2004, the total activity was slightly below normal, with three cyclones occurring compared to the annual average of four to five. One tropical cyclone developed within the central Pacific and the other two, Darby and Estelle, moved into the area from the eastern Pacific. Estelle was the strongest of the three systems and the only one of tropical storm intensity. There were no deaths recorded or property damage reported in the central North Pacific due to these three tropical cyclones.

With three systems in 2005, the total activity for the tropical cyclone season was below normal. One tropical cyclone developed within the central Pacific and the other two, Jova and Kenneth, moved into the area from the eastern Pacific. Jova was the strongest of the systems, maintaining category 3 strength for 36 hours, and the strongest tropical cyclone in the central Pacific since 'Ele in 2002. Although two storm remnants contributed to some locally heavy rainfall across Hawai'i, there were no deaths recorded or significant property damage reported in 2005.

In 2006, the central Pacific tropical cyclone season had near-normal activity despite the development of a relatively weak El Niño in September 2006. A total of five tropical cyclones occurred during the season, including two hurricanes. One of the hurricanes, Ioke, reached category 5 intensity and set several central Pacific tropical cyclone records. The other three systems only reached tropical depression intensity.

In 2007, Tropical Depression Cosme and Hurricane Flossie were tracked by the Central Pacific Hurricane Center in Honolulu. West-moving Cosme entered the region on July 19, but it passed well south of the Big Island, and with maximum winds of about 30 miles (48 km) per hour it caused little impact. A few weeks later, Hurricane Flossie entered the central Pacific on August 11. Flossie was an impressive hurricane with a distinct eye embedded within a solid eye wall. The hurricane maintained maximum sustained wind speeds of 130 to 140 miles (209 to 225 km) per hour, causing a hurricane watch to be issued for the Big Island early on the morning of August 13. Even though the center of Flossie passed about 100 miles (161 km) due south of the Big Island, it generated very large waves along the southeast-facing shoreline of that island. The height of the largest wave faces was estimated to be near 20 feet (6 m). In fact, coincident with the passage of Flossie, a 44-acre (18-ha) lava bench slipped into the ocean during the night on August 13 (see Figure 3.6). According to the U.S. Geological Survey (USGS), it is possible that this loss of shoreline was due to the large pounding surf generated by the hurricane, a 5.4-magnitude earthquake that occurred around the same time, or a combination of both events.

In 2008, only one storm materialized in the central Pacific. Tropical Depression Kika formed approximately 900 miles (1,448 km) to the southeast of the main Hawaiian Islands and caused no local impacts.

The 2009 Pacific hurricane season produced 3 tropical depressions, 12 tropical storms, and 8 hurricanes. In the central Pacific, tropical storm Lana was the first system to materialize; it passed south of the Big Island in early August and never gained hurricane strength. Hurricane Felicia approached the state from the southeast on August 8 but became disorganized and lost strength before crossing into state waters from the east. It released abundant rainfall among the Islands and produced localized flash flooding, especially in windward O'ahu. On August 16 Hurricane Guillermo entered the central Pacific as a category 1 hurricane but quickly weakened to a tropical storm due to very high wind shear. Ultimately it passed to the northeast of the Islands without local impact. Tropical storm Hilda traveled from east to west several hundred miles (km) south of the Big Island in late August, also without local impact. The last central Pacific storm of the 2009 season, Hurricane Neki, originated to the south of the Big Island on October 20 and took a northwesterly route whereby it avoided the main Hawaiian Islands. However, before it died completely, Neki passed through the Papahānaumokuākea Marine National Monument near French Frigate Shoals on October 23. It hit two small natural-habitat islands, Round and Disappearing islands. These were substantially affected; the former lost some land area due to erosion, and the latter was completely washed away. This made Neki the first tropical cyclone to directly impact the Hawaiian Islands since Hurricane ʻIniki in 1992.

Globally there is an average of about 90 tropical storms every year. Although the 2007 IPCC-AR4 concludes that there is no clear trend in the global number of tropical cyclones, in the North Atlantic there has been a clear increase in the frequency of tropical storms and major hurricanes. From 1850 to 1990, the long-term average number of tropical storms was about 10 annually, including about 5 hurricanes. For the period of 1998–2007, the annual average is about 15 tropical storms per year, including about 8 hurricanes. Some researchers want to tie this increase in frequency to the observed rise in North Atlantic sea-surface temperature and establish a causal relationship. For the North Atlantic, this may indeed be a valid correlation. However, despite a large number of hurricanes in the Atlantic region between 1995 and 2006,[26] the central Pacific region did not experience a high number of storms, and there has not been a long-term increase in hurricane frequency overall. Instead, there have been periods of high hurricane activity that last for several decades, followed by decades of low activity.[27] What is the last word on the connection between global warming and hurricanes? Leave it to the experts, and they haven't figured it out yet.

HURRICANE ʻINIKI: THE WAKE-UP CALL

Hurricane ʻIniki engulfed Kauaʻi on September 11, 1992, with 145 mile (233 km) per hour winds and gusts up to 175 miles (281 km) per hour.[28] The highest recorded wind speed from Hurricane ʻIniki was a reading of 227 miles (365 km) per hour from the Navy's Mākaha Ridge radar station. According to the *Honolulu Advertiser,* that remarkable figure was recorded at a digital weather station whose wind-gauging equipment blew off after taking the measurement during the storm. After crossing the island, 'Iniki rapidly accelerated north-northeastward, weakening as it went. 'Iniki's winds, waves, and flooding destroyed 1,421 buildings and damaged another 6,292, totaling 90% of all houses, buildings, and structures on the island. An additional 607 buildings on other islands were damaged or destroyed by the pounding surf.[29] 'Iniki caused around $2.4 billion in damage and six deaths. At the time, 'Iniki was among the costliest U.S. hurricanes, and it remains one of the costliest hurricanes on record in the Pacific. This level of damage proved to be crushing to the economy of Kaua'i, which was not seen to recover for another decade.

The hurricane's devastating waves penetrated well beyond the shoreline, smashing buildings, ripping up roads, tearing down power lines, filling swimming pools with sand so that they disappeared from view, and spreading boulders, gravel, and debris across properties and roads. Survey parties measured damage higher than 15 feet (4.5 m) above

Fig. 5.8. The eye of Hurricane 'Iniki passed directly over the island of Kaua'i on September 11, 1992, at high tide, as a category 4 hurricane with sustained winds of 145 miles (233 km) per hour. (National Oceanographic and Atmospheric Administration)

sea level on interior walls of demolished homes, trees, and among the vegetation in open fields. At eye level, the trunks of coconut trees were bruised, smashed, and had been laid open to their core by cobbles and coral heads suspended in the swirling storm surge and heaving waves. However, the most severe cause of property damage from 'Iniki was its turbulent wind.

The wind sucked roofs and walls off houses, leaving the interiors exposed to torrential rains and contributing to the windborne debris hurtling into adjacent buildings. Many houses were uprooted and pulled from their footings by the strong winds or floodwaters. In some cases, a house blew to shreds, and the next-door neighbor survived with minor damage—the product of proper engineering and careful construction. In some cases, the margin between total structural failure and only moderate damage was determined by whether a single line of roofing nails was well aimed and found the underlying joist or missed and failed in its job of anchoring a plywood roofing sheet to the frame of the house.[30]

'Iniki was the most costly natural disaster in Hawai'i's recorded history. At $2.4 billion in losses, it ranks as one of the top 10 costliest hurricanes in U.S. history.[31] Thousands of people were left homeless, 100 injured, and many lives were disrupted for years after, as Kaua'i struggled to recover. For a decade following the event, Kaua'i remained in the grip of a recession as tourism shifted to the less-damaged islands.[32] 'Iniki was neither a particularly large hurricane nor unusual in terms of frequency of occurrence, yet its impact was still devastating over an extended period of time. The shock and destruction are indelibly printed in the memory of those who experienced the storm and suffered through its aftermath.

HOW MUCH DAMAGE CAN A HURRICANE CAUSE?[33]

If a category 1 storm as strong as Hurricane 'Iwa, with sustained winds at 74 miles (119 km) per hour, strikes any of the Islands in the state, we can estimate from past experience that about 12% of the houses and apartments will be destroyed or heavily damaged and about 18% will experience minor damage.

If a category 4 storm strikes any island with the force of 'Iniki, with winds raging at over 130 miles (209 km) per hour we can estimate that about 38% of the homes will be heavily damaged or destroyed. An additional 40% will probably have minor damage.The information in the accompanying table was extrapolated from Kaua'i damage in 1982 and 1992.

TABLE 5.3   POTENTIAL HURRICANE DAMAGE IN HAWAI'I
(BILLIONS OF $)

Storm Type	O'ahu	Maui	Hawai'i	Kaua'i
'Iwa-strength storm	$4.5–7.5	$0.8–1.4	$0.8–1.4	$0.3–0.6
'Iniki-strength storm	$13.9–23.3	$2.7–4.5	$2.6–4.4	$2.1–2.9

WHAT DETERMINES THE LEVEL OF DAMAGE DURING A HURRICANE?

Although warning systems and evacuation procedures have improved over the years, the best way to mitigate damage is to build in areas less susceptible to high winds and floodwaters and their associated hazards, and to upgrade building codes as new construction materials and techniques become available. The single most important step to mitigate hurricane damage is to exercise avoidance; that is, to build away from the coast, ridgelines, and exposed hillsides; to build homes with continuous load paths from the foundation to the uppermost components of the roof; and to minimize external features such as eaves, balconies, and other structural components that are easily ripped off by high winds. These breach the envelope of a building and render it vulnerable to further destruction by the high gusts associated with winds near the eye wall of the storm as it passes across an island.

The greatest building damage during a hurricane occurs in the area of highest winds and strongest storm surge. However, because the track of any hurricane may vary in a number of ways, it is not always obvious where highest winds and storm surge will occur. The damage a building may experience depends on several factors: the track of the storm, position to the left or right of the eye, exposure to the open ocean and free or accelerated airflow, the topography of the surrounding landscape, and the construction and design of the building.

There is no way of knowing the chances of a hurricane damaging a particular home. We can only go on past history. Almost no one expected a hurricane as powerful as 'Iniki to strike Hawai'i. The best we can guess is that hurricanes in the future will probably hit Hawai'i as frequently as they have in the past. Unfortunately, this is like driving down the highway looking in the rearview mirror, not the most desirable way to predict what lies ahead—especially since the conditions of Pacific climatology may be changing with global warming.[34] In addition, increased development, increased alteration of watersheds, and increased loss of beaches add up to the likelihood of increased property losses in future storms.

Builders emphasize two important concepts in hurricane preparedness for your house.

- Build a continuous load path connection that ties your roof to your house foundation. If you are building a new house, this is something you want to ask your builder to ensure, and if you want to retrofit your home, a competent builder can perform this for you. A continuous load path will prevent the roof from blowing off your house and causing catastrophic damage.
- Create a rain- and wind-resistant envelope. This is done by strengthening your windows and doors to prevent a breach in the envelope of your house. Windows especially need your attention. One shattered window can be the first step to the failure of the whole house in hurricane-strength winds. You need shutters (there are several types: colonial shutters, Bahama shutters, and others), impact-resistant glass, and strengthened doors that will not blow in.

STAY OUT OF HARM'S WAY

Recommendations[35] on what to do before, during, and after a hurricane:

Stay or leave?

- When a hurricane threatens your area, you will have to decide whether you should evacuate or whether you can ride out the storm in safety at home. If local authorities recommend evacuation, you should leave!
- If you live on the coastline, on an offshore island, near a stream/ river, in a floodplain, in a high-rise, or in a mobile home, plan to leave.
- If you live in a sturdy structure and away from coastal and rainfall flooding, consider staying.

At the beginning of hurricane season:

- Learn the elevation and the flooding and wind damage history of your area.
- Learn safe routes inland if you live near the coast.
- Learn the location of official shelters.
- Determine where to move your boat in an emergency.

- Trim back dead wood from trees.
- Check for loose rain gutters and downspouts.
- If shutters do not protect windows, stock tape and boards to cover glass.

When a hurricane watch is issued:

- Check often for official bulletins on radio, television, or National Oceanic and Atmospheric Administration (NOAA) Weather Radio.
- Fuel your car.
- Moor small craft or move them to safe shelter.
- Stock up on canned provisions.
- Check your supplies of special medicines and drugs.
- Check batteries for radio and flashlights.
- Secure lawn furniture and other loose material outdoors.
- Tape, board, or shutter windows to prevent shattering.
- Wedge sliding glass doors to prevent their lifting from their tracks.

When a hurricane warning is issued:

- Stay tuned to radio, TV, or NOAA Weather Radio for official bulletins.
- Stay home if your home is sturdy and away from the coast.
- Board up garage and porch doors.
- Move valuables to upper floors.
- Bring in pets.
- Fill containers (bathtub) with several days' supply of drinking water.
- Turn up refrigerator to maximum cold and do not open unless necessary.
- Use phone only for emergencies.
- Stay indoors on the downwind side of the house, away from windows.
- Beware of the eye of the hurricane.
- Leave areas that might be affected by storm tide or stream flooding.
- Leave early, in daylight if possible.
- Shut off water and electricity at main stations.
- Take small valuables and papers, but travel light.
- Leave food and water for pets (shelters will not take them).
- Lock up the house.
- Drive carefully to nearest designated shelters using recommended evacuation routes.

After the all clear is given:

- Drive carefully; watch for dangling electrical wires, undermined roads, flooded low spots.
- Do not sightsee.
- Report broken or damaged water, sewer, and electrical lines.
- Use caution when reentering your home. Check for gas leaks. Check for food and water spoilage.

MITIGATION ACTIVITIES[36] AT HOME

Homes can be destroyed by high winds. Flying debris can break windows and doors, allowing high winds and rain into your house. High winds can also cause weak places in your home to fail. Strengthen these areas in your house: roofs, exterior doors, windows, and garage doors. Other mitigation activities include clearing debris from the area and building a safe room in your home.

Roof

Gable roofs need additional truss bracing to make your roof system stronger. Truss bracing consists of 2 by 4s running the length of your roof, and gable-end bracing consists of 2 by 4s in an X pattern. Hurricane straps and clips can also keep your roof attached to your walls and are very inexpensive. For a 1,200-square-foot (111-m²) house, hurricane clips might cost between $400 and $600. Learn more about protective measures from your local home improvement store.

Exterior doors

Most double-entry doors have an active and inactive door. Check to see if the fixed door is secure enough. Some door manufacturers provide reinforcing bolt kits, or you can buy and install door-bolt materials. Check with your local home improvement store.

Windows

There are many types of manufactured storm shutters, or you can make your own plywood shutters. Plywood shutters should be made of 5/8 inch (1.9-cm) exterior plywood and installed using bolts and masonry anchors. Remember to cover all exposed windows and glass, even French doors and skylights. Hurricane shutters cost between $1.67 and $50.00 per square foot (1 square foot = 0.009 m²) for motorized roll-up shutters. Check with your local home improvement store.

Garage doors

Two-car garage doors pose a problem because they wobble in high winds and can blow out of their tracks and collapse. Some garage doors can be strengthened with retrofit kits. Installing horizontal bracing on each can reinforce some garage doors. Backing a car up against the inside of the garage door in the event of a hurricane can also help resist strong winds. Check with your local home improvement store.

Clear debris

Clear debris away from structures, so that they do not become airborne missiles. Check with local officials for additional information about tree pruning and clearing.

Safe rooms

There are several options for building a safe room. A safe room made of concrete would have 8-inch (20-cm) mortar-filled tile walls and a 4-inch (10-cm) concrete ceiling, all supported with iron reinforcing rod. A wood-framed safe room would have doubled 2-by-4 studs, set 16 inches (40.6 cm) on center, faced on one side with 3/4-inch (1.9-cm) plywood and on the other with 12-gauge steel, and sheathed inside and out with 1/2-inch (1.3-cm) gypsum board. Go to your local library for additional information.

CHAPTER 6

Climate and Water Resources

Hahai nō ka ua i ka ulu lāʻau

"The rain follows after the forest"

" Born of the heavens, embraced by Mother Nature, entrusted to us to preserve for future generations, water for life—*Ka Wai Ola*." This is the opening message on the Web site of the City and County of Honolulu, Board of Water Supply.[1] *Wai* (water) is the lifeblood of the Hawaiian people. Woven into their existence and etched into daily life, this precious resource enabled ancient culture to thrive. Without fresh water life would

Fig. 6.1. Rainfall among the peaks of our high volcanic islands is an important source of water. (Photo by C. Fletcher)

simply not exist; it is essential to all living beings and becomes ever more critical on land surrounded by a salt-saturated sea.

Water's natural character and flow has historically been widely altered for our use. Beginning with the first Hawaiians and amplified 150 years ago by industrial agriculture, engineering projects diverted the natural flow of *wai* for irrigation projects. The greatest changes were made for the irrigation of sugarcane crops, but pineapple and others were also thirsty dependents. Some irrigation projects took water directly from stream channels, lowering their flow. Downstream of numerous fields, many streams simply dried up. In other cases, groundwater embedded in volcanic rock among the high ridges was tapped, impacting the natural environments fed by this source. This so-called "dike water" (named because groundwater was trapped in the mountain interior behind "dikes" [dense volcanic rock that is impermeable]) nourishes streams and wetlands on windward slopes. As dike water was drained for agriculture, these streams, wetlands, and estuaries dried up, and the taro fields that fed local communities were permanently affected. The estuaries where these streams enter the ocean have become shrunken slivers of their former selves, degrading the delivery of nutrients and the ecological health of reef communities. In arid regions, wells pumped groundwater for irrigation and lowered the water table. All these projects did, however, spread diverted water across porous fields, thus allowing at least some of it to return to the ground.

Today, a building and tourism boom is replacing the vast sugar plantations of yesterday. Thousands of acres (ha) of impervious concrete surface have paved over the soil. Runoff, that previously recharged the water table and was cleansed by infiltration, now flows directly into the ocean. In two significant ways this extraordinary change to *wai* increases the stress on our major water sources: (1) less water is available to recharge our groundwater stores, and (2) groundwater is being more heavily used. Groundwater is a vital resource for all Hawai'i communities. On the island of O'ahu, for instance, groundwater provides more than 90% of the drinking water and about half of the water used for modern irrigation. Adding to the water stress is a long-term decrease in rainfall,[2] and persistent pumping at a rate that exceeds full, natural recharge. Water tables are falling around the state. Meanwhile, millions of gallons of clear, fresh, no longer used irrigation water continue to be wasted. Islandwide calls for restricted water use are broadcast, and we endure yet another multiyear drought. Hawai'i does not have its water-house in order.

The need for more potable water increases yearly. In 1990, a resident population of 1,108,229 used nearly 136 million gallons (514 million liters [mld]) per day (mgd) or 123 gallons (465 liters) per person per day. By 2015, population projections show an increase of the resident population, and the Hawai'i Department of Health anticipates that 168 mgd (636 mld) will be needed if the 1990 per-person consumption does not change. Population growth with commensurate increases in water demand continually challenges the ability of groundwater to sustain Hawai'i's needs. The Commission on Water Resource Management[3] has stated that we are approaching the limits to developing our water resources in some parts of the state. For example, on O'ahu, it is estimated that about 415 mgd (1,571 mld) of groundwater is available where water allocations (of all types, potable and other uses) total about 340 mgd (1,287 mld). Within 25 years the unallocated amount of 75 mgd (284 mld) will be committed. But this does not take into account the potential for reduced rainfall that is already occurring and may worsen as global warming increases in the Hawaiian Islands. Other problems facing the management of groundwater supplies include inadequate data to estimate water availability, overestimation of water supplies, and inadequate planning to meet increasing water demands.

Although some former irrigation water is used in our homes and resorts, it is eventually returned to the ocean via sewage-treatment plants. Hence, it does not replenish our aquifers. Water tables around the state are dropping for many reasons, but among these is that in many places no one has changed the diversion system now that sugarcane has left the Islands.

The state of Hawai'i needs a management plan that comprehensively evolves us out of the days of sugar and other industrial crops, and into the days of sustainable development and climate change. It is reckless to allow millions of gallons of diverted fresh water to pour unused into the ocean every day while West Moloka'i, Kona, Honolulu, central Maui, windward Kaua'i, and other locations suffer water shortages. In an environment that includes some of the wettest places on Earth, to suffer such shortages should act as a wake-up call for improving our water management skills.

WAIWAI

In the Hawaiian language, *waiwai* signifies an abundance of fresh water and embodies the meaning of wealth. Though blessed with annual rainfalls much greater than in many parts of the world, Hawai'i nonetheless experiences localized water shortage problems. These are partly the product of inadequate water resources where we have chosen to live and partly the product of inadequate water distribution and management.

With water, as with most limited essential commodities in contemporary life, we attach rights, rules, and regulations to govern its fair distribution. The current management approach follows a fundamentally different design than the one established by early Hawaiians.[4] Today's model is one of engineering strength, where natural boundaries are overridden when inconvenient and water is regarded as an abundant resource to be moved where we see fit. The consequences of this system are frequently felt not only by humans but also by a range of organisms from bacteria to gobies and the downstream plants and animals impacted by water withdrawal.

The single most important factor contributing to the success of the Polynesian settlers was an abundance of pure water, and unfortunately, through our current negligence, we put at risk the very resource that enabled life on the Hawaiian Islands to flourish. By unfolding the Native Hawaiians' relationship to water and the process through which they allocated this precious resource, we may discover lessons to better manage today's supply.

For ancient Polynesians who arrived on Hawaiian shores, water was freely available to everyone and was divided among the people in a spirit of mutual dependence, the single stipulation being that because everyone used the water, everyone must help build and maintain the Islands' expansive irrigation systems. The word *kānāwai*, literally "belonging to the waters," is the Hawaiian word for water rights. Water, like land, did not belong to any individual, and with its use came the responsibility of care and respect. The *ali'i nui* (high chief) administered *wai* along with the *konohiki* (headman of the *ahupua'a*).[5]

Ancient residents paid close attention to the shifting moods of the climate, developing an eloquent vocabulary to describe its countless nuances. Rain was generically termed *ua,* but poetic descriptions depict the intricacies of rainfall. *'Awa* falls as cold mountain mist, a continuous rain is *ua ho'okina,* a downpour *ua lanipili,* and *uakoko* is a "blood rain," one that falls so hard the streams run red with eroded dirt.[6] The Hawaiians studied the detailed movements of *ua,* selecting their dwellings and communities based on its presence or absence. Through their resourcefulness and deep connection to the land they managed to live in Hawai'i's more challenging climates, including the desert regions.

Troughs made from the wood of the 'ōhi'a, koa, and *kukui* trees were carved to collect water inside blackened lava tubes, and milk from coconuts was extracted to sustain life during periods of drought. This thoughtful attention to the behavior of water and the seasonal shifts that molded their daily activities led early Hawaiians to recognize two broad seasons: *kau wela,* the high sun period corresponding to warmth and steady trade winds, and *ho'oilo,* the cooler period when trades are less frequent and rainy-stormy days may be more common. We now know that weather in Hawai'i follows a more complex pattern than originally contemplated, but that fundamentally these two seasons still hold true today.

CLIMATOLOGY OF *WAI*

In a study of stream flow characteristics at long-term gauging stations in Hawai'i,[7] the U.S. Geological Survey reported that year-to-year changes in stream flow are related to the El Niño–Southern Oscillation (ENSO) phenomenon as well as the Pacific Decadal Oscillation (PDO). It is not surprising that trends in rainfall are also linked to these phenomena; Hawai'i tends to be dry during most El Niño events, but low rainfall may also occur in the absence of El Niño.[8] Similarly, rainfall tends to be low during "warm" phases of the PDO; such as Hawai'i experienced from mid-1970 to the start of the twenty-first century. The state was previously characterized by high rainfall lasting for nearly three decades in the preceding "cool" phase of the PDO through the late 1940s to mid-1970s. ENSO and PDO are regional climate patterns that affect the distribution of rainfall in Hawai'i and therefore exert important influence on the discharge of streams and the recharge of groundwater stores from one year to the next.

ENSO is a large-scale meteorological pattern characterized by two conditions: the so-called La Niña and the El Niño.[9] These govern temperature and rainfall trends in the Pacific Ocean as well as exert a global influence on weather patterns. ENSO is related to the atmospheric pressure difference between a body of dry air (a high-pressure system) located in the Southeast Pacific over Easter Island and a body of wet air (a low-pressure system) located over Indonesia in the Southwest Pacific. Under normal conditions in the Southern Hemisphere air flows from the high pressure to the low pressure and creates the trade winds. These blow east to west across the surface of the Pacific and drive a warm surface current of water into the western Pacific. This water is replaced in the east by nutrient-rich cold ocean water that rises from the deep sea, a process called upwelling. The upwelling current is loaded with nutrients, fueling an important fishing industry off the coast of South America.

On occasion, the pressure difference between the two centers decreases and the trade winds (largely in the Southern Hemisphere)

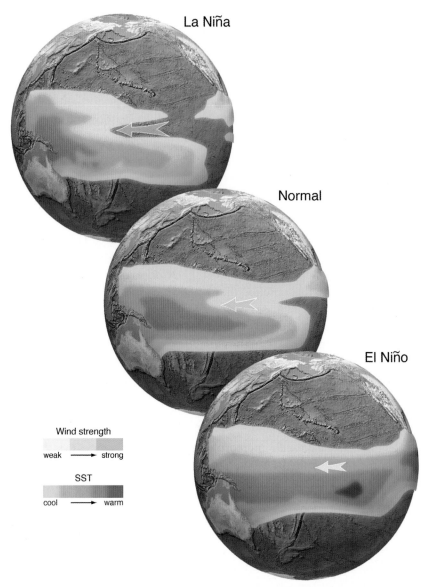

Fig. 6.2. The El Niño–Southern Oscillation is a large-scale meteorological pattern characterized by two conditions: the La Niña and the El Niño. These govern temperature and rainfall trends in the Pacific Ocean as well as exert a global influence on weather patterns. Winds blowing to the west are weak during El Niño, and sea-surface temperatures rise in the central and eastern Pacific. (SOEST Graphics, Matthew Barbee, Perspective Cartographic LLC, temperature patterns modified from: http://www.cpc.noaa.gov/products/analysis_monitoring/ensocycle/ensocycle.shtml [last accessed November 5, 2009])

respond by weakening. This condition is known as El Niño. As a result, the warm water of the West Pacific surges to the east and heats up the ocean surface in the central and eastern Pacific. Precipitation in the east increases because the warmer water evaporates more readily. Upwelling temporarily stops. Torrential rains and damaging floods across the southern United States have resulted, as well as faltering of the Peruvian fishing industry, wreaking economic hardship.

During an El Niño, the Hawaiian Islands usually experience a decrease in rainfall. In fact, the 10 driest years on record are all associated with El Niño years. This happens because El Niño causes a southerly shift in the atmospheric circulation system of the North Pacific, a feature called the Hadley Cell.[10] The Hadley Cell is a large continuous belt of air that rises, moisture-laden, from the warm waters north of the equator at about 8° latitude (Hawai'i lies between 16° and 23° N latitude) and moves north across the Hawaiian Islands, raining as it goes. During its journey to the north the air cools and moisture condenses, which produces abundant rainfall. Eventually airflow descends back to the ocean surface as a column of dry, cool air and creates a pressure system known as the Pacific High to the north of the Islands at around 30° to 35° N latitude.[11] Under normal conditions, the Hawaiian Islands experience a wet climate, but to the north and northeast the Pacific High creates a dry climate. However, during El Niño the surface waters at the equator become significantly

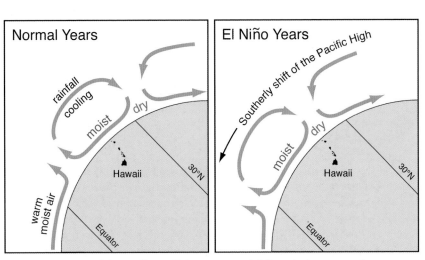

Fig. 6.3. Location of the Hadley Cell during normal (left) and El Niño conditions (right). (SOEST Graphics)

warmer, and the rising motion of the Hadley Cell shifts to the south. This brings the Pacific High south as well, and the Hawaiian Islands experience a decrease in rainfall as we fall under the influence of the dry high-pressure center.

The other important phenomenon that impacts rainfall in Hawai'i is the Pacific Decadal Oscillation.[12] The PDO, like ENSO, is a pattern of climate variability in the Pacific caused by shifting surface waters. During a "warm" or "positive" phase of the PDO, the western Pacific above 20° N latitude cools and part of the eastern Pacific warms; during a "cool" or "negative" phase, the reverse pattern occurs. The underlying cause of this oscillation has not yet been identified by scientists, but it is recognized that shifts in phase occur approximately every 20 to 30 years.

Researchers have reconstructed that the PDO displayed a strong warm phase beginning ca. 1905, changed to a cool phase around 1946, and reverted back to a warm phase in about 1977.[13] Beginning in 1998 the PDO showed several years of cool behavior but did not remain consistently in that pattern. A cool phase is characterized by a wedge of lower

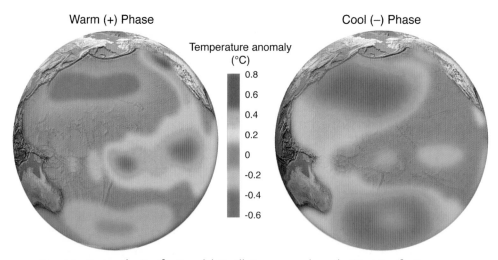

Fig. 6.4. During the Pacific Decadal Oscillation warm phase the West Pacific Ocean becomes cool and the tropical eastern Pacific warms. The cool phase is characterized by a wedge of cool ocean temperatures in the eastern equatorial Pacific and a warm horseshoe pattern of ocean temperatures connecting the Northwest, West, and Southwest Pacific. (SOEST Graphics, Matthew Barbee, Perspective Cartographic LLC, modified from Joint Institute for the Study of the Atmosphere and Ocean, University of Washington (PDO Web page): http://jisao.washington.edu/ [last accessed on June 9, 2009])

than normal sea-surface heights and cooler than normal ocean temperatures in the eastern equatorial Pacific and a warm horseshoe pattern of
higher than normal sea-surface heights connecting the Northwest, West,
and Southwest Pacific. Hawai'i tends to receive more rainfall in a cool
phase. In the warm phase, which appears to have lasted from roughly
1977 to 2000, the West Pacific Ocean becomes cool and the wedge in the
east warms. Hawai'i tends to receive less rainfall in a warm phase.

The PDO can modulate aspects of the ENSO pattern. When the
PDO is in a cool phase, surface-layer water temperatures are colder than
normal over a large area of the Pacific and La Niña events predominate
in Hawai'i, usually leading to more rainfall in the Islands. The reverse is
true for the warm phase (i.e., El Niño events tend to predominate and
drought may persist in the Islands).

The 2004 U.S. Geological Survey study of stream flow in the Hawaiian Islands found that during the months of January to March the average
total stream discharge tends to be low under El Niño conditions and high
under La Niña conditions.[14] This pattern is accentuated during positive
or warm phases of the PDO.

WINDS OVER HAWAI'I

Winds have always been an important part of the Hawaiian climate. There
are hundreds of names for localized winds in Hawai'i that are formed by
the interplay of the topography, the orientation to the ocean, and their
relationship to rain and the weather. Historically, the names of winds
were as familiar to Hawaiian communities as the name of the local beach
or mountain peak. Winds reveal current and impending characteristics of
the weather. They play a major role in the energy future of Hawai'i, driving the shift from carbon fuels to renewable energy. Winds, additionally,
are the primary source of our fresh water. They are, in many ways, the
most important climate element in Hawai'i.

The Hawaiian Islands are located in the subtropics, embedded in
a persistent band of winds known as the trade winds that were the
chief power behind traditional Polynesian and Micronesian navigation.
The trade wind, called Moa'e or A'eloa, is generated by the Pacific
high-pressure system generally located to the north or northeast of
the Islands.[15] Hawai'i, situated toward the bottom edge of this system,
experiences these clockwise winds where they blow easterly. Not only
did trade winds fill the sails of ancient navigators, but in the search
for new islands by early Micronesian and Polynesian explorers, when
trade winds temporarily died it was time to explore upwind. This was

a safe strategy because the trade winds would eventually return you to home.

On average each year, winds blow from each of the southeast, southwest, and northwest compass quadrants 10% of the time and from the northeast 70% of the time. These persistent northeast winds are the most common winds over Hawaiian waters. They became known as "trade winds" centuries ago when trade ships carrying cargo depended on the broad belt of easterly winds encircling the globe in the subtropics for fast passage.[16] In the summer, trade winds prevail more than 90% of the time, often persisting throughout the period of May, June, and July, and only weakening in the heat of August and September when their absence turns the weather in Hawai'i uncomfortably muggy. In the winter, November through March, trade winds may blow only 40% to 60% of the time.

Kona winds, named after the district on the leeward side of the Big Island, is a meteorological term for the stormy, rain-bearing winds that blow over the Islands from the southwest. When Kona winds blow, the western or leeward sides of the Islands then become windward, as the predominant wind pattern is reversed. Kona winds occur when a low-pressure center is within about 500 miles (800 km) northwest of Hawai'i and has an unusually low central pressure, below about 1 atmosphere (1,000 millibars). Strong Kona winds may uproot trees, strip the shingles off roofs or the roofs off houses, but they usually don't last for more than a day. When reinforced by mountainous topography, downslope winds can gust over 100 miles (160 km) per hour, causing destruction in low-lying areas. On O'ahu, the Schofield area below Kolekole Pass in the Wai'anae Range and the Kāne'ohe-Kahalu'u area below the Ko'olau Mountains have experienced extensive wind damage due to strong Kona winds.

Thunderstorm winds are accompanied by lightning and thunder, and may additionally bring heavy rains, strong downdrafts, hail, and waterspouts. Thunderstorms are uncommon in Hawai'i and are usually associated with cold fronts, Kona low-pressure systems, or upper-level low-pressure systems (known as "upper-level troughs") between October and April. During the summer, tropical cyclones may have thunderstorms embedded in their westward-moving cloud bands. The frequency of "thunderstorm days" in Hawai'i is low. The monthly average is 1 to 3 days during the winter, and 0 to 1 day during the rest of the year. Thunderstorms are most hazardous on the ocean. Strong downdrafts may gust up to 70 miles (112 km) per hour and are capable of knocking down sailboats, swamping motorboats, and wreaking havoc among canoes and other small watercraft.

Waterspouts and tornados are the same: small-diameter localized storms formed by winds rotating at very high speeds, usually in a counterclockwise direction. The funnel is called a waterspout if it is over water and a tornado if over land. When a waterspout moves from water to land, it becomes a tornado. These severe storms of short duration are more common in Hawai'i than many people realize. Waterspouts occur mainly in the winter months, with about 30–40 sighted annually in Hawaiian waters, but this is probably a small fraction of the true number of occurrences.

Hawaiian funnels do not usually pack the devastating wind of midwestern tornadoes. However, the few that have been studied in Hawai'i had destructive wind speeds over 100 miles (160 km) per hour. The path of destruction is usually narrow, 50 to 300 feet (15 to 90 m) wide and less than a mile (1.6 km) long. However, its speed of travel can exceed 30 miles (48 km) per hour (faster than many boats). Waterspouts have

Fig. 6.5. In Hawai'i dozens of waterspouts are spotted each year. These are small-scale tornados. (Photo by C. Heliker, U.S. Geological Survey)

been known to pick up a six-person canoe (and paddlers) and toss it like a matchstick. These are dangerous phenomena that may cause significant damage if they run into developed areas. Even though most waterspouts in Hawai'i spin away harmlessly over the ocean without causing trouble, every waterspout should be treated with extreme caution.

Hurricanes are an important wind event in Hawai'i, and they have been given their own chapter in this book (chapter 5) because they warrant your full attention. Hurricanes don't strike Hawai'i often. Most of them weaken before reaching Hawai'i or pass harmlessly to the west and south of the Islands. However, strong winds and rain are always a potential threat from these storms, which can occur from June through November. Additional information on hurricanes can be found by visiting the Pacific Disaster Center home page.[17]

Clouds, harbingers of *WAI*

Men and women who spend their time on the sea have made a fine art of interpreting clouds to foretell the weather, and Polynesian navigators are chief among them. This account is from A. Y. Varela, who visited Tahiti in 1774.[18]

> What took me most in two Indians whom I carried from Otahiti
> to Oriayatea was that every evening or night, they told me,
> or prognosticated, the weather we should experience on the
> following day, as to wind, calms, rainfall, sunshine, sea, and
> other points, about which they never turned out to be wrong: a
> foreknowledge worthy to be envied, for, in spite of all that our
> navigators and cosmographers have observed and written about
> the subject, they have not mastered this accomplishment.

Meteorologists classify clouds by their height and shape: high clouds appear above 20,000 feet (6,100 m), middle clouds are found between 6,500 and 20,000 feet (2,000 to 6,100 m), and low clouds form near the ground up to 6,500 feet (2,000 m). The names of clouds are not commonly known to most of us, but they form the basis for recognizing these phenomena. The word cumulus means to "heap up," and it describes a dense, sharply outlined cloud with a mostly vertical configuration, like a rising dome or tower. The upper part of a cumulus cloud often looks like a cauliflower. A stratus (meaning "spread") cloud has a more horizontal configuration. One or the other of these two terms is found as the root in most cloud names.

Nimbus means "rain producing," and a nimbostratus is a low, rain-bearing cloud that tends to be broad and flat. A cumulonimbus is a rain-producing, middle to high cloud that towers several thousand feet (about 1,000 m) in vertical length. The word alto means "high" and is used to refer to middle or high clouds. Hence there are altocumulus and altostratus clouds. The roots cirri, cirro, or cirrus ("curl of hair") refers to high, wispy clouds that may be stratus (cirrostratus), cumulus (cirrocumulus), or very wispy (cirrus) in their shape.

In 1983 retired Rear Admiral W. J. Kotsch wrote a comprehensive weather guide for mariners that interprets cloud meanings.[19] Among high clouds, scattered cirrus that are not increasing indicate that the weather is stable. Cirrus in thick patches, associated with the tops of thunderstorms, means that rain showers are close by. Cirrus shaped like hooks or commas indicate that warm weather is approaching and that a continuous rain will follow, especially if the cirrus is followed by cirrostratus. Cirrostratus in a continuous sheet (and increasing) signify approaching warm weather

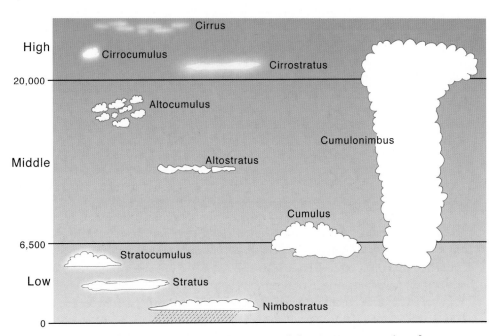

Fig. 6.6. Clouds are classified by their elevation and shape. (SOEST Graphics, from W. P. Crawford, 1992, Mariner's Weather, Norton, New York; illustration modified from Polynesian Voyaging Society Web site: http://pvs.kcc.hawaii.edu/navigate/winds.html [last accessed November 5, 2009])

with rain and stormy conditions. If these are not increasing and are not continuous, the storm is passing to the south.

Altostratus (middle clouds) can be the most reliable weather indicators of all the clouds. They signify warm air flowing above colder air, signaling that continuous rain is imminent, especially if the cloud layer progresses and thickens. These indicate the development of a new storm at sea with poor visibility, large waves, and heavy swells. When altocumulus are followed by thicker high clouds or cumulus low clouds they designate an advancing warm front with steady rain. Altocumulus in the form of turrets rising from a common flat base are typically the forerunner of heavy showers or thunderstorms.

Among low clouds, nimbostratus are rain clouds; when they are overhead the bad weather is already at hand. If they are at some distance and heading your way, you can expect persistent high winds, rain, and hazardous sea conditions. Stratus clouds do not signify much danger. In true stratus clouds there is little or no vertical circulation, hence moisture is not actively condensing and only a fine drizzle is expected. Stratocumulus look like altocumulus clouds but at a lower level. These produce light rain and are usually followed by clearing at night and fair weather. Isolated, puffy cumulus clouds indicate that the weather is fine, and nothing hazardous is in the offing. However, when cumulus swell vertically, heavy showers are likely, associated with gusty surface winds. Massive and forbidding, cumulonimbus look like a huge tower with a flattened top. They are thunderheads, and there will be gusty winds and heavy rain. It is these clouds that can bring lightning, hail, and occasionally a tornado or waterspout.

AHUPUA'A REVITALIZED

The concept of *ahupua'a* evokes a sacred image imbued with the principles of stewardship, cooperation, and respect. Unrecognized by many, it also evokes a system of management based on scientific principles grounded in observations of natural processes. *Ahupua'a* concepts embody the approach favored by citizen action groups[20] and are beginning to guide government management of natural resources. For example, the revised Hawai'i Ocean Resources Management Plan, the official state effort to unify and coordinate coastal and ocean-related activities and missions among government agencies, is guided by *ahupua'a* principles.[21] The updated plan replaces the earlier "sector-based approach" (meaning that it dealt with problems one by one based on the categories they fell into, rather than integrating the categories to manage the origin of the problems) with an integrated, place-based approach to management of ocean resources. The plan is framed around three perspectives: connecting land and sea, preserving our ocean heritage, and promoting collaboration and stewardship. Each of the framing perspectives is accompanied by concrete management goals and strategic actions to address them in 5-year implementation phases over the next 30 years.

As described in chapter 2, the basic definition of the *ahupua'a* is a land division that stretches from the mountain ridges to the deep ocean, tracking the movement of water within the confines of the topography. The term arises from the practice of marking the coastline of each district with an altar (*ahu*) on which a sculptured wooden head of a pig (*pua'a*) is offered to the *ali'i* as a ceremonial gift. Each *ahupua'a* was designed to take into account the ecological characteristics of its specific location, and boundaries were set to ensure subsistence shares in food, fishing rights, firewood, housing timbers, and water. People living in one *ahupua'a* were free to use whatever grew wild within their own borders, except for the revered portions high in the watershed. They were forbidden to take from a neighboring *ahupua'a* without permission, but trade between *ahupua'a* was widespread.

Ahupua'a were also political subdivisions. Individuals were viewed as part of their unit and subject to taxation through tributes to the *ali'i* in the form of *kapa* woven mats, food, clothing, and other products. Health and orderly governance was assured through a rigid regulatory system that assigned duties, roles, and consequences for disobedience. For example, an unauthorized diversion of *wai* was punishable by death.[22] Political boundaries coincided with agricultural and natural resource boundaries, affording immense power to the chiefs. Still, no person claimed ownership of the land. Instead, the Native Hawaiians embraced the notion of territorial custody.

During the reign of the *ahupua'a* management system, Hawaiian agriculture ranked among the most productive in Polynesia, mostly due to meticulous observations of how the flora and fauna responded to various weather conditions. A carefully conceived zoning arrangement demonstrates the truly integrated Hawaiian system of beliefs. The *'ilima*, the flower used for the lei of the *ali'i*, grew in a narrow band of soil just below the *wao*, the upland zones. From the edge of the rarely used rain forest to the sea spanned the *kula*, the dry plains. Within the *kula*, there were three sections. Farthest inland from the ocean lay the dry-forest uplands, *kula uka*, cultured with drought-tolerant sweet potato and yam, carefully nurtured by moisture-rich recycled mulches. The middle plain, *kula waena*,

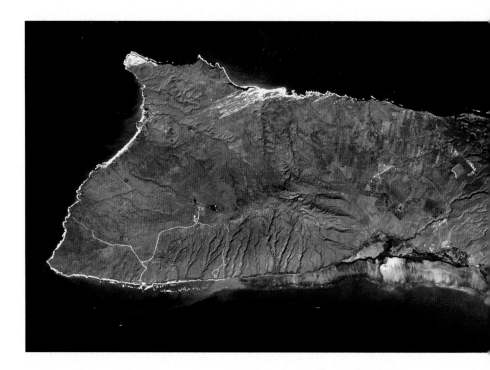

spanned between the *kula kai,* or sea plain, and *kula uka.* There fields of taro absorbed the diverted fresh cool waters through irrigation ditches. The *kahakai* (seashore), although inappropriate for cultivation, became the ideal region for Hawaiians to build villages. All of the major Hawaiian settlements developed along the coast. Within every *ahupua'a,* the *ali'i* discouraged residents from using certain areas to ensure a constant, untainted source of water.

Success as a civilization rested in large part on understanding water-flow dynamics, complex hydrological processes, and the ways to apply this knowledge to land management. This understanding enabled Hawaiians to complete feats that still intrigue researchers today, including the construction of waterways to channel fresh water and their layered agricultural terraces carved out of the mountain slopes. In many ways, the early Hawaiians merged with nature by accepting the limits imposed by the physical and ecological properties of their *ahupua'a.*

EARLY ENVIRONMENTAL IMPACT

Although the Polynesians transported strict conservation ethics to the Islands, their arrival and subsequent behavior would shape the future

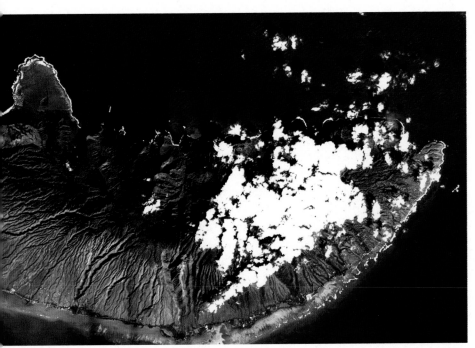

Fig. 6.7. A long history of land abuses on Molokaʻi, including poor soil management, deforestation, lack of cover crops, overgrazing, and impacts of feral ungulates, now leads to intense soil erosion during every rainfall. Much of this eroded mud ends up on the reef, causing severe ecological damage. (Photo provided by U.S. Geological Survey, Matthew Barbee, Perspective Cartographic LLC)

quantity and quality of water and the ecosystem where it collected. After traveling thousands of miles (km) in their seafaring canoes, the Polynesians introduced many hardy plants such as *kukui* (candlenut), *ʻulu* (breadfruit), *kalo* (taro), *kō* (sugarcane), *maiʻa* (banana), *ʻuala* (sweet potato), and others. These "canoe foods," all important staples to this day, had a demanding thirst and when introduced into the pristine Hawaiian ecosystem precipitated unforeseen changes in the movement and natural storage of *wai*.

Lush forests covering the mountain slopes acted as cloud nets before the land was settled, catching the updraft of air currents and capturing their moisture over the trees. But as villages blossomed across the landscape and the inhabitants cleared windward talus slopes with intentional fires, these steady rain cycles were disrupted. When the land was cleared for cultivation by Hawaiian and later Western settlers, rains caused the barren volcanic soil to wash down the now-craggy hillsides, turning previously clear streams brown.

The destruction of forests accelerated in the nineteenth century and triggered a progressive desiccating cycle. Lacking the protective upper canopy dominated by 'ōhi'a and a secondary canopy formed by *hāpu'u* tree ferns, the amount of solar radiation reaching the soil increased, robbing moisture from the earthen surface. Less water resulted in less vegetation, and a degrading cycle ensued, resulting in the loss of the "A" soil horizon, the primary nutrient-providing layer for forest growth. Water ran off the land surface rather than seeping into the subsurface and refreshing the water table. Increased runoff caused gully erosion, washing topsoil and nutrients off the slopes. With no protective ground cover, wind velocities blew unchecked, whisking even more soil from the slopes. Throughout this accelerated cycle the sun continued to beam strongly on the barren soil, enhancing the cycle of increased evaporation and land degradation.

In many areas of Hawai'i today this pattern of accelerated erosion is perpetuated by poor soil management, fallow ranch and farm lands, open construction sites lacking vegetative cover, and tilled earth exposed to heavy rains.

PLANTING THE LAND

Early Hawaiians replaced cleared forest with fields of nourishing crops. Subsistence agriculture demanded less environmental disruption than modern industrial farming, and, being a resourceful society, they added manure and blue-green algae to fertilize the crops. Many crops such as dryland taro, sweet potato, yam, and breadfruit thrived on the arid lava-coated slopes, relying solely on the rain for hydration. In these rocky areas, fields were allowed to lay fallow for periods to replenish soil nutrients. These simple yet wise methods yielded prolific crops. Even today, calls for untilled agriculture, mulching, cover crops, and crop rotation reverberate throughout industrialized nations, which Hawaiians have implemented from the beginning.

Nowhere in Polynesia was the cultivation of plants brought to a higher state of refinement than in Hawai'i, and of all that was grown *kalo* (taro) became the most revered. *Kalo* is the spiritual and nutritional centerpiece of Hawaiian culture, with its success depending heavily on steady flows of cool, fresh water. Elaborate ditch systems were built, diverting mountain streams to *kalo* fields to sustain wetland cultivation. These diversions stole a portion of the natural base flow of streams (water from the water table that feeds a stream), but not in the totality that the later Western plantations achieved. Three hundred varieties of *kalo* are named in the Hawaiian language, 85 of which are still known today.[23] In addition to being the source of the staple Hawaiian food poi, *kalo* was also used for fishing bait and as a dye. The slopes and valleys of old Hawai'i waved green with vast fields of *kalo,* and the number of *kalo* planters exceeded any other occupation. But things change, and so it was with the hydrology of Hawai'i.

As European ships unloaded their cargo in the late 1700s, goats, pigs, sheep, and cattle were part of the inventory. These grazers and grubbers roamed and reproduced in the native forests and lowland hills of Hawai'i and further upset the flow of water across the landscape. Their hooves dug into the soft earth creating new pathways for water and opening up the now-muddy slopes to erosion. Physical damage to the landscape is only part of the story with ungulates. The bigger problem, at least with goats, deer, sheep, and cattle, is the unchecked grazing that denudes the landscape. Ungulates also spread invasive plants and diseases that decimate native bird populations.

Captain Cook left behind the first three goats on the island of Ni'ihau. They proliferated and by the mid-1800s occupied all of the Hawaiian Islands. A few years later, Captain Vancouver delivered several cattle to Kamehameha I, who imposed a *kapu* to protect them from being hunted. Like the rats, pigs, and goats before them, the cattle population swelled beyond any natural balance. Adding insult to injury, the European boar, brought in by Captain Cook, was crossed with the smaller Polynesian pig, and the resulting new species has wreaked havoc on Hawaiian vegetation ever since. In these animals' wake remains a rototilled landscape, bearing little resemblance to the original forest. Many species of native Hawaiian orchids and other plants are being devoured to extinction by these pigs.[24]

BLESSED WITH *WAI*

Hawai'i's rugged topography and reliable trade winds are the critical factors in its hydrologic cycle.[25] The Islands' towering peaks and cliffs form a barrier to the moisture-laden trade winds, and as this air rises against the island, the falling temperature allows condensation to outpace evaporation, forming familiar rain clouds. Meteorologists call this the orographic effect. When rain bursts forth and gushes down the volcanic slopes, porous, permeable lava and cinder layers store the waters that seep underground. Solar energy and gravity constantly move the moisture

Fig. 6.8. The orographic effect is the primary source of the groundwater that we drink. (SOEST Graphics)

from the ocean to the atmosphere and back to Earth again. This great cycle converts saltwater to fresh, perpetually recycling the treasured liquid through a natural cleansing process. Within the ground is concealed rain from past decades, and today's sky is the source of our tap water a generation into the future.

To reach these underground coffers, a raindrop must undertake a long and slow journey. Condensing as mist caught by high forests, water drips off the canopy, seeps into the moist vegetative mat on the forest floor, and gains dissolved nutrients from the decomposing mat. Soon, it percolates into the porous soil. From soil to permeable bedrock, through pores, cracks, and vesicles, gravity pulls the trickle into and through the basaltic rock of the shield volcano. Deep within the roots of our volcanic home, rainfall accumulates as a lens of fresh water that we draw upon for our daily needs. The fresh water floats on an underlying layer of seawater that seeps into the core of the island through submerged fractures and other permeable routes from the surrounding ocean. Fresh upon salt, this hydrologic stratigraphy is the result of density—seawater is denser than fresh water by virtue of its dissolved salt and other compounds. The position of the boundary between the seawater and the freshwater lens is controlled by a combination of forces: the level of the sea around the island, the mass of fresh water pushing downward on the saltwater, the

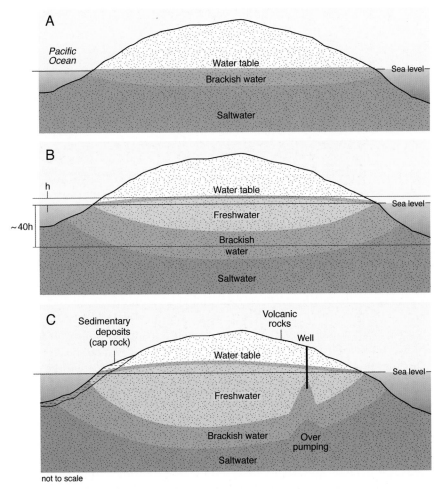

Fig. 6.9. Within islands, fresh water floats on saltwater. The transition zone between the two extends about 40 feet (12 m) below sea level for each foot (0.3 m) the water table stands above sea level. If overpumping occurs, brackish water will intrude into the aquifer. (SOEST Graphics, modified from Gordon A. Macdonald, Agatin T. Abbott, and Frank L. Peterson, 1983, *Volcanoes in the Sea: The Geology of Hawaii*, 2nd ed., University of Hawai'i Press, Honolulu, p. 234)

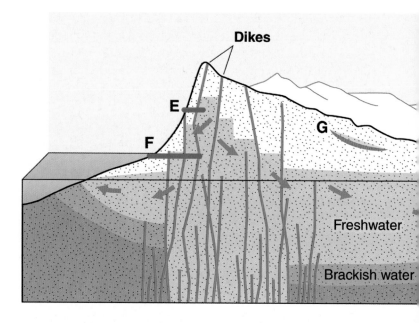

rate of fresh water flowing through the lens, and the amount of ground-water pumped for human consumption.

Over millennia a freshwater lens accumulates within an island. Fresh water displaces the underlying saltwater, but only in proportion to the difference in density. A ratio of 40:1 defines the midpoint of this water body; half fresh water and half saltwater. That may seem like a great deal of water, but, consequently, for every foot (0.3 m) the groundwater table is lowered by pumping for our use, the contact between fresh water and saltwater rises 40 feet (12 m), bringing the salty interface within reach of our drinking wells. The result is that if we carelessly overpump the aquifer, as has been done with the 'Īao Aquifer of Maui and the Pearl Harbor Aquifer in Honolulu, we run the risk of saltwater intrusion into the wells that supply human needs, destroying that essential resource as a water source. If pumping is stopped or sufficiently decreased, it will recover. But that can take a while, and in the meantime, considering that most Island populations are served from a single aquifer, where will you turn for your water needs?

Island water occurs underground in another form. If a saturated zone develops where solid layers of rock or ash retard percolating groundwater from moving toward the freshwater lens in the core of the island, a perched aquifer is formed. This water is less common but can constitute an impor-

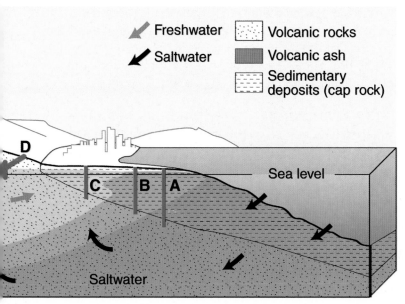

Fig. 6.10. On Oʻahu, fresh water occurs as an underground lens or as a confined body behind impermeable dikes of volcanic rock. Wells A, B, and C produce water from volcanic rock that is confined under sedimentary rock (mostly limestone, clay, and sand). Well A produces saltwater. Well B produces brackish water. Well C produces fresh water. Shaft D produces large amounts of fresh water by skimming groundwater just below the top of the water table. Shafts E and F produce fresh water from aquifers confined behind volcanic dikes; and shaft F probably gets more water because it penetrates through more dikes. At site G a layer of impermeable volcanic ash creates a "perched aquifer." Arrows indicate the generalized pattern of flow. (SOEST Graphics, modified from Gordon A. Macdonald, Agatin T. Abbott, and Frank L. Peterson, 1983, *Volcanoes in the Sea: The Geology of Hawaii*, 2nd ed., University of Hawaiʻi Press, Honolulu, p. 237)

tant resource for our use. Secured upon low-permeability ash beds, dense lava flows, or buried clays, perched aquifers can be substantial and they are known to feed many streams. Islands lacking the freshwater-lens type of aquifer are forced to rely more on these secluded, high-elevation water bodies as well as surface streams to provide for agriculture and industry. Perched aquifers may be detected by the presence of springs; however springs can also come from dike-confined aquifers and a vertically extensive lens in the core of the island. An island lens may also discharge at low elevation through coastal springs.

Groundwater is important for more than just drinking. Most Hawaiian streams are fed by groundwater that seeps into their channels and thereby sustains stream flow and a host of native aquatic ecosystems. This water is called base flow; it is the portion of stream discharge that is not attributable to direct runoff from precipitation. Unfortunately, according to the U.S. Geological Survey base flow to our streams is decreasing, as is the long-term annual trend in rainfall.[26] As a result, Hawai'i streams are slowly drying.

From 1913 to 2002, base flow generally decreased in streams for which there are data. This trend is consistent with a long-term downward trend in annual rainfall over much of the state over the same period. During the early part of this period, 1913 to the 1940s, monthly mean base flow to streams was generally above the long-term average. However, after the early 1940s, monthly base flow decreased to below the average. The primary factor responsible for this decrease is likely to be the decreased recharge to groundwater resulting from decreased rainfall (and potentially other sources). In addition, where pumping for human use has historically been important, base flow to streams may suffer from reduced groundwater storage because of human consumption.

Search for *WAI*

Though blessed with prolific rainfall, the search for water nevertheless intensified during the 1800s and continues to escalate today. The "communal" or "feudal" system of landownership enjoyed by Kamehameha III held lands in trust by a high chief, *ali'i 'ai moku*, who then distributed his land to subchiefs, *konohiki*.[27] They, in turn, allotted lands to the commoners, who grew food for the masses. Before the Mahele, the revolutionary episode of land distribution in 1848 that essentially ended the *ahupua'a* system, disputes concerning water were rare. But the Mahele sparked a hunger for private property rights, the outcome of which was the acquisition of large areas of land for the full development of commercial agriculture. By the late 1800s, the face of Hawaiian agriculture was undergoing a dramatic shift. As Hawai'i developed into a significant trade route, many prosperous entrepreneurs discovered Hawai'i's agriculture-friendly climate. The only complication facing these savvy business moguls was access to enough fresh water to transform gigantic plots of arid land into flourishing plantations.

Enter James Campbell, a California agricultural tycoon who time and again watched the dark waterlogged clouds hanging over O'ahu's east side while his 41,000 acres (16,600 ha) of parched leeward soils remained deprived of the elusive rain. The frustrated businessman eventually turned his sights downward and hired well-borer James Ashley to accomplish his vision. In 1879, equipped with only a hand-operated tool, Ashley struck subterraneous water on the 'Ewa Plain at Honouliuli. Overflowing the top of the pipe, a seemingly endless supply of buried water treasure was released from its geologic grasp.[28] Once these underground reservoirs were identified, misguided users began expunging the water faster than nature could replace it.

By the turn of the twentieth century, industrial agriculture companies learned the secrets of tapping into the groundwater reserves of the Hawaiian Islands. Sugarcane, pineapple, and cattle farming began a century-long consumption of fresh water that was essentially unregulated. Drilled wells, infiltration galleries, and drainage tunnels probed the volcanic canyons and valleys for water and found abundant quantities of the elusive liquid stored within. More than 1,000 substantial wells have been drilled on O'ahu alone, with most production wells 250–750 feet (76–228 m) deep.[29] The 4,000-foot (1,220-m)-deep Waiki'i Well located on the slopes of Mauna Kea is one of the deepest water wells in the world.[30]

Infiltration galleries are dug slightly below sea level with one or more tunnels extending outward to skim fresh water from the surface of an aquifer. O'ahu's most famous infiltration gallery, Hālawa Shaft, draws enough water each day for the needs of more than 100,000 people.[31] Horizontal shafts drilled into mountain areas to tap high-level perched or dike-confined water have been successful on only a couple of the islands, though they have been attempted on all. The largest operation statewide is the controversial Waiāhole system traversing a 3-mile (4.8-km) tunnel punched through the Ko'olau Mountain Range.[32] However, the longest transmission tunnel in the state brings water across 5 miles (8 km) of arid scrubland in western Moloka'i.[33]

By the time of annexation in 1893, Hawai'i's plantation economy was owned and controlled largely by five corporations: Castle & Cook, Theo H. Davies, C. Brewer, American Factors, and Alexander & Baldwin.[34] Sugar interests expropriated the irrigable lowlands and rain-drenched prime upland, and pineapple interests seized most of the remaining land suitable for cultivation. Plantations soon dominated the landscape and used the Islands' freshwater supply for the production of export crops, directly contradicting the self-sustaining model of the *ahupua'a* system. The sugar industry demanded a copious and secure supply of water, the right to own the waters that originated on lands they owned, and the right to transport water out of its original watershed, even if transport greatly depleted the stream flow to the detriment of the people downstream.

These demands, and the fact that they were met by a willing government, changed the fundamental principles of water management throughout the Hawaiian Islands.

COMMUNITY *WAI*

In a report dated 1926, the Honolulu Sewer and Water Commission stated:

> *The City is supplied with water through antiquated equipment which is by no means a system, but an unplanned patchwork of unrelated units. There are frequent water shortages in many parts of the City. There is an ever-present fire menace, growth is hampered, and, in general, the water system has lagged far behind the needs of the City.*[35]

In response, the Honolulu Board of Water Supply was formed with the intention of modernizing the system of water needed to run the growing city of Honolulu.[36] Under the leadership of its first manager, Fred Ohrt, the new board capped wasteful artesian wells, put casings inside leaky wells, and installed water meters for consumers and billed them at fixed rates.[37] The water table stabilized and watersheds critical to maintaining the natural flow that recharged the aquifer were placed under protected status.

By the end of World War II, land use shifts were imminent. Labor unions gained a major role in the Democratic Party, and profits by large plantation companies began to decline. In the 1950s, unionization increased wages and benefits, driving up the costs of labor, and competition from other tropical producers (Brazil, Puerto Rico, and the Philippines) magnified. Water use across Hawai'i became a focus of concern among multiple users: cattle ranchers, taro farmers, aquatic biologists, Native Hawaiians, industrial agriculture, environmentalists, and others. Water conflicts catapulted to the forefront of political battles in the state.

In 1987, the landmark State Water Code established the Commission on Water Resource Management and its authorities and responsibilities.[38] This law asserts that, "the waters of the State are held for the benefit of the citizens of the State."[39] Thus the state highlighted sovereign water rights and the concept of public trust. However, the water commission has not quite lived up to the promise of solving the problems associated with *wai*. A 1996 state auditor report criticized the commission for not moving fast enough to produce a plan for water use by Native Hawaiians, a plan to protect Hawai'i's streams, or a comprehensive water plan for the state.[40] Partially in response to these problems, the state administration in 2005 moved to transfer water management responsibilities to the counties and dismantle the commission.[41] Environmentalists objected to this plan and labeled it "an unraveling" of controls and a step backward in water resource management.[42]

Fears that local control of water management would be inadequate stem from numerous problems that persist in county water administration. There are many examples of water disputes that involve state authorities, county government, private corporations, Native Hawaiians, local watershed partnerships, developers, and others.[43] For example, in his 2004 annual "State of the County" address, Maui Mayor Alan Arakawa said that for years the county has performed "a delicate dance around the truth: that the 'Īao Aquifer—Central and South Maui's main domestic water source—was being tapped out."[44] The situation was so bad that in July 2003 the state water commission moved to take over control of the aquifer because of perceived problems in management by the county.[45] Further, Arakawa declared that there was a "dire" need for alternative water sources and had even researched the possibility of building a desalinization plant.[46] Contributing to the problem is that 50 million gallons (189 million liters) of West Maui surface flow each day remains under the control of private companies, a practice left over from the days of big sugar. Pledging to continue negotiations with water owners, the mayor declared, "Ultimately, one way or another, Maui County will soon have this major water source under public control."[47] Five years later this has yet to happen.[48] Overarching the entire question of public water management and "ownership" by corporate entities is the belief framework of Hawaiians that water is not to be owned: it is to be served and stewarded.

Although it is always possible to find problems, managers say that there are highlights in Hawai'i's water landscape. Water agencies are typically unified in predicting that we are not on the precipice of a water shortage. Central Maui and central O'ahu are growing population centers where water consumption and quality are concerns, but in general there are supplies enough for the next 20 years or so in most communities.[49] Managers always have the fallback of desalinization plants: "there is really no scarcity of water in Hawaii. If people are willing to pay for fresh water (made from ocean water), it certainly can be had." However, water issues in Hawai'i extend beyond mere supply; they represent our skill as stewards. Water is a yardstick by which we measure our proficiency as a community; and if there is one parameter that defines us, it is the fact that we are island-bound. As island stewards measured against the water yardstick, there is room for improvement.

Is there a problem with water? It seems that the answer is an unqualified "yes." When county authorities declare a dire water problem, are faced with takeover by the state, investigate the need for expensive desalinization while tens of millions of gallons (liters) of fresh water flow into the sea, and must negotiate with commercial concerns to secure the lifeblood of their booming economy, something is clearly amiss with the management of *wai*.

QUENCH OUR THIRST FOR *WAI*

The state of Hawai'i today has shifted away from the days of industrial sugar farming and into a mosaic of tourism and population-centered development activity. In this setting, water management issues promise to become even more pressing. Hawai'i's urban growth has been concentrated on O'ahu, where the population has doubled since 1950 and is projected to reach 930,000 people by 2010.[50] Population centers across the Islands tend to spotlight sunny, dry coastlines where fresh water is scarce, amplifying the uncertainty of their water supply. This point is worth emphasizing: most growth and increase in water demand has been in leeward areas. This requires a more expensive and elaborate infrastructure to move water farther, plus more is needed because per-capita water use is higher in leeward areas.

According to the Hawai'i Department of Business, Economic Development, and Tourism,[51] the total fresh water used in Hawai'i in 1995 averaged a little more than 981 mgd (3,713 mld). By 2000, the last year for which there are data, this had decreased to about 628 mgd (2,377 mld), a decrease resulting from the demise of sugarcane as a major producer in the state.

- Kaua'i with a population of 58,460 withdrew 45.2 mgd (171 mld) (down from 239 mgd (905 mld) in 1995).
- O'ahu with a population of 876,160 withdrew 216.91 mgd (821 mld) (down from 264.23 mgd (1,000 mld) in 1995).
- Maui County with a population of 128,090 withdrew 312.82 mgd (1,184 mld) (down from 368.83 mgd (1,396 mld) in 1995).
- Hawai'i County with a population of 148,680 withdrew 53.41 mgd (202 mld) (down from 108.52 mgd (411 mld) in 1995).

On Kaua'i, shifts in water usage have been dramatic. In 1995, more than 223 mgd (844 mld) of Kaua'i's water went to agriculture (93% of total freshwater use). By 2000, agriculture use had dwindled to only about 30 mgd (113 mld). On O'ahu, 1995 agriculture use was almost 90 mgd (341 mld) but by 2000 had decreased to only about 40 mgd (151 mld). Groundwater use on O'ahu went from 227.85 mgd (863 mld) and surface water use of 36.38 mgd (138 mld) in 1995 to 208.84 mgd (790 mld) of groundwater and 8.07 mgd (31 mld) of surface water use by 2000. In Maui County, 1995 surface water use was about 240 mgd (908 mld) and by 2000 had fallen to 164.13 mgd (621 mld). Groundwater use in Maui County rose over the same period from about 128 mgd (485 mld) in 1995 to 148.69 mgd (563 mld).[52]

The 'Iao Aquifer on Maui supplies more than half the municipal water supply on the island. Pumping from the aquifer increased tenfold from 1950 to 1995 but has since steadied out at about 15% less than its 1995 maximum. In parts of the aquifer, the persistent pumping caused the water table to fall to as much as half its prepumping level. However, because pumping rates decreased slightly at the start of the twenty-first century, the water table has stabilized.[53] Throughout that period the brackish transition layer underlying the freshwater lens has moved consistently upward, effectively shrinking the aquifer. Despite the decreased pumping, several freshwater supply wells continue to show increased salinity levels, a trend that has persisted since measurements began in 1986. Is the aquifer shrinking because of overpumping? The maximum sustainable rate of pumping for the 'Iao Aquifer has been estimated to be 20 mgd (75.7 mld). Average pumping during 2002 was only 17 mgd (64.4 mld); hence it was thought that the resource was being used sustainably. However, the sustainable pumping estimate is based on a relatively simplified method of calculating how much water can be extracted from the aquifer without causing it to shrink. Obviously the rate of water withdrawal is too high for the current distribution of wells, and the estimate of a threshold for sustainable withdrawal is off target. Improvements in this calculation are under way, and it is hoped that a clearly defined sustainable use for the resource will be defined.

The comparison between each island's supply of water and the amount of consumption suggests that there is generally an adequate amount of water to meet current needs. But when projected population growth and encroaching urbanization are added to the equation, the picture is less clear. A combination of location, population density, urbanization, and agricultural withdrawal set the stage for water shortages across the state;

add to this the historical decline in rainfall and stream flow,[54] a lack of knowledge about how global warming will impact the trade winds[55] (the primary source of our water), the possibility of a continued decrease in rainfall,[56] along with other future unknowns, and there is reason to be concerned.

A general shift to decreased surface water use related to the demise of the sugar industry and increased focus on groundwater sources related to growth in population characterizes the past decade of water use. Overall, the numbers showing water use island by island do not indicate a widespread problem because Hawai'i continues to be blessed with *wai* and the end of sugar has decreased the overall demand for water.

But a problem exists in local form: shifts away from streamborne water to groundwater as a major domestic supplier despite poor understanding of this resource, private corporate control of major stream resources and lack of public scrutiny over their use, salinization of Maui's 'Iao Aquifer, declines and salinization in O'ahu's Pearl Harbor Aquifer, community disputes over water on Moloka'i, decreased groundwater recharge due to growing urbanization and decreased irrigation. The Kona coast of

Fig. 6.11. The shift from sugarcane to housing tracts has changed land use. Impervious surfaces of asphalt and concrete divert more rainfall to the ocean rather than recharge the aquifer.

Hawai'i, the west portion of Maui, O'ahu's dry leeward coast, and even parts of Kaua'i and Moloka'i are rapidly developing or may be in the near future if a sustained growth of tourism continues. These many interconnected elements promise to propel water rights to the forefront of Hawaiian political debates in the twenty-first century.

Hawai'i does not have a water shortage problem; we have a water distribution problem. And despite the best intentions of managers, authorities have yet to provide a comprehensive plan for the husbandry of our water resources.

MOVING WATER

Hawai'i is facing mounting problems resulting from the extraction of too much groundwater, decreased groundwater recharge, and decreased rainfall (and decreasing flow in streams). Global warming questions about future trade wind flow[57] and the possibility of changes to ENSO[58] add additional uncertainty to this picture. As if this is not bad enough, global warming models also suggest that in a warmer atmosphere the precipitation events will be fewer but more intense[59] with the net result that less water is likely to infiltrate to the water table, with most of it simply running into the ocean in the short period of the storm. These dramatic changes threaten the organisms that dwell in aquatic ecosystems and impact the nearshore estuaries and bays where brackish water is important. By rerouting water where we need it and with climate changes afoot, Hawai'i has evolved into a massive experiment in the effects of moving water. But as water becomes scarce in aquatic systems, the stakes are getting higher.

At any moment in Hawai'i, water is on the move; with tunnels, irrigation ditches, water wells, and miles of buried pipes connecting homes, sewage plants, and the sea, we do not even know how much water is in transit at any given moment, though it exceeds 200 billion gallons (757 billion liters) a year.[60] On O'ahu more than 18 tunnels transport large quantities of fresh water from high-rainfall areas to dry agricultural and residential communities. Each island is riddled with miles of tunnels flowing with fresh water.

When water is diverted, it is channeled away from streams, extracted from high vaults of dike rock, or pumped out of the ground. On O'ahu, pulling water from the Ko'olau dikes has caused a reduction in stream flow on the windward side of the island. This is essentially a diversion of water from windward communities to leeward communities, which are more populous. This diversion can have numerous effects on the ecological and cultural functions of the windward community: streambeds go dry, agriculture is impacted, cultural taro-based practices decline, and estuaries and reefs lose critical ecology. Basically these tunnels are diverting windward waters from their original ecological and community functions into tourism and residential growth. As the windward side of the island experiences the stress of an interrupted water system, on the leeward side new neighborhoods, golf courses, hotels, and tourist adventures soak up the "extra" water. But in most instances this new use does not return the water to the water table (as irrigation partially did). It sends it into the ocean or into salty groundwater in the coastal plain after treatment.

ARE WE RUNNING OUT?

The answer to this question depends on your point of view. On West Maui the population has increased (and along with it the withdrawal of groundwater), aquifer recharge rates have declined, chlorinity (saltiness) of well water is up, and sustained drought has hit the region—all this despite decreased agricultural use of surface water over the past decade.

Because the combination of Hawai'i's geology and topography makes its surface water difficult to exploit and most efforts to divert stream flow have been intended for irrigating sugar, water used by the burgeoning population for drinking and manufacturing comes from the easily extractable stores underground. For over 100 years we have been siphoning from the water table. The last decade of the twentieth century and the first decade of the twenty-first century have witnessed the most intense water shortages as the extraction rate grew by millions of gallons (liters) in the midst of a prolonged drought and sustained expansion in population and housing. In some places, the water table has plummeted by more than 13 feet (4 m), allowing saltwater to contaminate coastal wells as it replaced the retreating fresh water. In Kona and portions of Maui, drinking water has become less enticing as saltwater finds its way into the lowered water table.

Although some areas have an abundant supply and have even been referred to as possessing "extra water," others have limited surface and groundwater supplies. The inequality of water resource distribution has serious implications. For example, water supply in Līhu'e, Kaua'i (mere miles from Mount Wai'ale'ale, among the wettest spots on Earth), is constrained by a lack of new groundwater sources, and the county is paying to use a new desalinization plant. The problem with groundwater near Līhu'e is that the aquifer has a low permeability. When a well is pumped, water levels drop, so only a limited amount of water can be pumped out of each well. When they analyzed the situation, Kaua'i authorities decided it was cheaper to just treat saltwater than to put in many small-capacity wells. Līhu'e is a major population center on the island and an area slated for new residential development, and the water shortage there is becoming a major problem. This not only limits future development around Puhi and Hanamā'ulu but also restricts current residential plans. Elsewhere on the island, unused fresh water runs through out-of-date irrigation ditches formerly used by the sugar industry.

What becomes evident when examining the water supply is that the Līhu'e situation is not atypical. In fact, several populous areas (Honolulu, central and West Maui) across the Islands are withdrawing amounts of groundwater dangerously close to the estimated availability level. And these estimates are just that—estimates—with margins of error existing that may be cloaking the fact that many towns are already flirting with the limits of sustainability.

Though very difficult to estimate, planners use the concept of maximum sustainable yield (MSY) to determine how much groundwater is extractable. This important planning guide means "the maximum rate at which water may be withdrawn from a water source without impairing the utility or quality of the water source as determined by the Water Commission."[61] Considerable disagreement exists over how to derive valid estimates for MSY because the outcome dictates the future fresh water supply for the entire state.

The Commission for Water Resources Management currently divides each island into "aquifer sections" and estimates the MSY for each. Withdrawal rates for municipal and agricultural demand are established in accordance with these MSY targets. For decades the state has used methods that tend to overestimate the MSY, and although they still use these liberal estimates, they have begun a process to incorporate more modern computer modeling approaches when such information becomes available. Improved estimates of MSY are critical because overpumping in one aquifer can pull water from an adjacent watershed. For example, water shortages in portions of the Pearl Harbor Aquifer may be due to overpumping from other sectors of the same aquifer. New models are being proposed by federal agencies that take into account water transport between aquifer sections.

As groundwater use in certain areas approaches the sustainable yield estimates and water tables begin to decline, refinements of MSY estimates are mandatory. On O'ahu, the MSY has been reduced for this reason. Before 1991, MSY was set at 495 mgd (1,873 mld) and then was reduced to 465 mgd (1,760 mld) in 1992 due to overstated estimates for the central aquifer.[62] Discrepancies between guidelines such as MSY and the ability of natural recharge to replace these withdrawals can spell problems on a local scale. Because of population growth, and the shift from agriculture to residential and tourism uses, rates of water usage are on a path of convergence with rates of sustainable yield. Given the high degree of uncertainty in our models of the groundwater resource, actual usage and MSY may be already overlapping in places, or soon will—an undesirable situation.

Following the path blazed by central Maui and Līhu'e, some hydrologists feel that O'ahu will soon face a water crisis. To a large extent, two factors influence how much water reaches the aquifer: groundwater recharge and future land use. In light of this, two questions should be present on all residents' minds: Are we willing to risk contaminating aquifers by withdrawing too much fresh water? What does climate change hold in store for our water resources?

Water management pivots on the overarching concept of sustainability. Does a commitment exist to preserve the integrity of the hydrologic system for future users, or is our concern only with the short term? If the word "sustainable" is interpreted to mean "without impairing the utility or quality of the water source," then we have already violated this standard in several places and now have some backpeddling to do.

WATER RECYCLING

Water recycling was implicit within the valley-type *ahupua'a* management system, with boundaries between adjacent land units following ridgelines to separate the drainage basin of one *ahupua'a* from the next. Today, water recycling and reuse (not for human consumption) have been successful elsewhere, but until the 1970s Hawai'i largely ignored this important opportunity. As water shortages become a reality, Hawai'i is scrambling for solutions to improve water sustainability. The Hawai'i Department of Health is in charge of regulating recycled water use in the state. Department of Health regulations specify logistics such as irrigation, handling, and hook-up procedures to make sure recycled water is used safely.[63]

The potential users of recycled water in the state are large tourism complexes, large-scale farms and ranches, residential and commercial landscaping, and golf courses. Water recycling has been practiced for many years[64] in Hawai'i. On O'ahu, recycled water is used at the Polynesian Cultural Center, Turtle Bay Golf Club, Brigham Young University campus, and the Kāne'ohe Marine Corps Base Hawai'i. It is also used on other islands, including on Maui at golf courses, public parks, schools (again, not for human consumption), libraries, resort areas, and for growing seed corn. Not only is treated wastewater a reliable source of water because it flows evenly throughout the year, but it can be treated to a level that ensures it is safe for use, therefore saving millions of gallons (liters) of fresh water each year. At the same time, it allows natural potable aquifers to be replenished when recycled water is spread on the land. Of the more than 157 mgd (594 mld) of wastewater generated in the state, at least half of it is reusable, which is enough, for example, to water more than 100 golf courses each day. However, in 2004 only 24 mgd (91 mld) of the 157 mgd (594 mld) (approximately 15%) of treated wastewater was put to reuse.

While we dispose of millions of gallons (liters) a day of potentially reusable "waste" water, parts of Hawai'i face the prospect of freshwater shortages. Yet, a solution based on water reuse remains in the shadows of acceptance. Notably,[65] Los Angeles County, home to 10 million people, treats its wastewater to a tertiary level, pumps it to natural recharge areas, and allows it to infiltrate back into the potable groundwater system in a massive (and successful) display of efficient water reuse.

The Hawai'i State Department of Health released the Guidelines for the Treatment and Use of Reclaimed Water on May 15, 2002.[66] These guidelines provide specifics for evaluating reuse projects to safeguard public health and preclude environmental degradation of aquifers and surface waters while seeking to maximize the potential of the relatively new resource. Although water reuse is on the upswing among consumers such as golf courses and other forms of water use not directly tied to human consumption, as a state we remain behind the times in terms of water recycling. For instance, no large-scale groundwater recharge effort exists in Hawai'i to utilize recycled water. Global warming, and the climate change that will accompany it, may force us to have a higher appreciation for the value of water reuse.

The three main stumbling blocks to recycling water are demand (public acceptance), money, and governmental organization. Demand for reused water is quite insignificant, in part because Hawai'i users already have access to low-cost, high-quality water. The initial start-up expenses for water recycling are considerable. It is understandable that customers are hesitant to assume the added financial cost of water reuse. The state of Hawai'i's water-management structure must change to attract users and make it economically and politically desirable to incur the initial hassle and cost of switching over to reused water.

Using the threat of fines as an incentive, the federal government goads state and county agencies into promoting reclamation efforts. At issue is who should pay and whose responsibility is it? This lingering obstacle of responsibility pits those who create potentially recyclable wastewater against those who reuse it. Managing the environmental risks is costly and potentially hazardous because improperly treated wastewater can contain fouling nutrients, toxins, and disease-causing organisms. Unfortunately, the lines of liability are unclear, leaving considerable opportunity for inaction.

To date, the recycling strategy of Hawai'i has embodied local rights and responsibilities rather than creating a vision for the entire state. Consequently, sewage facilities are not accountable for how much treated water is recycled, no single agency has a clear mandate or authority for managing water, and the supply agencies are not forced to worry about how treated effluent gets to potential recycling users from treatment plants.

Fig. 6.12. Intense rainfall combined with expanding paved land surface results in polluted runoff to the ocean and decreased recharge to groundwater resources. (Photo by Carl Berg)

However, shifting sentiments are beginning to redefine the way Hawai'i manages water, and a suite of solutions is being discussed.

- New sewage treatment plants can be located close to agricultural users to reduce distribution costs.
- For many of the mega hotel complexes the reuse and treatment facilities are integrated from the beginning. Wastewater treatment plants can be designed with reuse as the objective, supplying all the needed water to irrigate their golf courses, lawns, and open space.
- Merging water supply offices with wastewater managers under one roof may also be a step in the right direction to promote more urgent examination of the long-range costs that effluent reuse can avoid.

ON THE HORIZON: CHANGING CLIMATE AND SHIFTING LAND USE

Climate change has the potential to move the discussion of water use from its current footing of concern to one of anxiety. The orographic effect that supplies much of our rainfall is the primary source of recharge to the aquifers we use. However, scientists are unsure exactly how our local system of water production will respond to global warming. As noted earlier,[67] it is unclear how anthropogenic warming will impact the trade winds and cloud formation. Will this lead to a related decrease in recharge? The answer is currently unclear.[68]

Researchers are groping to understand the impact of global warming on rain production in Hawai'i. One hypothesis proposes that increased drought will result as surface heating in coming decades drives the cloud base higher, limiting orographic rainfall to less land area. Another posits that higher surface temperatures will enhance evaporation and offset this trend. Cloud base elevation today is approximately 2,000 feet (610 m). If it rises, former groundwater recharge areas will receive less rainfall, aridity in leeward regions will increase, and the overall amount of rainfall falling on mountainous lands will decrease. Only monitoring of our environment will reveal the emerging pattern among these critically important processes. Yet monitoring our water resources is decreasing. For example, the number of stream gauges in Hawai'i, easy to cut out of the budget in tight financial times, has decreased 300% from a high of nearly 200 in the late 1970s to fewer than 50 today.

The Intergovernmental Panel on Climate Change hypothesizes that global surface temperatures will rise between about 4° and 10°F (2.2° to

5.6°C) by 2100 relative to 1990.[69] How high this heating will raise the cloud base, or sea-surface temperature, and what will be their potentially offsetting influences on Hawai'i water production is the subject of ongoing research.

Above the clouds lies the upper-elevation inversion. This is a region of dry, still air that limits the altitude of the top of orographic clouds. Those who have visited the summits of Mauna Loa or Mauna Kea (or looked out the window on their last airplane flight) have experienced the inversion firsthand as they looked down on the tops of Hawai'i's clouds. Climate researchers are uncertain if warming will raise the inversion, or if it will hold its current elevation and constrict the zone of orographic rainfall between a rising base and an immovable ceiling. It is even conceivable that the inversion could descend—further squeezing our water source in the sky.

One pattern of future rainfall in Hawai'i that should be of particular concern is the possibility of increased extreme rainfall. In fact, between 1958 and 2007, the amount of rain falling in the very heaviest downpours (defined as the heaviest 1% of all events) has increased approximately 12% in Hawai'i.[70] Climate models suggest that extreme rainfall events, such as the type that caused over $80 million in damage to the University of Hawai'i library in 2004,[71] will become more frequent in a warmer climate.[72] Researchers used a combination of direct meteorological patterns and advanced computer models to reveal a distinct link between rainfall extremes and temperature, with heavy rain events increasing during warm periods and decreasing during cold periods. Notably, they found that rainfall extremes were larger than the events predicted by global circulation models, implying that predictions of extreme rain events in response to future global warming may be underestimated. With Hawai'i's steep topography, heavily developed and cemented watersheds, and natural proclivity for flash floods and heavy rainfall, weather patterns with more extreme behavior pose a dangerous future for both steep housing tracts and low-lying coastal plains where floods fan out and stagnate.

Though water supply has certainly been a major issue for the state over the course of history, new dilemmas heighten concern. The closure of large-scale agricultural operations is leaving future use of the land uncertain.[73] The fate of our water supply may well be linked to the fate of fallow fields. Many hydrologists agree that converting agricultural land into impervious residential sprawl will cause groundwater supplies to be depleted more rapidly, especially if residential withdrawal rates increase. In Maui, Kaua'i, and Honolulu at least, this pattern is well under way and threatens to hasten the already declining water levels in those communities.

Combine these shifts in land use with potential decreases of our main source of water due to climate change, and add to this the fact that no one is integrating these disparate issues and sources of information into an overarching water management plan—and you have the basis for continued, potentially accelerated, water shortages. Here again is another reason why Hawai'i needs a state science agency rather than a disparate collection of federal, state, and local offices that operate without integration.

WATER'S ROLE IN THE FUTURE OF OUR SOCIETY

There are no easy answers to the questions concerning growth management, climate change, and water rights in Hawai'i. The state needs to address land use changes and the privatization of water. We also need to devote resources toward improving our understanding of the potential impacts of climate change on local water availability. With a resource as

Fig. 6.13. 'Iao Stream (flowing toward the camera) on West Maui is a sparkling clear body of water before it hits an intake for the commercially owned 'Iao-Waikapū Ditch. Nearly the entire stream is diverted to the old Wailuku Sugar Company irrigation system for use in commercial agriculture. (Photo by D. Oki, U.S. Geological Survey)

finite and essential as water, solutions to these issues are emerging as the foundation for Hawai'i's future. The time has arrived for the people of Hawai'i to decide, as the Polynesians did upon arrival, water's role in the future of our society. The Hawaiians agreed that water was a right, a resource entitled to all living things, and a life force that needed to be protected, respected, and stewarded, because existence depended on it.

Let us remember the lessons of the *ahupua'a* and the basic tenets that underlie this management system: stewardship, service to the resource, cooperation, and respect. All of these concepts must be incorporated within our water management system if Hawai'i is to choose a sustainable future.[74]

CHAPTER 7

Stream Flooding and Mass Wasting

ʻĀwīwī! ʻĀwīwī! O pea oʻe I ka wai
"Quick, quick or the waters will stop you"

Fig. 7.1. Rainfall in Hawaiʻi is largely produced by northeast trade winds or by south and southwest winds associated with storms that approach from the northwest. (From NASA Earth Observatory "Image of the Day" for May 27, 2003: http://earthobservatory. nasa.gov/IOTD/view.php?id=3510 (last accessed November 7, 2009)

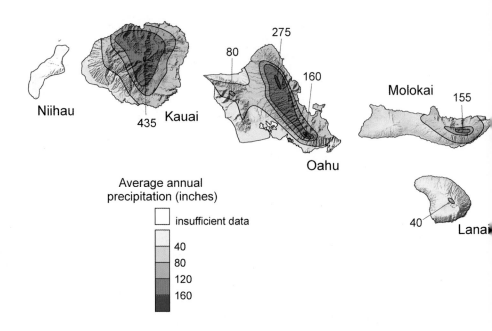

Niihau

435 Kauai

80

275

160

Oahu

Molokai 155

Average annual
precipitation (inches)

☐ insufficient data

40
80
120
160

40 Lanai

I have sometimes sat on the high bank of a streamlet, not more
than fifteen to twenty feet wide, conversing with natives in the
bright sunshine, when suddenly a portentous roaring, like the
sound of many waters, or like the noise of the sea when the
waves thereof roar, fell upon my ears, and looking upstream, I
have seen a column of turbid waters six feet deep coming down
like the flood from a broken milldam. The natives would say to
me, 'Awiwi! Awiwi! O pea o'e I ka wai'—"Quick quick, or the
waters will stop you."[1]

It is no fluke that one can usually predict Hawai'i's daily weather: "Today's
weather will be partly cloudy, with passing showers windward or *mauka*,
and temperatures ranging from the mid-70s to mid-80s." This balmy
climate provides one more reason why Hawai'i's moniker is "paradise."
At sea level the warmest daytime temperatures in summer infrequently
exceed the mid-90s, and the chilliest nighttime temperatures in winter
rarely fall below the high 50s. The difference in average daytime temperature at sea level throughout the year is only around 11°F (6°C), making
the Hawaiian Islands home to Earth's most temperate climate. But as you
have learned by now, weather in paradise has a volatile side.

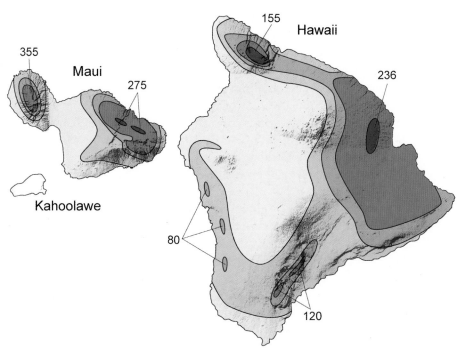

Fig. 7.2. Trade winds deliver moisture to the northeastern sides of the Islands, making these the wettest regions in Hawai'i. Large-scale storm systems passing near the Islands, usually from the northwest, generate winds that provide rainfall to southern shores that are otherwise dry. (SOEST Graphics, modified from T. W. Giambelluca, D. Nullet, and T. A. Schroeder, 1986, *Rainfall Atlas of Hawai'i*: http://pubs.usgs.gov/ha/ha730/ch_n/N-HItext1.html (last accessed 7 November 2009)

Two seasons dominate the Hawaiian climate: summer (*kau wela*) and winter (*ho'oilo*). Summer extends from May through October, accompanied by the misnamed "gentle" trade winds. On any other shore, consistent 25 miles (40 km) per hour winds gusting to 35 miles (56 km) per hour are hardly considered gentle. These persistent air currents blow toward the equator from the northeast in all Northern Hemisphere oceans and toward the equator from the southeast in all Southern Hemisphere oceans. They are so reliable that for centuries the global trading industry under sail relied on their consistency. In Hawai'i, they usually provide constant, natural air-conditioning throughout the late-winter and summer. Yet beyond the winds and seasonally tranquil north-shore waters, the summer also ushers in threatening tropical storms (occasionally growing into hurricanes) that develop in the eastern Pacific.

From October through April, Kona conditions often interrupt the trades (and can do so any time of the year), bringing the frequent rain and cool, cloudy weather of winter. Kona storms derive their name from the typically sheltered west coast on the Big Island where local southerly winds or stagnant air dominate, resulting from the shadowing effect of the high volcanoes. Occasionally a Kona storm, originating from the northwest but generating winds that arrive from the south or southwest, will stall over Hawai'i. These bring persistent islandwide rain, strong winds, and high surf that can cause flooding and property damage. These wet events can lead to remarkably dangerous and unpredictable flash flooding in usually tranquil streambeds.

Not only do wind patterns significantly control the Hawaiian archipelago's weather and microclimates but so also do high-elevation mountains. In areas where onshore trade winds are obstructed by tall mountains, the moist warm air rises up the mountain slope, cooling on its ascent and condensing into heavy cloud cover and rainfall on the windward side. Hawai'i's preeminent example of this orographic rain is Kaua'i's Mount Wai'ale'ale, which receives an average drenching of 460 inches (38 feet) (11.5 m) of rain a year.[2] In sharp contrast, the dry air descending Kaua'i's leeward side creates local semiarid conditions. Polihale Beach on the west side of the island receives on average a mere 8 inches (0.2 m) of rain a year.[3] This pattern is similar for most of the Islands, with the windward side receiving frequent rain squalls and the leeward side boasting eternal sunshine.

STREAMS IN URBANIZED HAWAI'I

On windward sides, the locale of streams that flow year-round, many channels hold dozens of large boulders. Sure-footed hopping from one large rock to another can get you across most streams with flowing water. If you travel to the leeward side of an island, you will notice a similar phenomenon, dry streambeds that contain a population of alien massive boulders seemingly dropped off by a county work crew. What force of water does it take to roll a 5-foot-high (1.5 m) solid block of rock and carve the sharp corners off to make it as round as a *manapua* (a local food made of stuffed, rounded dough)? These boulders are the telltale signs of powerful floods that characterize our watersheds. Tranquil streams can become frothing rapids, 10 feet (3 m) deep, moving fast

Fig. 7.3. In October 2004 flash flooding in Mānoa Stream on Oʻahu caused over $100 million of damage. (Photo by K. Kodama, National Weather Service)

enough to quickly drown the strongest of hikers unfortunate enough to be in the way.

Because of potential flooding, streams flowing through urban areas have been contained by concrete-lined channels. Block these channels during a storm with debris such as tree trunks, sediment, or other refuse and the wave of backed-up water that is created can jump out of a channel into a neighborhood and flow with sufficient force to tear a house from its foundation—and has on more than one occasion.

Most (but not all) flooding in Hawaiʻi is caused by Kona storms, which drift slowly over watersheds and release torrents of precipitation that arrive too fast to be absorbed by the ground. Flash floods can roar down a stream valley in minutes.

FLASH FLOODS: A RIVER OF SURPRISE

Flash floods in Hawaiʻi can occur during any month of the year. There have been 461 flash floods in the past 48 years, an average of nearly 10 per year throughout the state.[4] However, these events are most frequent during *hoʻoilo*, the wet season of October through April. As the

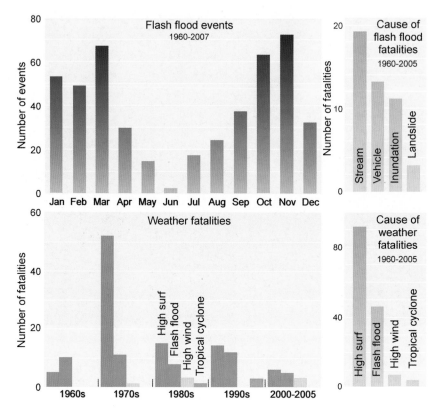

Fig. 7.4. Flash floods are the leading cause of direct weather-related deaths in Hawai'i. High surf is the leading cause of indirect weather-related deaths. (SOEST Graphics, National Weather Service: http://www.prh.noaa.gov/hnl/pages/weather_hazards_stats. pdf [last accessed November 7, 2009])

graphs show, March and November, followed closely by October, have the highest flash-flood event frequency. Flash floods are the leading cause of direct weather-related deaths in the state of Hawai'i, far exceeding the toll caused by high wind events and tropical cyclones. Although high surf causes the most "weather-related deaths" in the Hawaiian Islands, surf is considered to be an indirect weather impact because high swells are usually formed far from the island chain.[5]

Particularly notable flash floods occurred in February to April 2006 when 28 flood events throughout the state caused seven deaths and more than $50 million in damages. In Mānoa, the flash flood in October 2004 resulted from an extreme rainfall of 8.7 inches (22 cm) in only 5 hours and led to more than $100 million in damage through Mānoa Valley and on the University of Hawai'i campus. November 2000 was accompanied by flash flooding on the Big Island and resulted in $70 million in damage due to 37 inches (94 cm) of rainfall in a 24-hour period. These and other similar events provide clear warning that in Hawai'i when it rains, water levels in streams can change rapidly.

FLOODING ON O'AHU

Flooding on O'ahu[6] most frequently occurs where steep hillsides meet low-lying coastal plains, such as those found along the windward side: Kahuku, Lā'ie, Ka'a'awa, Kualoa, Kāne'ohe, Kailua, and Waimānalo. On the leeward side, usually the site of dry and sunny weather, Māpunapuna, portions of Waikīkī, Mānoa Valley, and the watersheds of East Honolulu all have flooding histories.[7] Heavy damage from flooding occurs a couple of times every decade, most recently in 1987, 1991, 1993, 1994, 2006, 2007, and 2009. The heaviest rainfall in Kāne'ohe occurred in October 1991, when 15 inches (38 cm) fell in 48 hours, leading to intense flash flooding.[8]

In the spring of 2006, Lā'ie and Ka'a'awa experienced a failure in their storm drains during a 6-week-long drenching.[9] A combination of forces overwhelmed the drainage system including heavy runoff onto the flat coastal plain from steep watersheds, high seas driven by sharp trade winds that flooded drainage outlets with saltwater, extensive debris that blocked channels and pipes designed to carry floodwaters into the ocean, and the refusal of saturated ground to absorb any more precipitation after weeks of rain. Communities along Kamehameha Highway, at the base of the steep mountain slopes, and throughout the flat coastal plain all flooded as high as 5 feet (1.5 m) within homes. Coastal homes are susceptible to a double whammy: stream mouths and storm drains that usually funnel rainwater to the sea are most vulnerable to having their capacities exceeded at high tide, during onshore winds, and in high-wave events. This is because channeled flow reaches a sea that is unusually high, blocking drainage paths. As a result, a stream channel can fill with water when it is unable to discharge its liquid load to the sea. The incidence of this type of condition is going to increase with time as sea level continues to rise in response to climate change (see chapter 9 for more on climate change and sea-level rise).

The most severe flooding on O'ahu occurs where low-lying lands are heavily developed. The land surface is armored with impermeable concrete, forcing waters to run along the surface rather than soak into the ground. The lands of greatest concern are those that lie below, or near, sea level. Wells show that the water table rises and falls with the tides and even rises and falls with groups of high waves breaking in the surf zone.[10] Hence, when the ocean rises at high tide and during high-wave events, and as sea level continues to rise due to global warming, the groundwater table reaches and often breaks through the land surface. When this happens simultaneous to heavy rainfall, flooding in the coastal zone is nearly impossible to avoid. If you have a house that is built with post and pier architecture and sited several feet above the land surface, you stand the best chance of avoiding the hazard on low-lying lands.

During the first 15 days of November 1996, record-breaking rainfall occurred along the Wai'anae coast. Twenty-one inches (53 cm) fell in an area usually subject to annual rainfall of only 2 inches (5 cm). In 'Ewa, 12.5 inches (32 cm) fell in 7 hours on the fifth of that month, leading to flooding of the low limestone plain and its wide neighborhoods and paved surfaces.[11] When undeveloped, limestone is filled with pits, holes, and subterranean tunnels that easily funnel away heavy rains. To a geologist, the idea of floodwaters occurring on limestone seems as unlikely as an ocean without waves. The fact that 'Ewa is capable of flooding is stark testimony that the natural geology has been ignored by drainage specialists when instead it could have been efficiently incorporated into mitigation plans. Buried under yards of impermeable cement that direct surface waters to a few insufficient storm drains (often flooded by high tide), an entire landscape of permeable water-absorbing limestone lies in dry impotency under the impervious cement of the 'Ewa Plain. A simple and inexpensive solution would have been to work with the natural characteristics of the geology by using permeable cement on roadways and sidewalks. Permeable cement, invented over two decades ago, allows runoff to seep into the underlying substrate rather than flood in torrents into a few inadequate canals that run to the sea.

Perhaps the most severe flooding in recent memory was the New Year's Eve flood of 1987, which, in less than 24 hours, caused over $34 million in damage to some of O'ahu's most densely settled residential areas. The Ko'olau Range separating windward Waimānalo and Kailua from Hawai'i Kai and East Honolulu received a heavy drenching that caught meteorologists and Civil Defense authorities by surprise. High clouds masked the event from satellite observation, and the area had sparse radar coverage at the time. This extreme event was equivalent to a 200-year storm that accumulated as much as 20 inches (51 cm) of rainfall over a 24-hour period in some areas.[12] Torrential rain led to flash flooding in Waimānalo and Hawai'i Kai, fluidized mudflows in Niu Valley, levee breaching in Kawainui Marsh in Kailua, road closures, and damage to over 1,200 homes. A population of 40,000 people in East Honolulu was isolated by road closures; stranded with little public aid for hours as the storm developed at the height of New Year's revelry. Compounding the problem was difficulty rounding up emergency personnel due to the celebratory date, inability to reach stranded families due to ongoing flooding and landslides in the most affected areas, and the unrelenting intensity of rains that made rescue operations dangerous to all involved. Drainage systems blocked by rocks and debris caused unanticipated diversions of floodwaters, resulting in extensive damage to many ridgeline neighborhoods not accustomed to flooding. Meanwhile, in Waimānalo, a low-lying region, floodwaters inundated homes with up to 5 feet (1.5 m) of swirling water at the peak of the runoff.

Other severe events on O'ahu include the October 1981 flooding of Waiawa Stream after heavy rains that led to $786,000 damage, and the January 1968 flooding in Pearl City, which caused $1.2 million in dam-

Fig. 7.5. Urban flooding results from heavy rains on impervious surfaces. (Photo by Morphing Puppy, Flickr: http://www.flickr.com/photos/morphingpuppy/120971429/ [last accessed January 6, 2010])

age.[13] Worst of all, $100 million of damage took place at the University of Hawai'i when Mānoa Stream jumped its carefully channelized banks after almost 10 inches (25 cm) of rain fell in only a few hours on October 30, 2004.[14] Careful maintenance of debris buildup in the channel might have helped prevent this expensive catastrophe.

FLOODING ON KAUA'I

On average, Kaua'i receives between 8 and 80 inches (20 to 203 cm) of annual precipitation along the coast, and more than 450 inches (1,143 cm) at the higher elevations of Mount Wai'ale'ale.[15] Because of this abundant rainfall among the headwaters, stream flow is characterized by frequent flash floods as well as prolonged flooding associated with slowly passing rainstorms that saturate the soils. The tendency to flood is exacerbated when development ignores nature's cues—such as the wetlands that were filled with soil and cement to allow for the construction of the Kalapakī Bay Marriott on Kaua'i. On March 11, 2006, when heavy rains and debris blocked a culvert upstream of the hotel, two dozen rooms were flooded and sinkholes appeared in the parking lot.[16]

Flash floods resulting from a storm on December 14, 1991, that dropped over 20 inches (51 cm) of rain in 12 hours over windward Anahola tragically wiped out an entire family, causing five deaths, intense flooding, stream bank failures, erosion, and landslides, totaling more than $5 million in property damage.[17] Such events are not uncommon. On January 24–25, 1956, 42 inches (107 cm) of rain fell in 30 hours on the northeast side of Kaua'i, leading floodwaters to rise 10 feet (3 m) above normal in the streams between Kīlauea and Anahola.[18] The Hanalei River, which most directly drains the wettest region of Mount Wai'ale'ale, overflows its banks at the coast nearly every rainy year. Some years are considerably more damaging than others (for example, November 1955, January 1956, April 1994, September 1996, March 2006, and November, 2009).[19] In September 1996, for instance, 9 inches (23 cm) of rain were recorded in 12 hours along the coast, and an undetermined amount fell in the uplands. This event led to flooding of Hanalei town and temporary closure of the Hanalei Bridge, the residents' sole access to the rest of the island. In mid-November 2009, 17 inches (43 cm) of rain fell over 24 hours on the north shore. The Hanalei River flooded, forcing closure of the Hanalei Bridge and Kūhiō Highway at the Kalihiwai Bridge. Power and clean water were lost, stranding thousands until floodwaters retreated more than a day later. When waters finally receded, the town of Hanalei was buried in a layer of fine red silt.

Fig. 7.6. The Hanalei River valley on Kaua'i floods every few years. (Photo by Carl Berg)

In the western and southern portions of Kaua'i, the flooding hazard is primarily due to standing water that will not drain, especially after Kona storms. Waimea River, for example, has a long record of flooding dating back to 1916, which includes numerous occasions where its channels overflowed after storm-fed precipitation in Waimea Canyon at the head of the watershed.[20] The challenge to improving public safety and reducing damage from stream flooding is largely one of obtaining adequate warning in the case of flash floods and in improving the style and location of development in areas of known flood history.

Unfortunately, a new flooding concern has entered our consciousness in Hawai'i—dam failure. The deaths of seven people due to breaching or failure of the Kaloko Reservoir in northeastern Kaua'i on March 14, 2006, revealed a hidden killer.[21] With the demise of plantations has also come the end of careful dam maintenance. Reservoirs built in the last century to ensure that the thirst of sugar fields was adequately met have since fallen into neglect and decay, and a state of potentially criminal abuse. In the past, the state and landowners were careful to monitor, measure, and maintain the multimillion-gallon (tens of millions of liters) holding ponds that used to ensure adequate irrigation. Throughout the last century state engineers had routinely performed safety checks on

Fig. 7.7. Searchers with dogs looked for bodies in the mud and debris after the Kaloko Dam breach, March 14, 2006, on Kaua'i released a roaring, tree-snapping torrent of water and raised fears about the safety of dozens of similar dams across Hawai'i. (Photo by Casey and Cyndi Riemer, Jack Harter Helicopters; provided by Honolulu Star-Bulletin: http://archives.starbulletin.com/2006/03/15/news/story15.html [last accessed February 11, 2010])

Hawai'i's dams. But in the past decade, dams throughout the state have gone unchecked for flaws, proper maintenance, and other basic steps to ensure public safety. As records indicate, it has been years since some dams have been inspected.[22]

With no one watching, the Kaloko Reservoir was neglected to the point of tragedy; it became a deadly trap waiting to be sprung by an unfortunate combination of hydrologic and meteorologic conditions. What led to the dam-breach flood is unclear. Was it unauthorized changes to the dam? Poor maintenance? Lack of government oversight? Or another example of human engineering that was underdesigned to withstand the extremes of nature? Whatever the root cause, heavy rains finally pushed the dam beyond its breaking point and a wall of water 10 to 20 feet (3 to 6 m) high rushed like a tsunami through the valley below. Snapping trees like twigs, the churning cauldron cut a 3-mile (4.8-km) swath of destruction to the sea, wiping out homes, roadways, forests, and entire families.

FLOODING ON MAUI

Stream flooding on Maui is not only common but is the very agent responsible for making it famous as the Valley Island. The deep V-shaped valleys of West Maui have been carved by over a million years of stream flow. Along the eastern half of Maui, the mountains and valleys are much younger, and as a result the channels are not as well developed. Most of the streams cut steeply down to the narrow coastline of Hāna, often in cascading waterfalls. Annual rainfall is greatest (360 inches [914 cm]) at the summit of West Maui and nearly as high (240 inches [609 cm]) along the flanks of East Maui just below the trade-wind inversion.[23] Rainfall tapers off dramatically toward West Maui and Haleakalā and is lowest (<15 inches [38 cm]) in the vicinity of Kīhei and Lahaina.

Despite the general trend of fewer historic stream floods along the arid south slope of Haleakalā and more frequent severe floods in the wetter regions of Maui, flooding in dry areas such as West and Southwest Maui is common. Flooding in areas around Lahaina and Kīhei is in part a result of the abrupt transition in slope at the base of the highlands and the behavior of flash flooding. Many historic floods in those two areas occurred after heavy precipitation at higher elevations that fed narrow stream channels, and channelized drainages near the arid coast, to the point of overflow. Flash floods due to heavy precipitation, in some cases equaling the average annual maximum, have occurred throughout the historical record.

During the week of January 14–22, 1990, over 20 inches (50 cm) of rain fell on many parts of Maui, causing significant flooding in the coastal zone.[24] The north central portion of Maui and the Hāna coast, however, have the greatest stream-flooding histories. Nearly once a decade, a major flood emanates from 'Īao Valley bringing sheets of water down into the urban centers of Kahului and Wailuku. Events such as on November 30, 1950, and November 2, 1961, produced enormous volumes of stream discharge out of the 'Īao Stream Valley and generated sheet flows (open sheets of water flowing across the land) on the coastal plain below.[25]

In addition to flooding from stream channels, portions of Maui, notably the Lahaina region and Kīhei, are vulnerable to standing surface water flooding that refuses to drain into the ground. This may interrupt transportation and damage low-elevation buildings. Standing surface water develops after intense rainfall events where poor soil permeability and urbanization prevent adequate drainage. Along the road to Hāna, temporary road closures are common due to flash floods and mud slides from the steeper slopes of East Haleakalā.

FLOODING ON THE BIG ISLAND

Stream flooding in the coastal zone of the Big Island of Hawai'i results from heavy precipitation on the steep mountain slopes of Mauna Kea, Mauna Loa, and Kohala, as well as flash flooding from extraordinary rainfall events on the coastline itself. Kīlauea and Hualālai volcanoes are located in more arid regions but occasionally do receive intense rainfall that causes flash floods. Annual rainfall ranges between 300 inches (762 cm) on the slopes of Mauna Kea above Hilo, to below 10 inches (25 cm) on the arid regions of Kawaihae and South Point.[26] Soils that can absorb precipitation are more developed on the older volcanoes of Kohala and Mauna Kea, so mud slides and landslides are more common along the coastal cliffs of the Waipi'o and Hāmākua coasts.

The young lavas that compose the coastal terraces of Mauna Loa, Kīlauea, and portions of Hualālai are very porous and have less soil development. Often, heavy precipitation simply infiltrates into the rock and flows toward the sea in underground lava tubes, fractures, or in the highly porous young lava beds. As a result, stream flooding is generally less of a hazard on the younger coastlines. Nonetheless, many occurrences of islandwide stream flooding have been reported that are associated with precipitation from passing tropical storms and hurricanes or their remnants.[27]

Flooding along the wet, windward side of the island has been common and rather expected due to the large input of rainfall. Much of the

windward coastline is relatively steep, and so runoff occurs in deep channels that reach the shore below precipitous cliffs. Most of the flooding
that has caused damage is associated with extreme rainfall events that
bring about sheet flows between stream channels.

The Hilo and Puna areas are probably the most frequently flooded
and hardest hit by flash floods on the Big Island and perhaps in the state.
Severe flooding occurred in 2000 and again in 2003, 2004, and 2006.
On November 18–20, 1990, 30 inches (76 cm) of rain fell in the region,
bringing about intense flooding of the low-lying coastal areas.[28] What is
more surprising is the degree of flooding that the more arid regions of
the Big Island have sustained. The Kohala and Kailua-Kona coasts have
a long and active history of flooding largely due to flash flooding and
intense storms. From 1997 to 2001, the South Kohala and Waikoloa areas
experienced intense flash flooding that caused considerable damage.[29]

According to the data from the last 50 years, on average a damaging
flood event occurs on the Big Island every 2 years.[30] During this past
50 years, however, the threat due to stream flooding has increased dramatically because of the risk taken to develop extensively in flood-prone
areas.

Fig. 7.8. Flood damage on the Big Island occurs on average every 2 years. (Photo by K.
Kodama, National Weather Service)

OFFICIAL NOTICES

The National Weather Service is responsible for providing us with warnings when flash floods are imminent.[31] They issue bulletins to the media, who in turn make you aware of potential weather problems. The first warning is a "Flood Potential Outlook." This is issued 2–4 days ahead of a possible flooding event. Within 0–2 days of a potential event the National Weather Service will issue a "Flash Flood Watch." This means that it is time to prepare for flooding; families should account for members and modify planned activities to achieve maximum safety and protect against exposure to a dangerous situation.

A "Flash Flood Warning" will be issued when flooding is imminent or already occurring. This means that there is a direct threat to life and property, and immediate action should be taken to avoid the threat and protect oneself, one's family, and one's property. Flash Flood Warnings are usually issued for a portion of a county; hence the public can assess the degree of applicability to their particular situation and location. "Flash Flood Statements" will also be issued to alert the public about specific occurrences such as roadways that are not usable or other types of site-specific information.

FLOOD INSURANCE AND FLOOD DAMAGE AT HOME

For information on flooding risks in your neighborhood, refer to a Federal Emergency Management Agency (FEMA) Flood Insurance Rate Map. Flood Insurance Rate Maps are produced by the National Flood Insurance Program. These maps depict potential floodwater surface elevations as a probability of annual occurrence. They provide information on areas subject to flooding and are used by your insurance company to calculate your flood insurance premiums. They are used to guide future development away from flood-prone areas and to regulate development that is proposed to occur within such areas. The National Flood Insurance Program is a federal program that was established to allow property owners in participating communities to purchase insurance against losses from flooding. Participation in the National Flood Insurance Program is based on an agreement between local communities and the federal government stating that if a community will adopt and enforce a floodplain management ordinance to reduce future flood risks to new construction and substantial improvements in Special Flood Hazard Areas, the federal government will make flood insurance available within the community.

Unfortunately, updates to flood maps have not kept up with development in Hawai'i. Each year, numerous properties not officially mapped as "flood-prone" are flooded. Call your State Flood Coordinator (808-587-0248) for information about flooding. Or ask your local librarian to show you the Flood Insurance Rate Map for your community.

Besides purchasing flood insurance, you can strengthen areas in your home. Proactive mitigation activities include:

- clearing debris from your gutters and storm drains and alerting proper authorities (usually your county public works department) to debris in flood channels and storm drains in your neighborhood
- elevating the main circuit breaker or fuse box in your home as well as the utility meters above the anticipated flood level
- adding a waterproof veneer to the exterior walls of your home and sealing all openings, including doors, to prevent the entry of water
- anchoring external water or propane tanks so that a flood does not wash them away (an unanchored tank outside your house can be driven into your walls, and it can be swept downstream, where it can damage other houses)

Flood insurance covers the value of your home and/or its contents; whether you live in a flood zone or not, you should purchase flood insurance as an additional rider on your principal insurance coverage. Don't think that you are immune from flooding because you live in a high location. In Hawai'i, it is important to know that even homes on steep streets and high ground are vulnerable to flash flooding, intense rainfall, and stream overflow onto streets. If you have a flood insurance policy, you can be reimbursed for all your covered losses, even if a disaster is not federally declared. In contrast, if you do not have flood insurance and your home is damaged in a federally declared flood, you may get federal disaster assistance in the form of a loan—repayable in full—with interest. An average policy from the National Flood Insurance Program costs about $300 per year for $100,000 of coverage in high-risk areas. In low- and moderate-risk areas, the cost is about $100 a year.[32] Compare that low cost to paying back a $50,000 disaster home loan that will cost an average of $300 a month—for an average repayment of 20 years! Keep in mind that as many as 25% of flood insurance claims come from nonhigh-risk areas. In addition, rain damage is covered under flood insurance but not homeowners or hurricane insurance. To learn more about insurance call FEMA's National Flood Insurance Program at 1-888-CALL-FLOOD, or contact your local insurance agent. It pays to get flood insurance for your home.

HEAVY RAINS CAN DO MORE THAN CAUSE FLOODS

A sometimes-overlooked aspect of heavy rains in Hawai'i is the damage they produce that is not related to flooding. For instance, during the lingering Kona conditions in spring 2006, sewage pipes burst in Waikīkī, Waimānalo, Enchanted Lake, and elsewhere across the island of O'ahu.[33] In most cases, events such as these result from high water pressure due to rainwater seepage into aged and perforated sewer pipes. In Waimānalo, the overflow was delivered directly into the ocean, closing 3 miles (4.8 km) of polluted beaches to the public. Ten thousand gallons (37,854 liters) of sewage flowed into Enchanted Lake, which in turn delivered the pollution into the ocean through the center of Kailua Beach Park, closing the entire length of 2.5-mile(4 km)-long Kailua Beach.

In Waikīkī, work crews repairing a burst pipe diverted raw human sewage from dozens of hotels directly into the Ala Wai Canal for 5 days. Forty-eight million gallons (182,000,000 liters) of untreated sewage emptied into the ocean.[34] Sadly, the highest bacterial concentrations were recorded at popular surf sites Ala Moana Bowls, Rock Piles, and In Betweens. Worse, beaches were closed along the length of Waikīkī, the gateway of Hawai'i tourism, to prevent bacterial infections among the many swimmers and surfers. Captain Paul Marino with the City lifeguards said of the mess, "the beaches from the Royal Hawaiian Hotel to Sans Souci are now yellow, smelly with a brownish tint."[35]

The dumping of raw sewage and the delivery of sediment-laden waters containing the dissolved wastes of our roads, yards, and sidewalks have a largely unknown, but likely negative effect on the animal life of our reefs. Debris on beaches, layers of silt, and fresh water delivered at unnaturally high rates through our system of concrete stream channels all foul coastal waters every time a Kona storm settles in for a stay among the Islands. Although Kona storms are perfectly natural events, the products that their rainwaters deliver to the ocean reflect the waste of our society. Like flushing the toilet, heavy rains send human waste of all types into our coastal playground. This problem is discussed in more detail in chapter 8.

CHANNELIZING WATERSHEDS

Flooding is one of Hawai'i's most frequent, damaging, and dangerous hazards. Have we responded to this threat in an acceptable manner? The

Fig. 7.9. Channelized streams give sediments, nutrients, and other pollutants from developed watersheds a rapid ride directly into the ocean. Aquatic ecosystems and coastal marine environments, including reefs, are negatively impacted by this modern phenomenon. (Photo by R. Richmond)

Mānoa channel that backed up with debris and flooded residential properties and the campus at the University of Hawai'i in 2004, causing over $100 million in damage, was built to prevent flooding, not be the cause of it. In addition to occasionally not doing their job as planned, these channels destroy the native aquatic ecosystems of our streams.

Ideally, streams in Hawai'i should have natural beds and banks. Cool, clear water should meander through features such as pools and riffles that also provide habitat to native species. Naturally overhanging vegetation should provide shade from the hot subtropical sun for the indigenous ecology. Instead, dozens of cement causeways have been built by authorities to contain the raging torrents that now develop after hard rains. These replace the natural soft substrate of the streambed with a slick concrete surface and armored banks that offer no habitat to native species, and prevent normal stream behavior. Flood-control channels also replace nature's own version of a flood mitigation system—the floodplain. Normally, floodplains lie on either side of a stream channel and carry excess water and sediment during flood conditions. They are as much a part of a stream environment as the beach is part of an ocean. Channelization projects replace floodplains and encourage neighborhoods to be built on lands that have been repeatedly flooded throughout history.

Channelized streams tend to be free of overhanging vegetation, so they are exposed to the full heat of the sun, significantly increasing the water temperature. Each of these modifications from natural conditions (accelerated water flow, decreased habitat, and increased water temperature) has a negative impact on the ability of native species such as the o'opu (the native goby) to survive in the stream. Introduced species such as tilapia tend to be more robust than native species and are better able to survive in channelized streams. As we channelize more and more streams, we not only reduce the habitat available for native species but we also increase the amount of habitat for their competitors.

Channelization alters the natural rate at which a stream transports sediment and fresh water and delivers them to the shoreline. Silts, clays, and sands that would usually be stored by a stream along its floodplain and in its channel are instead carried to the sea. This leads to sediment pollution in the estuary that occupies the stream mouth and a deluge of fresh water with every rainfall. Brackish-water species adapted to tolerate estuary conditions cannot take this chemical and material stress and ultimately disappear. Many of these species and the organisms they feed on are marine fishes, so with the decline of estuaries comes an impact to our coastal fishery, the reef, and even offshore species.

Often forgotten is the fact that huge flood channels destroy wetlands that nature originally designed to mitigate flood hazards. But because we now build houses, streets, and entire communities in former marshy floodplains, their original function is forgotten and we cut these same communities in half with cement channels. A more rational approach is to avoid the flood hazard by mapping the geologic telltales of floods, look for floodplain deposits, statistically predict the frequency and water height of flooding, and build our communities out of the way. Channelization should be discouraged in most cases in favor of simply avoiding the floods by not building our homes in their path. Homes can be engineered with post and pier construction methods to lift them above flood elevations, and property owners can be required to manage precipitation that falls on their property rather than seeing it diverted into storm drains.

Is it possible that some communities might be redeveloped to recover and restore former floodplains? Yes, they have done just that in Sacramento and other communities trying to restore the natural ecology of their streams.[36] These places exercise flood avoidance by removing poorly sited homes and streets. In urbanized watersheds we can engineer a solution: retrofit a natural streambed component in channelization projects, use permeable surfaces when paving in a watershed, design no-build corridors along stream banks, buy the most hazardous properties and return them to an undeveloped state, and construct catchment areas on undeveloped lots and open lands. Floods are a natural event and we have much more to learn about coexisting with them.

ROCKFALLS, LANDSLIDES, AND HAWAI'I'S UNSTABLE HILLS

The watersheds and slopes of Hawai'i are the setting for a process geologists call mass wasting, where heavy clusters of rock and soil unhinge from a hill and slide, tumble, and ooze downward under the unending influence of gravity. Technically speaking, mass wasting includes rockfalls, landslides, slumps, debris avalanches, mudflows, and debris flows. But regardless of these details, they all refer to the perpetual work of two unseen culprits among the Hawaiian hills: gravity and time.

The Islands combine several essential components that lead to mass wasting: steep hillsides, a layer of loose rock and sediment, heavy rainfall, clay-rich soil, and strong demand for residential development (that may destabilize slopes) in upland areas. Intense rainfall triggers moving masses of soil, vegetation, and loose weathered bedrock that flow rapidly down the slopes and stream channels of watersheds. In one study[37] at the University of Hawai'i, 1,779 mass-wasting scars were identified in the hills of Honolulu between 1940 and 1989. Studies of slope stability in the Honolulu region indicate that about 55% of the area is highly unstable, highlighting a critical island problem.

- Heavy rains in spring 2006 spawned a dozen mass-wasting events that backed up traffic and closed lanes during peak rush-hour traffic on Kailua's Pali Highway. It was fortunate that no one was hurt by the various mudflows, debris slides, and rockfalls that rained down on the highway that season.[38]
- On March 6, 2000, rock and debris fell onto Kamehameha Highway at Waimea Bay. A new, $4 million stretch of highway has been constructed 38 feet (11.5 m) *makai* (toward the sea) to mitigate future risk to motorists there.[39]
- A mud slide in Mākaha swept away several cars and bikes and left rocks and mud in the lobby of the Mākaha Valley Towers condominium in November 1996.[40]

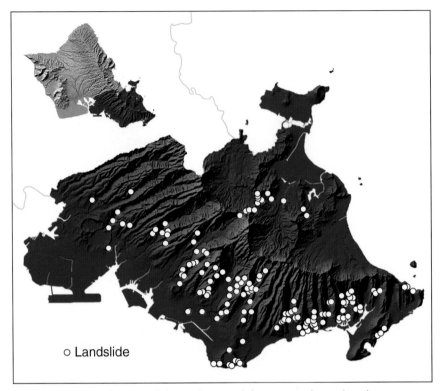

Fig. 7.10. Studies of slope stability in the Honolulu region indicate that about 55% of the area is highly unstable, highlighting a critical island problem.(SOEST Graphics, modified from S. K. Deb and A. I. El-Kadi, 2009, "Susceptibility Assessment of Shallow Landslides on Oahu, Hawaii, under Extreme-Rainfall Events," Journal of Geomorphology 108:219–233)

- On Woodlawn Street in Mānoa Valley a whole series of houses is slowly sliding down the clay-rich hillside.
- In ʻĀina Haina and Niu Valley several houses are slowly sliding downhill. The homeowners patch fractures in the walls, only to see them open again 6 months later.[41]

A PERSISTENT PROBLEM

In 1999, tragedy struck one of paradise's most admired destinations, Sacred Falls. This 80-foot-high (24 m) waterfall, popular with hikers, is located in the lush, upper Kaluanui Valley on Oʻahu's windward side. On Mother's Day, May 9, while dozens of tired hikers were sunbathing around the deep blue pool at the base of the waterfall, a mass of rock and soil separated from the valley upper wall and rocketed down the sheer cliff onto the unsuspecting families.[42] Free-falling boulders the size of cars and masses of dirt, cobbles, vines, roots, and sharp-edged rock crashed to the ground in the narrow confines of the valley. Eight people were killed instantly, another 32 were injured. Those who survived this shocking disaster did not notice any discernable event that would trigger such a catastrophe—no earthquake, no drenching rain, nothing specific to dislodge the rock.

Regrettably, the horrifying event at Sacred Falls was not a unique occurrence. Many examples of slope failure have been documented across the island chain although, unlike the catastrophe at Sacred Falls, there is often an obvious perpetrator, a trigger to start the moving slope. Two of the most common triggers for mass wasting are earthquake vibrations that reduce soil adhesion to the ground, and heavy rains that saturate unattached debris sitting on bedrock. Both vibration and saturation destroy the particle-to-particle packing that stabilizes loose sediment and rock. Saturated debris also gains the enormous weight of water; this alone is often enough to overcome slope stability.

Hawai'i's violent earthquake of April 1868 followed a period of torrential rain that triggered several slides, and a mudflow that buried 31 people and 500 livestock in the Wood Valley of the Ka'ū District on the Big Island.[43] The November 1983 earthquake on the Big Island generated landslides on precipitous slopes all around the island and destabilized hillsides that later collapsed from heavy rainfall.[44] Driving rain prompted several slope failures on O'ahu in November 1965.[45] One, the Pali Highway slide, released 20,000 tons (18,000 metric tons) of rock and mud over a 4-day period. Although each of these catastrophes was unpredictable, the damage they caused could have been diminished through specific mitigation strategies based on an analysis of vulnerable areas.

Potential damage from rockfalls and small slides can be avoided or reduced by reinforcing the ground with steel fencing, steel-mesh netting, and slope stabilization with grouting and the use of anchored gabions. These techniques are most commonly used to stabilize artificial cuts into bedrock and are only employed as a final option because they have to be maintained regularly. Another solution for stabilizing soils is to remove the threat by blasting away overhangs or by grading steep slopes to lower, more stable angles. Grading and vegetating the steep dirt cliff that threatened daily commuters on the Pali Highway was a successful maneuver that greatly increased safety on that mountain road.

But modifying slopes to improve safety can be unpopular. Residents

Fig. 7.11. Mass wasting is often started by heavy rainfall and flooding. (Photo by K. Kodama, National Weather Service)

do not like to see heavy engineering in the pristine hills of Hawai'i. Open views are important attributes to the beauty of our daily Hawaiian life-style—who wants to look at heavy-gauge steel fences among the placid greenery? Caves holding artifacts and other sacred sites dot the hillsides in hidden abundance and it is disrespectful, even sacrilegious, to alter the character of these legacies. Of course, cost is always a concern. Shoring up a hillside that has displayed no immediate impact to our safety, at a cost of millions, is often far down the "to-do" list of local legislators and agencies.

WHEN ISLANDS FALL APART

Among the awe-inspiring natural features of the Hawaiian Islands are the dramatically beautiful *pali* that are found on every island. These sheer, vertical rock faces of ancient basalt lava flows are hundreds to thousands of feet (m) high and often drop directly to the coast. They are the product of prodigious landslides that happen on an islandwide scale.[46]

The geological history of these immense cliffs is a tale of mass wasting on an extraordinary scale. The cliffs are initially formed by massive landslides that take away whole pieces of an island; they later evolve and retreat under the forces of erosion that continue to assault their steep slopes. To understand this process, it is important to understand that the Hawaiian Islands are composed of what are essentially piles of glassy volcanic rock. This rock (produced by basaltic magma) solidified within only a few minutes of reaching the seafloor or atmosphere where it cooled. With so little time to crystallize, the lava has the strength of glass and is very brittle because it has no crystalline structure.

Although no one was present to witness when these *pali* formed, it is hypothesized that swelling of a volcano associated with magma intrusion before an eruption causes the flank of a shield volcano to break off and slide into the sea. This is possible because from the floor of the sea to the surface the Islands are composed of glassy talus—literally a pile of jagged broken basalt.

Evidence in support of the prodigious landslide hypothesis is found in several observations. Maps of the seafloor show topography that is consistent with a landslide source. Submarine fields of chaotic debris can be traced back to the Islands where some of the highest *pali* are located. In addition, the main shield volcanoes composing the Islands have incomplete outlines. Rather than the expected broad oval shape, each of the two shields forming the island of O'ahu show only half their base—each forms an open crescent rather than the closed oval of a complete volcano.

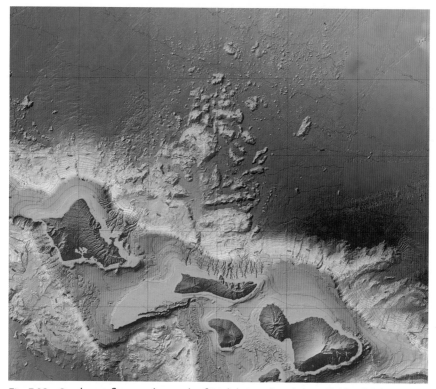

Fig. 7.12. On the seafloor to the north of Moloka'i and O'ahu are fields of debris resulting from prodigious landslides of the Ko'olau and East Moloka'i shield volcanoes. (Matthew Barbee, Perspective Cartographic LLC, U.S. Geological Survey, Western Coastal and Marine Geology, "Giant Hawaiian Underwater Landslides": http://walrus. wr.usgs.gov/posters/underlandslides.html [last accessed November 7, 2009])

The same is true of the island of Moloka'i. Missing pieces can be found some distance offshore where they have presumably slid in the course of an extraordinary mass-wasting event.

Scientists have identified over 15 massive landslides surrounding the Hawaiian Islands.[47] Surveys[48] show that submarine slumps and debris avalanche deposits are exposed over about 38,000 square miles (98,000 km²) of the seafloor from Kaua'i to Hawai'i, covering an area more than five times the land area of the Islands. Some of the individual debris avalanches are more than 120 miles (193 km) long and about 1,200 cubic miles (5,000 km³) in volume, ranking them among the largest on Earth, and most have taken place throughout the past 4 million years. The youngest landslide, south of Kona, is estimated to have occurred only 100,000 years ago. Today, geologists consider a phase of island mass wasting to be an important chapter in the formation of all Hawaiian Islands and perhaps other islands of similar origin around the world. Where will the next prodigious landslide occur? No one knows for sure, but the volcanically active shores of Mauna Loa and Kīlauea on the Big Island are being closely watched by researchers around the world as a likely candidate.

DEBRIS FLOWS AND HOW TO AVOID THEM

Hopefully the next landslide will not occur anytime soon. Debris flows, however, are happening all around us. Debris flows are fast-moving slope failures that begin as a shallow landslide on a steep incline during intense rainfall. The sliding mass of clayish soil, weathered bedrock, and vegetation turns fluid as it pours downhill. A debris flow may carve its own path or follow shallow drainage channels down a hillside, dragging along anything in its path. If the avalanche turns soupy from gathering lots of soil, the term "mudflow" may be more fitting, although its thinner texture does not diminish a mudflow's power to carry boulders or cars. An ancient mudflow deposit on Maui's south coast, near Kaupō, is 325 feet (99 m) thick, over 2 miles (3 km) wide, and contains blocks of rock over 40 feet (12 m) across![49] Debris flows are frequent visitors on the heels of prolonged rainfall that saturates hillsides.

With over 55% of Honolulu hillsides classified as "highly unstable," mass wasting is a critical islandwide problem.[50] Government concern reached a peak when the New Year's Eve storm of 1987–1988 dumped 23 inches (58 cm) of rain and triggered more than 400 debris flows in the Niu, Kuliʻouʻou, and Hahaʻione valleys. The torrential rain also contributed to slow-moving landslides that persisted long after the storm in the Kuliʻouʻou, Moanalua, and Mānoa valleys, as well as in ʻĀina Haina. However, these events were not completely surprising because many pre-1930s deposits and topographic features associated with earlier slides have been found in all of these valleys, and in many cases entire subdivisions have been built upon them.

Debris flows are surprisingly common, but there are ways to avoid them:

- Avoid construction on or at the base of steep slopes.
- Avoid slide run-out areas at the mouths of gullies, gulches, and narrow stream valleys.
- Keep away from regions where scars of previous events can be identified.

- Identify areas where other indicators of mass wasting exist, such as leaning trees or structures and hummocky topography.
- Do not construct in areas where native vegetation that anchors the soil has been cleared.
- Get advice from a consulting geologist or engineer if questions persist.
- A detailed report on landslides, debris flows, and rockfalls is available from the City and County of Honolulu.[51]

If a building or home is already located on an unstable slope, only a few options exist to reinforce the soil. One can revegetate the hillside with native plants, grade slopes to a more stable angle, install drain systems to dewater slopes, anchor one's house to bedrock, and avoid increasing the load on the slope. Of course, one could always consider relocation if these options are not effective. Much of the information on how to avoid a debris flow applies to slow-moving landslides as well.

CREEPY HILLS

Slow-moving landslides encompass relatively large areas of ground that gradually and sporadically move downhill. As the soil and loose rock on an unstable hillside slowly creep downward over months to years, moving like a massive earthworm, part of its bulk may stretch and fracture while other areas compress and fold due to its sheer size. If any structures such as houses, streets, sidewalks, or utilities are carried on top of the slow landslide, they too will suffer the irregular buckling that can result. The resulting damage ranges from minor flaws to total structural collapse.

Shortly after the Wai'ōma'o Subdivision in upper Pālolo Valley of O'ahu had been completed, the first signs of slope instability appeared.[52] Cracks in walls and poured slabs, framing out of square, displaced landscaping: these were the first sure signs of unstable ground. In the second year of its existence, Wai'ōma'o experienced a 9-inch (22-cm) rain that aggravated these instabilities. Within 3 years the side effects of slow creep were becoming alarmingly obvious as water mains and utilities that suffered damage were relocated in an attempt to compensate for the ground's inexorable downhill creep. This continued for several years, until the landslide had consumed 40 homes and thwarted every attempt to stabilize the subdivision. Several years later the subdivision was completely abandoned, and the city purchased the land and converted it into a park—a memorial to the costs of ignoring the geology when developing the land.

There are two lessons to be gleaned from the Pālolo example: (1) Avoid this situation at all costs by integrating a thorough geohazard analysis into all development plans. (2) If such an analysis is not possible before development, the only safe and cost-effective option once the hazard is identified is to relocate. Slow-moving landslides move a very small fraction of an inch per day, creating an essentially unobservable process. Any building caught in this kind of motion can be damaged slowly enough to mislead the owner into thinking that a few repairs will solve the problem. But the problem is irresolvable, and damage can continue to the point of endangering the residents of the dwelling. Fortunately, there are signs that can reduce the risk of a homeowner buying a residence that will take him or her for a ride.

Trying to predict future sites of slow-moving landslides is not 100% accurate and can be extremely tricky. Knowing the factors associated with such slides at least provides guidelines for evaluating an area's risk. The key lies in knowing how to decode the evidence.

- First, slow landslides occur in areas along the margins of valleys on relatively low slopes of 5° to 25°, near the base of steeper valley walls.
- Second, beware of dark, clay-rich soils, characterized by clay minerals that expand and swell when wet, and shrink and crack when dry. These clay-rich soils exist within the region of moderate rainfall (39 to 79 inches per year [100 to 200 cm/year]) and can act as slick surfaces for the slow-moving slides.
- Third, it is heavy rains that initiate the first movement of slow-moving landslides and periodic reactivation of the slippage. Once a slide begins to move, evidence of the shift can be spotted in the form of scarps (a steep slope or small cliff formed by erosion or faulting) soil-surface hummocks (irregular rises or humps); offsets in curbs; cracked driveways and foundation walls, windows, and doorframes and other linear features; uniformly curved trunks among a stand of trees; and bulges of soil in the middle or lower reaches of a slope.
- The City and County of Honolulu has a map of unstable slopes subject to debris flows. Check this before considering any real estate purchase.[53]

AVOIDANCE IS ALWAYS EASIER—BUT WHERE DO WE FIND THE INFORMATION?

The watersheds of Hawai'i are the birthplace of our waters. But they are not the tranquil, cloud-shrouded regions they may appear from afar. Nestled between valley walls are flash floods, rockfalls, debris flows, creeping hillsides, and torrential rains. Knowing the geologic clues that Mother Nature always leaves for our discovery can spell the difference between catastrophe and serenity on these verdant slopes. Avoiding hazards means knowing how to find them in the first place. Geologic analysis should be part of every development and the final factor in siting everything from homes and roads to pathways and parks.

Information pertinent to landslides is still limited, despite the fact that landslides, especially rapid shallow landslides, pose severe threats to life, infrastructure, and property. Fear of possible future landslides has become a major constraint on further property development. Annual landslide inventories are rarely done, and there is almost no information on slope stability with respect to landslide occurrence. Greater efforts are needed to understand mass-wasting processes, to analyze threatening

Fig. 7.13. Flood debris on the Big Island; there is little public information and no centralized resource providing expertise on natural hazards in Hawai'i. The public needs a lead science agency to fill this gap. (Photo by K. Kodama, National Weather Service)

landslide hazards, and to predict future landslides as accurately as possible to minimize damage.

Unfortunately, the state of Hawai'i has no lead science agency. In fact, being the only state in the union without a state geologist and lacking an office of the state geological survey, there is literally nowhere for citizens to turn for geologic information. Would you like to see a map of unstable slopes? A map of streams prone to flash flooding? The latest report on mass wasting in Hawai'i or regions vulnerable to rockfalls? None of this information is easily or immediately available to the public. Some of this information exists, but "who ya gonna call?" the state? the county? the feds? The answer is yes, yes, and yes—in each case different agencies handle different hazards (if at all). In fact, neither is there a collated body of literature on rock and mineral resources, sand and gravel reserves, soil erosion, coastal erosion, torrential rains, flash floods, landslides, or any number of geological issues of critical importance to public safety and resource management.

CHAPTER 8

Sewage Treatment and Polluted Runoff

'O ka mea ukuhi ka i 'ike i ka lepo o ka wai 'o ka mea inu
   'a'ole 'oia i 'ike

"He who dips knows how dirty the water is, but he who
   drinks does not. He who does the work knows
   what trouble it takes, he who receives it does not."

Fig. 8.1. The Kahului wastewater treatment plant on Maui puts nearly all its treated effluent into injection wells. (Photo by Z. Norcross-Nu'u)

Human waste is not a subject discussed with ease or finesse, yet it is an unavoidable by-product of human existence that can contaminate the surrounding environment and water systems if not disposed of in a safe and deliberate fashion. Within traditional Hawaiian culture, the disposal of human waste was treated with extreme care, associating it with a number of *kapu* to protect the purity of Hawai'i's water. It was forbidden to relieve oneself in any of the natural water sources, including streams, wetlands, and the ocean. So, on long outrigger canoe voyages, detailed ceremonies were performed to absolve the travelers for tainting the respected resource.

THE POINT ABOUT POINT SOURCE

Today, the City and County of Honolulu treats about 113 million gallons (428 million liters) a day of wastewater.[1] When this effluent is emitted from a pipe or some other identifiable point of discharge, it is termed point source pollution. Although the term often conjures up ugly images of brown sewage pouring into the ocean from a large, rusty pipe, this is rarely the case. Wastewater is generated from many sources including agricultural processing, industrial facilities, military bases, and county treatment plants, each with unique properties of contamination. Most of Hawai'i's wastewater comes from industrial sources, generated from cooling water, metal cleaning waste, and power generation stations. Sewage and, to a lesser extent, agricultural wastewater make up the rest.

Hawai'i has several major wastewater treatment plants that discharge into coastal waters. Sand Island with an upgraded capacity of 90 million gallons (340 million liters) per day and Honouliuli at 38 million gallons (144 million liters) per day are the two largest, and they serve the urban metropolitan region on O'ahu's south shore. These two offer advanced primary treatment, below the required secondary treatment mandated by the U.S. Environmental Protection Agency. Sand Island and Honouliuli, as well as Wai'anae, Kailua, and Hilo wastewater plants all discharge in waters below 130 feet (40 m) depth. It is widely believed by scientists that this is sufficient to prevent sewage from returning to the coastal area, though there have been challenges to this. Three outfalls are located in water shallower than 130 feet (40 m) and these are located at East Honolulu, Fort Kamehameha, and Wailua (Kaua'i). Other than Sand Island and Honouliuli all sewage is treated to the secondary level. These levels of treatment are discussed in this chapter. But first, let's start at the beginning—"What happens after you flush?"

WHAT HAPPENS AFTER YOU FLUSH?

After you flush, you have liquid waste, solid waste, toilet paper and other paper products, various types of medicines, cleaners, and other synthetic chemical compounds all carried by gravity into a pipe buried under your street. This toxic brew combines with the sewage of your neighbors and is carried by gravity, aided by pumps, to the closest sewage treatment plant. There, your tax dollars pay for it to be processed into a form that can be disposed of. This process consists of primary treatment, secondary treatment, and in some cases tertiary treatment.

All of these methods of treatment are used in Hawai'i. The City and County of Honolulu, responsible for the largest portion of the state's population, has six plants using secondary treatment: Kailua, Wahiawā, Kahuku, Wai'anae, Waimānalo, and Pa'ala'a Kai wastewater treatment plants. At these locations, treated wastewater is injected into the ground, discharged into a lake, or discharged into the ocean. These discharges return the water and remaining impurities for recycling back into our

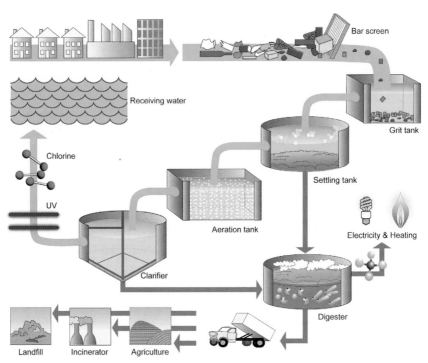

Fig. 8.2. Liquid sewage is treated to remove pollutants. (SOEST Graphics)

ecosystem. The Sand Island and Honouliuli wastewater treatment plants process sewage for the urban core in Honolulu and the ʻEwa Plain. These two plants treat most of the wastewater (more than 85%) generated on the island of Oʻahu.[2]

When your treated sewage is discharged into the sea, it is assumed that it is harmlessly diluted in the immense volume of the Pacific Ocean. Ground disposal assumes that by flowing through island bedrock, our wastewater is adequately purified. Behind these steps lies the recognition that in the ideal world of a closed ecosystem there is no such thing as waste. Just about everything organic (meaning a carbon compound made by a living organism) is eventually recycled. For example, animals exhale carbon dioxide, and plants use this gas and water in the process of photosynthesis to make their food. Plants thrive on natural fertilizer from the droppings of animals and in return provide food to the animal kingdom as well as oxygen to breathe. Nutrients, water, and carbon are constantly being recycled through the ecosystem.

The treatment of wastewater from humans borrows this natural recycling. The most popular wastewater treatment method (secondary treatment, mandated nationwide by the federal Environmental Protection Agency under the Federal Clean Water Act) is simply an amplification of what occurs naturally.

WASTEWATER TREATMENT

Treating wastewater consists of several steps, and most treatment facilities employ a unique blend of several methods to best suit their needs. All sewage goes through preliminary treatment when it enters a facility. Preliminary treatment is a mechanical process with the goal of removing physical objects and grit from the wastewater. Sewage arriving from neighborhood pipes is run through large screens, grinders, and grit channels to catch physical materials such as rags, large food items, sand and stones, and other garbage.

The next step is primary treatment, which relies on physical separation to remove solids from the wastewater. This mostly involves letting water move slowly through large sedimentation basins to permit fine solids and organic compounds to settle to the bottom or float to the top. Mechanical scrapers and belts clear the "sludge" from the bottom of the tank while slow-moving rakes cross the surface and remove floating "scum." Scum includes grease, oils, plastics, and soap (compounds lighter than water, hence they float). Scum is thickened and pumped along with the sludge to enclosed heated tanks called digesters. There, solids are kept

Fig. 8.3. "Scum" at the Kahului wastewater treatment facility on Maui is removed during the treatment process. (Photo by Z. Norcross-Nu'u)

for 20 to 30 days while bacteria digest organic material, reducing its volume and odors, and getting rid of some of the organisms and compounds that may cause disease. The finished product is mainly sent to landfills but sometimes can be used as fertilizer.[3]

Secondary treatment employs mostly biological processes to remove much of the remaining dissolved and microscopic organic matter. Several different techniques are used to achieve this. One common type is the activated sludge system that consists of two parts, an aeration tank and a settling tank, or clarifier. The aeration tank contains a culture of bacteria, protozoans, fungi, and algae that is constantly mixed and aerated with compressed air or large mechanical mixers. Wastewater enters the tank and mixes with the culture, which uses organic compounds in the sewage for growth—producing more microorganisms—and for respiration, which results mostly in the formation of carbon dioxide and water. Both growth and respiration destroy organic compounds in wastewater using the natural metabolism of the bacteria. The process can also be set up to provide biological removal of the nutrients nitrogen and phosphorus. After sufficient aeration to reach the required level of treatment has taken place, wastewater flows into the clarifier. Once there, sludge settles to the bottom of the tank producing a reasonably clear upper layer of water that is removed and discharged or subjected to further treatment.

Other techniques can be employed to further purify the water, including chemical treatment, filtering, and radiating the water in ultraviolet light to kill bacteria.[4] Clarified wastewater may flow into a "chlorine contact" tank, where the chemical chlorine is added to kill bacteria, just as is done in swimming pools. The chlorine is mostly eliminated as the bacteria are destroyed, but sometimes it must be neutralized by adding other chemicals. This protects fish and other marine organisms that might be harmed by the smallest amounts of chlorine when the treated wastewater is discharged.

Some treatment plants in the United States also use filtration in sewage treatment. After solids are removed, liquid sewage is filtered through a substance, usually sand, by the action of gravity. This method gets rid of almost all bacteria, reduces turbidity and color, removes odors, reduces the amount of iron, and removes most other solid particles that might remain in the water. Water is sometimes filtered through carbon particles, which is also effective at removing organic matter.

Fig. 8.4. Water-quality tests near the Honouliuli wastewater treatment plant's outfall near Barbers Point on Oʻahu revealed bacteria counts exceeding national standards for recreational uses, according to the Environmental Protection Agency. (Photo by F. L. Morris)

Combined primary-secondary treatment removes about 97% of suspended solids, 95–97% of organic waste, 70% of most toxic metal compounds and organic chemicals, 70% of phosphorus, 50% of nitrogen, and 5% of dissolved salts.[5] However, only a small percentage of pesticides (such as dieldrin and chlordane) and other types of particularly persistent chemical compounds are removed.

LESS THAN PERFECT

The federal government plays a significant role in regulating the discharge of point source pollution through the Clean Water Act of 1972.[6] The act's primary goals are to ensure that all waters in the United States are clean enough for fishing and swimming and to end all discharges of pollutants into navigable waters. The National Pollution Discharge Elimination System—part of the Clean Water Act—requires all dischargers to have permits and mandates all publicly owned treatment facilities to achieve secondary treatment of effluent to protect public health. Honouliuli and Sand Island treatment plants do not meet this standard.

In 1977, Congress amended the Clean Water Act to allow municipalities to discharge at less than secondary standards if they can establish that it would not harm the environment.[7] This stipulation has received widespread attention across the Hawaiian Islands because our two major sewage plants, Sand Island and Honouliuli, seek to remain at the level of "advanced primary" treatment. This is because of the great expense of upgrading (Honolulu Mayor Mufi Hannemann has estimated that upgrading the Sand Island and Honouliuli plants will cost more than $1.2 billion)[8], and because it is thought that dilution in the deep ocean water adjacent to our Islands is sufficient to mitigate potential health and environmental impacts.[9] Experts, and the court of public opinion, continue to debate whether the level of nutrients, suspended solids, ammonia production, organic matter, and industrial chemicals in Honolulu effluent poses a threat to sensitive marine habitats. Likewise, opinions vary with regard to the potential human health impacts of effluent dumped in the deep waters offshore of island beaches. In the meantime, Sand Island and Honouliuli treatment facilities are slowly improving the quality of water that makes its way into the oceans as they add components such as disinfectors, larger digestion facilities, and upgrades to sewage delivery pipes beneath our streets.

Poised for battle is the U.S. Environmental Protection Agency (EPA), which is pressuring Hawai'i to treat all sewage to the secondary level, while the City of Honolulu struggles to convince them that appropriate treatment standards and disposal guidelines specific to its unique midocean environment should be exempt from the Clean Water Act. Because of the Clean Water Act, most U.S. cities have combined primary and secondary sewage treatment plants.

To reduce the environmental and public health risks associated with primary wastewater effluent, managers have three basic options at their disposal: improve the treatment level, change the effluent discharge location, and upgrade the facilities' operation to minimize leakage and emergency bypassing of untreated water associated with flooding during rainstorms. Each, or all, of these steps will lessen the negative environmental impacts related to sewage discharge; and each has been employed in the Hawaiian Islands.

The advanced primary treatment being performed at Sand Island and Honouliuli are controversial. Both plants were formerly granted waivers under the Clean Water Act to legally discharge sewage treated at the primary level. Upgrades at the Sand Island plant allow it to operate with increased capacity, as much as 90 million gallons (340 million liters) per day, and to provide ultraviolet or chlorine disinfection, improved filtering, and recycling of treated sludge into fertilizer pellets. Extension of sewage outfalls at Sand Island and Kāneʻohe Bay from shallow water to deeper water in the 1970s and 1980s has resulted in significant improvements with regard to seafloor ecology and human health concerns.[10] Outfalls, placed far enough out to sea, benefit from oceanic currents that carry sewage offshore, trap it below a thermocline (temperature boundary), and otherwise mix the waste in the vast cauldron of the ocean. But is this sufficient?

The EPA says "no," citing studies done in March 2007 at both Honouliuli and Sand Island that found that bacteria levels were higher than national standards established to protect swimmers, surfers, and others from gastrointestinal diseases. Effluent from both plants was tested to determine impacts on sea urchin fertility. EPA results indicate the primary-treated wastewater produces excess ammonia that may harm aquatic organisms. Samples from both plants also contained excess levels of two pesticides, dieldrin and chlordane.[11]

The City and County of Honolulu says "yes," citing the enormous cost of retrofitting to secondary treatment as unaffordable and pointing to Hawaiʻi's unique position in the Pacific Ocean that renders secondary treatment unnecessary. Mayor Hannemann said in August 2007, "To upgrade these two plants to full secondary treatment, the city will need to expend an estimated $1.2 billion for construction at the facilities. This expense doesn't include the increased cost to operate and maintain the expanded plants nor does it include the additional energy consumption and greenhouse gas emissions of secondary treatment."[12] City Hall fought the EPA with a battalion of scientists who testified that the use of sea urchins as a telltale was inappropriate, pesticide measurements fell within acceptable guidelines, and the ammonia measurements were contradicted by real-world evidence. Over the course of four public hearings, not a single scientist or engineer familiar with the system testified that denying the Clean Water Act waiver would result in any environmental benefit. The City also argued that the money would be better spent upgrading the county's sewage-collection pipes.[13] Holes in this delivery network are a main source of nonpoint source pollution that enters our coastal ocean and groundwater system.[14]

FACILITY IMPROVEMENTS

Though treatment standards for wastewater and the location of where wastewater is ultimately released to the environment are both important factors dictating the environmental impacts of treated sewage, facility operations and maintenance problems have been at the heart of many water-quality violations in Hawai'i. The Clean Water Act played a key role in improving facility-level operations across the United States, beginning in the 1970s. The Clean Water Act not only imposes financial penalties when violations occur, but the federal government also offers a carrot of financial assistance for facilities that desire to improve their operations.

During the 1970s, federal funds were made available, under a Construction Grants Program, to build municipal wastewater treatment plants.[15] Through this program, the net cost of building or upgrading facilities in Hawai'i significantly decreased from their original price tag. Maui capitalized on the opportunity, and as a result, all of its sewage is treated to the secondary level, and nearly all of the effluent is disposed of using injection wells located in coastal areas below (down gradient from) potable aquifers, rather than through ocean outfalls. The Kahului wastewater treatment plant puts about 5 million to 5.5 million gallons (19 million to 21 million liters) per day of its treated effluent into injection wells. The county's Lahaina and Kīhei plants reuse from one-fourth to one-third of their effluent.[16] Kaua'i also seized the opportunity to upgrade its treatment facilities, recognizing that the island's increasing population could overburden its aging systems. The County of Hawai'i used federal Construction Grant money to expand and even overbuild their plants, allowing them to operate considerably below their capacity.

However, wastewater treatment on the Big Island is not without its problems. The Kealakehe wastewater facility in North Kona has no disposal unit. It discharges a daily load of 1.5 million gallons (5.7 million liters) of wastewater directly into the surface waters of Honokōhau Harbor on the Kona coast. This is a major problem that has no parallel on the other islands.

WRESTLING WITH OUTFALLS

In the last decades of the twentieth century, heightened public awareness, a rash of federal water-quality violations, and several notorious and lengthy lawsuits prompted the placement of outfalls deeper in the ocean and farther offshore. Studies associated with these lawsuits have highlighted Hawai'i's deep-water location, ideal in many ways for rapid diffusion of pollutants provided it can be established that effluent-carrying currents are not delivering waste to bathing waters or sensitive ecosystems.

The problem is that environmental politics sets different standards for different people. No scientific study is fail-safe and absolute. All have built-in assumptions and undersampling of the desired information. Depending on their inclination, two people may view the same study in a very different light. One person's thorough study is another person's flawed research. Hence in the politics of environment, environmentalists remain skeptical that our outfalls are safely located and not impacting nearshore water quality. They continue to call for wastewater treatment to the secondary level at all facilities, and there are experts that back their opinions.

With fewer people inhabiting the outer islands, their wastewater dilemmas have not reached the same boiling point as they have on O'ahu. In the past, treatment plants have been built only when population density has reached a critical mass and public concern erupts. On Kaua'i, there are only four treatment facilities, all of which treat their sewage to the secondary level, and the majority of this water is used for golf-course irrigation now. Due to the island's former character of intense agricultural production and sparse population, farming waste composes a larger proportion of Kaua'i's wastewater than that of any other island—but that is changing in the twenty-first century. The more rural islands of Moloka'i and Lāna'i have a combination of facilities shared between commercial resort and ranching operations authorized by the state Department of Health and the County of Maui public facilities. Settling ponds, injection wells, and water recycling characterize wastewater treatment on these islands with little to no direct ocean disposal.

IMPROVEMENTS TO SAND ISLAND

In the late 1990s and early this century the City and County of Honolulu spent hundreds of millions of dollars on improvements to its largest sewage treatment plant, Sand Island. The Sand Island wastewater treatment plant services an area stretching from Kuli'ou'ou Valley on the east side of Honolulu to Āliamanu on the west side of Honolulu. The facility diverts sewage from the city of Honolulu into the sea with advanced primary treatment. However, with deteriorating infrastructure, upgrades to the plant equipment have been necessary to correct federal Clean Water Act violations. These include the following:

- expansion of capacity from 82 million gallons (310 million liters [mld]) per day (mgd) to 90 mgd (340 mld)
- increased wet-weather capacity from 210 to 270 mgd (794 to 1,022 mld)
- introduction of an ultraviolet disinfection system to reduce pathogenic organisms
- construction of new main lines, pumps, and delivery/receiving components for sewage
- refurbishing clarifiers, new screening channels, and new aerated grit chambers
- installation of biofilter trickling filters and carbon scrubbers to reduce hydrogen sulfide gas emissions[17]

Despite these corrections, however, the plant continues to operate at the advanced primary level, and, in the past, the EPA has threatened to revoke permits to operate the plant if it is not upgraded to the level of secondary treatment.

Offshore discharge of treated sewage has improved since 1972 when sewage from Honolulu, amounting to about 62 mgd (234 mld), was being discharged off Sand Island at a depth of only 38 feet (11.5 m).[18] The discharge was untreated raw sewage, and thick sludge deposits accumulated on the seafloor with measurable impacts to the reef community. The ocean surface was marred by an ever-present thick, grayish brown plume usually heading in the direction of 'Ewa Beach and Barbers Point. During periods with calm winds the sewage was carried toward the shore and could be found at Ala Moana Beach Park. Studies revealed that viruses from the sewage discharge were being carried into recreational waters.

By 1976, Sand Island had installed a new 78-inch (198-cm) -diameter outfall that extended about 1.5 miles (2.4 km) offshore where the ocean was 225 to 240 feet (68 to 73 m) deep.[19] The new outfall was designed with a long diffuser section to discharge sewage over a greater length of the seafloor. The combined effect of extending the outfall, the diffuser head, and upgrades in treatment level at the plant have considerably improved water quality and, according to City and County scientists and engineers, virtually eliminated impacts to the seafloor ecology.

The diffuser heads distribute wastewater across a wider area of the seafloor. Instead of discharging sewage in one big mass from the end of a pipe, the outfall at Sand Island has a 3,400-foot-long (1,036 m) diffuser section with 282 openings, ranging from 3 to 3.5 inches (7.6 to 9 cm) in diameter and spaced 24 feet (7.3 m) apart. This way sewage is discharged in small amounts from each port and spread out over the length of the diffuser. The City and County monitors bacteria levels, micromollusk populations, over 64 chemicals in the flesh of fish caught in the area, abnormal growths in the livers of fish caught in the area, nutrient levels (nitrogen, phosphorus), chlorophyll levels (related to algae growth), turbidity, temperature, and properties of the sand and sediment from the seafloor. The data from these tests help to determine how the ocean environment is changing and whether the wastewater discharge is having any effects.[20]

EAST HONOLULU AND KĀNE'OHE

In 2002, a study of the Sand Island and East Honolulu sewage outfalls by researchers at the University of Hawai'i found the following: (1) an area off Sand Island that was essentially destroyed when the discharge was located in about 35 feet (10.7 m) of water has fully recovered to the same level as exists at nearby natural ecologies now that the outfall has been relocated below 200 feet (61 m) depth, and (2) discharge from East Honolulu at Sandy Beach has negligible effect on reef communities, but the communities are continually oscillating in coral abundance as a result of periodic large storm surf from the east.[21]

During the 1970s, Kāne'ohe Bay's marine organisms suffocated under a deluge of sewage being dumped daily into the nearshore environment. In response to the community's outrage over such environmental negligence, a new deep-water outfall was completed in 1977 to divert sewage into the open ocean off Mōkapu Peninsula on the outskirts of Kailua Bay. It took 2 years for visible improvements in the water quality to appear at the old site in Kāne'ohe Bay, and evidence suggests that the reef-based community there is recovered and thriving.[22]

INJECTION WELLS: A COMMON WAY TO GET RID OF WASTE

Still the most common method for disposing of wastewater in Hawai'i is through the injection well. Wastewater injected into the ground becomes part of the groundwater system and eventually discharges into the ocean. Along the way dilution and chemical processes can reduce the nutrient load. On Maui, from 1997 to 2008, the wastewater plants at Kahului, Kīhei, and Lahaina injected an estimated combined total of more than 51 billion gallons (193 billion liters) of secondary wastewater into the ground. Even treated to the secondary level, this effluent could be a source of pollutants if it gets into the natural ecosystem. The problem is, no one is quite sure if it is causing any problems.

Managing the location of injection well sites is based on a simple map (an underground injection control map, or UIC) of each island that out-lines the periphery of salty groundwater levels around the Islands and the location of drinking-water aquifers. Most injection wells are drilled close to shore, based on the assumption that treated effluent will mix with the salty groundwater and not threaten potable aquifers. The basic operating

Fig. 8.5. The County of Maui owns and operates 18 injection wells: 8 in Kahului, 4 in Lahaina, 3 in Kīhei, and 3 in Kaunakakai on Moloka'i. Seventeen of the injection wells in Maui County range in depth from 80 to 385 feet (55 to 117 m); Moloka'i has one that is 29 feet (8.8 m) deep. (Photo by Z. Norcross-Nu'u)

principle behind an injection well's location is to protect drinking water from any possible contamination. This responsibility belongs either to the wastewater plant using the well or to the county, which oversees wastewater treatment. Impacts to the natural environment are another issue; one may question whether we have sufficient understanding of the details of each island's subsurface geology to know where to place these wells to have minimal environmental impact. Over the years, compliance with federal and state regulations has decreased, which has led to discussions on implementing more-stringent regulations of injection well locations.

Subsurface injection wells can pose problems for Hawaiʻi's environment. First, a percentage of injection wells experiences problems such as clogging, with over half of Oʻahu's monitored injection wells periodically or continuously overflowing. Second, if the well's effluent migrates far from the site without a sufficient amount of dilution, coastal ocean waters might become contaminated. Groundwater flows from the mountains to the shore carrying a background nutrient load from the adjacent watershed. As this water moves beneath urbanized lands along the coast it picks up additional nutrients from nonpoint sources such as fertilizers, septic systems, and leaks in the sewer system. Wastewater injection at treatment plants adds to this load that potentially enters the ocean. With this possibility in mind, the U.S. Geological Survey conducted a study of the effluent from one injection well on the Kīhei coast of Maui.[23] Although they are careful to point out that the study focused on only one well, results indicated that the nutrient load was considerably attenuated (reduced) during flow through the permeable geology of the coastal zone. They found that despite this attenuation nutrient movement was 3.5 times greater than the background movement of both nitrogen and phosphorus. The USGS scientists compared their measurements with nutrients measured in coastal groundwater at golf courses on the Big Island. Compared with Big Island golf courses, the nitrogen movement was one-half to one-third less in the Kīhei study.

Use of injection wells in Maui has become controversial in recent years largely because they have been pointed to as a source of nutrients contributing to noxious algae blooms on beaches. However, there are many sources of nutrients that could contribute to the algae problem. The Maui County position on wastewater injection and algae is that studies do not prove that nutrients from these wells are the sole or most significant source of blooms or reef damage. Although independent studies detected injection-well discharge in some areas of algae blooms, other sources from rainfall runoff, reef siltation, agricultural fertilizer, overfishing, and human interaction must also be analyzed as contributing causes. It is important to thoroughly consider and identify all sources so that efforts to eliminate damage are successful.[24]

Crumbling U.S. sewage system

The mainland has about a million miles (1.6 million km) of sewage-collection pipes and pumps designed to carry over 50 trillion gallons (190 trillion liters) of raw sewage daily to some 20,000 treatment plants. But parts of this complex and aging infrastructure are crumbling, resulting in spilled sewage entering neighboring groundwater, aquifers, watersheds, and even homes. Hawai'i is included in this pattern, and many observers are of the opinion that fixing this aging infrastructure in the Islands is the most important sewage treatment priority. Accidentally spilled, discharged, and leaked sewage costs Americans billions of dollars every year for medical treatment, lost productivity, and property damage. This pollution poses a health risk to communities and ecosystems across the nation.

A statement by the Association of Metropolitan Sewage Agencies says that the Congressional Budget Office, the Government Accounting

Fig. 8.6. Repairing the system of pipes that carry sewage to treatment plants is considered a high priority. (Photo by D. Oda)

Office, and the EPA all agree that there is a national funding gap estimated to be as high as $1 trillion that is needed for upgrading water infrastructure.[25] Researchers found that in 2001 there were 40,000 sanitary sewer overflows and 400,000 backups of raw sewage into basements.[26] The EPA estimates that 1.8 million to 3.5 million individuals get sick each year after swimming in waters contaminated by sewage overflows.[27] A large part of the problem is one of aging infrastructure; some pipes still in use are almost 200 years old, although the average age of collectionsystem components is between 30 and 40 years. Federal officials predict that without substantial investment in the nation's sewage infrastructure, by 2025 U.S. waters will again suffer from sewage-related pollutant loadings as high as they were during the record year of 1968.

HAWAI'I'S SEWAGE TRANSPORT SYSTEM

Hawai'i's sewage transport system mirrors problems across the nation. Although the 1980s were spent attempting to resolve many sewagedisposal problems by relocating outfalls, the 1990s verified that Hawai'i's antiquated delivery and transmission system was one of the biggest contributors to sewage-based pollution. And in early 2006 this was brought to appalling relevance with the largest sewage spill in state history—48 million gallons (182 million liters) of raw untreated human sewage were dumped directly into the Ala Wai Canal to prevent it from surging out of toilets and sinks into hotel rooms and apartments throughout Waikīkī.[28] During that same week another 1.85 million gallons (7 million liters) of raw sewage spilled into Hawaiian waters at half a dozen other sites located around O'ahu. One man who fell into the sewage-laden Ala Wai Canal died a week later of massive infections that included a form of flesh-eating bacteria, although his case was compounded by additional factors.

Raw sewage spills, leaks in pipelines, treatment-plant problems, sewer-line blockages, structural failures from corrosion, sewage infiltration to groundwater, groundwater infiltration into overwhelmed treatment facilities, limits in sewer line capacity, and poor maintenance programs, together with Hawai'i's moist and salty environment, have created a persistent and pervasive water-quality hazard on O'ahu. As more sewer lines are added and the entire system continues to age, the likelihood of additional problems increases. The majority of the state's sewer lines were installed between the 1930s and 1950s, and some of O'ahu's most decrepit pipes are approaching 100 years in age.

Each county owns, operates, and maintains their sewage collection system, and budget constraints commonly limit the amount of attention

Fig. 8.7. In 2005, raw sewage poured from a manhole cover on Kalaniana'ole Highway on O'ahu due to heavy rains leaking into corroded pipes and built-up debris that clogged the sewer lines. (Photo by Rebecca Breyer)

the counties give to these systems. With more than 2,000 miles (3,200 km) of sewer lines dug beneath the soil of Oʻahu and lesser amounts on neighboring islands, the challenge of monitoring and maintenance is daunting. The flurry of bypass and overflow events that accompany every heavy rain are proof enough of the need for vigorous and unending upgrading of facility-level operations. Indeed, this refrain is not new to the ears of county managers because sewerage upgrades perennially top the list of major expenditures in county budgets. However, the combination of making up for past neglect, keeping up with ever-expanding housing growth, and getting out in front of the ever-aging infrastructure proves to be a daunting task.

The federal government ordered a major overhaul of Honolulu's ailing sewer system, blamed for at least 200 spills, overflows, and bypasses, in the 1990s.[29] In 1999, Oʻahu embarked on the ambitious mission of upgrading 1,800 miles (2,900 km) of sewer pipes. This is an expensive endeavor and requires the year-by-year expenditure of hundreds of millions of dollars that the city can barely afford. However, the 2008 call by the EPA to upgrade the Sand Island wastewater treatment facility may cost $1.2 billion, threatening further progress on replacing the aging in-ground infrastructure. There is good news however; Honolulu's Department of Environmental Services announced in mid-2008 that sewage spills were down for the second-straight quarter. Records show 20 spills from main pipes and pump stations from April 1 through June 30, 2008, a 51.2% reduction compared with 41 spills recorded during the second quarter of 2007. In addition to the second-quarter decline, there was a 40.8% drop (42 to 71) in sewage spills during the first 6 months of 2008 as compared with the same stretch in 2007. The majority of spills were attributed to grease, roots, and debris clogging pipes, as well as broken or sagging pipes and wet weather.

As the number of tourists visiting the state skyrockets past 7 million per year, however, most of the existing systems are simply not adequate to handle the increased loads. Most counties are locked into playing an expensive game of catch-up that incrementally improves their sewage system with each passing year, but is it fast enough?

UNSEWERED LAND

An area of concern in Hawaiʻi are the numbers of sources whose sewage is not treated in municipal facilities. In the United States, 75% of sewage is treated in municipal treatment plants and 25% is treated in household septic systems, mostly in suburban and rural areas.[30] A large proportion of the private residences in Hawai'i, perhaps as many as 40%, are not attached to centralized sewer systems; instead these sources count on the ability of organic-munching microbes, and the filtering effect of soil and rock, to cleanse their sewage. Sewer systems are only installed in areas where the population reaches a level high enough to justify the cost and effort of trenching for lines, hooking individual homes and properties to pipelines, and building treatment facilities to handle the transported sewage.

In much of Hawai'i, particularly in rural areas, residents are forced to deal with sewage on an individual basis. Oftentimes individuals simply dig a cesspool, nothing more than a hole in the ground, and divert raw, untreated sewage directly into the earth. Septic tanks are also employed. The term "septic" refers to anaerobic bacteria in the tank that decompose or mineralize the waste discharged into the tank. The environmental risk

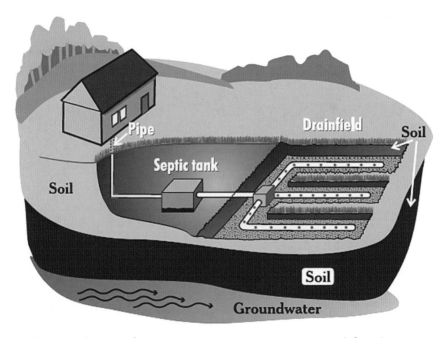

Fig. 8.8. The objective of a septic system is to retain sewage in a tank for at least 30 hours, allowing solids to settle to the bottom and grease to float to the top. The remaining liquid flows by gravity to a drain field where it percolates into soil and rock. (U.S. Environmental Protection Agency, 2005, A Homeowner's Guide to Septic Systems: http://www.epa.gov/owm/septic/pubs/homeowner_guide_long.pdf [last accessed January 6, 2010])

of cesspools and septic tanks depends on what lies beneath the property, and varies from island to island. Hawai'i has the largest number of cesspools in the nation, approximately 100,000, and the greatest number per capita. However, cesspool construction has been banned on the islands of O'ahu and Kaua'i, and on Maui, Moloka'i, and Hawai'i cesspools are only allowed in certain areas.

The alternative to septic and cesspool systems is for communities to hook up to a nearby sewer system. Unfortunately, these costs are always high, particularly in sparsely populated areas where sewer lines and centralized systems have yet to be built. In new developments, property owners incur the cost of hooking up to a public system and are also charged a monthly service fee. Communities that have enjoyed free services up until now by relying on self-supported inexpensive methods are hesitant to assume the financial responsibility of hooking up to a county system.

Monitoring of cesspools and septic systems has been a difficult and costly dilemma for the state. Ideally, the state would monitor the safety and condition of each cesspool and septic field, but rarely are funds allocated specifically for that purpose. Monitoring is made even more difficult by the fact that there are no records of how many private systems are in place or of how much wastewater they generate.

Over time wastewater from cesspools and septic tanks seeps into the underlying substrate and frequently overflows during periods of high rainfall, generating public health risks. Studies by the University of Hawai'i affirmed the suspicion that increased levels of nitrogen and phosphorus are leaching into surrounding environments from cesspools and septic systems.[31] The problem is that wastewater engineers still disagree on how much of a threat cesspools and septic tanks pose to Hawai'i's environment and its people. Additional problems may result from septic systems located in coastal areas. Some feel that noxious algae blooms on the Maui coast (*Hypnea musiformis*) are at least partially feeding on nutrients coming from household systems. However, scientists studying the problem[32] have determined that natural nutrient levels in coastal water are sufficient to support the algae explosion and that this is a case of a new species (introduced to Maui ~20 years ago) finding a niche in the ecosystem where it enjoys relatively little competition, a low wave environment, and a broad fringing reef that supports algae growth. The problem with the noxious algae is that it should never have been introduced to Hawai'i in the first place.

Currently, septic and cesspool complaints are handled on a case-by-case basis, and the only mechanism for improving the situation occurs when individual building permits are requested. The state still has very few courses of action in place to promote centralized sewage for proposed subdivisions. It seems glaringly obvious that Hawai'i will need to dedicate a worthy portion of its financial resources to this problem, because it is potentially affecting the purity of Hawai'i's groundwater, stream, and coastal water, and our public health and safety.

PAVING PARADISE: NONPOINT SOURCE POLLUTION

Most of Hawai'i's lowland forests are gone, having been felled by Native Hawaiians and Westerners to support intensive agriculture and urban development. Thousands of acres (ha) of soil that once supported the native vegetation, absorbed the rain, and naturally filtered water runoff are now buried and vaulted beneath paved impermeable surfaces. As asphalt replaces forests and fields, and rainfall is forced to flow across denuded landscapes, the water degrades into polluted runoff. Once referred to as "any pollution that doesn't come out of a pipe," nonpoint source pollution is sinister in that it is diffuse and difficult to trace. Nonpoint source pollution is simply a term for polluted runoff.

As water runs across paved surfaces, it absorbs whatever lays in its path: oil, grease, and toxic chemicals from vehicles; pathogens from leaking sewage systems and cesspools; chemical spills from dry cleaners, paint stores, and automotive shops; fertilizers and pesticides from lawns; animal feces; petroleum products; heavy metals; and many other undesirable compounds that make their way onto the ground surface. Unlike natural landscapes where rainwater pools in gullies and filters slowly into the ground, urbanized streets, parking lots, driveways, rooftops, and gutters divert polluted runoff into channels, culverts, and storm drains all headed at high velocity into our streams and coastal waters. This artificial maze of conduits alters the natural course of the runoff, accelerating its pace and increasing its volume, thereby eroding stream banks, damaging streamside vegetation, and widening channels while carrying the polluted discharge into the ocean. The potential for polluted runoff to affect coastal and marine environments is immense in Hawai'i because most of the population and the waste they generate are located within only a few miles (km) of the shoreline.

In places as diverse as Kona, Kā'anapali, and Pearl City, paradise is disappearing under a labyrinth of paved surfaces. Outlet malls, shopping plazas, and miles (km) of roads are connecting spillways, storm drains, and culverts, and diverting the natural flow of *wai*. As prime real estate becomes increasingly scarce, remnant parcels on the arid coastal plains are acquired, and cities are created that depend on stressed water resources. Access to new development requires more roads. Water quality is diminishing in direct proportion to the pace at which land is developed. The fate of the Islands' water quality is closely linked to the choices that will be made in coming years about how the vast agricultural fields will be used.

POLLUTED COMPONENTS

One big contributor to polluted runoff is sediment. Derived from eroded soils, hillsides, construction sites, excavation pits, and, most important, abandoned or fallow agricultural fields, sand, silt, and clay find their way into aquatic and marine ecosystems. Through storm drains, channels, streams, and ultimately into sensitive nearshore marine communities including reefs, estuaries, and other restricted shallow-water environments, abundant sediment accumulation is harmful. Water turbidity increases where sediment loading is intense, suffocating delicate marine organisms or filling in wetlands and waterways. Although sediments are present naturally in water, any type of land-disturbing activity generally increases sediment delivery to surrounding water bodies. Few disagree that construction projects and unvegetated agricultural fields accelerate erosion, but best management practices can be adopted and enforced to minimize the damage.

Another major component of polluted runoff is derived from industrial and urban sources. Combined pollutants from roadways, vehicles, industrial facilities, gas stations, and manufacturing plants create a lethal soup. Over the past few decades, many hazardous substances from industry have found their way into the environment through a combination of routine releases, leaks, and spills. Although regulations exist that govern the use and disposal of such substances, it is difficult to keep pace with the proliferation of new chemical compounds.

With more than 4,000 miles (6,400 km) of roads and 1 million vehicles registered statewide, the residue of heavy metals from cars and trucks alone contributes significantly to nonpoint source pollution. In Hawai'i, tons of lead were released into the environment until 1971 from the use of leaded gasoline, and it took 17 years until all alkyl-leads were finally eliminated from measured runoff. Studies suggest that lead, possibly from brake pads, is still being added to the environment. Zinc and cadmium are contributed by the wear and tear of car tires. Chemical runoff from homes is also significant: copper from plumbing and gutter linings, droppings from pets and birds, chemicals in asphalt driveways and

Fig. 8.9. Coastal waters are the receiving environment for polluted runoff, disposal of treated sewage, and groundwater discharge. How clean are the waters we play in? (Photo by C. Kojima)

used on lawns, and spills of common home chemicals all seep into storm drains and ultimately the natural environment. The impacts of urbanization are growing daily.

Maps of impaired water bodies are available from the Hawai'i Department of Health.[33] These areas, known as "water quality limited segments," are mapped so that practices within these regions can be monitored and regulated. According to the EPA's National Water Quality Inventory 2000 Report, in Hawai'i only 32% of surveyed rivers have daily quality that met their designated uses, and 69% of surveyed rivers are impaired; all of the surveyed estuaries and bays in Hawai'i fully support aquatic life, but only 86% fully support fish consumption and shellfish harvesting, and only 74% fully support swimming. The most significant pollution problems in Hawai'i are siltation, turbidity, nutrients, organic enrichment, and pathogens from nonpoint sources including agriculture and urban runoff.

MANAGING WATERSHEDS

Within a watershed the land is characterized by various types of water bodies: runoff, stream channels and lakes, wetlands and estuaries, and marine coastal waters. In Hawai'i, these water bodies are increasingly found to violate water-quality standards set by the EPA. A list of impaired water bodies in 1998 included 19 coastal sites and 3 streams. In 2002, this had risen to 59 streams and 139 coastal stations exceeding federal regulations.[34] By 2004, the list included 70 streams and 174 coastal stations.[35] From 2002 to 2004 no stream or previously listed coastal station was delisted despite funding for mitigation from the State Department of Health and the EPA. The 2006 report of impaired water bodies listed 136 coastal sites, 95 streams, 56 bays and harbors, and 21 estuaries. Among the causes of impairment reported, 42% were impaired by nutrients, 30% by turbidity, and 11% by algal growth. Other causes included pathogens (7%), ammonia (3.6%), and trash, pesticides, metals, and PCBs.[36] Although watershed management, it appears, is losing ground to pollution every year, there are examples of individual stream systems and coastal sites where the level of degradation has been reduced due to community and state efforts.

Historically, Hawaiians managed watersheds by emphasizing sustainability. The key to their success was that they managed the watershed as a whole system, not piece by piece. Watersheds today are managed by separate agencies primarily concerned with achieving isolated and uncoordinated mandates. The result is that over time, water quality declines, watersheds decay, and coastal water bodies accumulate pollutants. However, taking a cue from the *ahupua'a* practice, the Department of Health now recognizes that the fragmented management system employed in the Hawaiian Islands is failing to achieve advances in coastal water quality. Hence, the Department of Health has combined with the Hawai'i Coastal Zone Management Program to offer watershed planning guidance to agencies, nonprofits, industries, and neighborhood groups seeking help to recover and sustain watersheds. The goal of the effort is to improve coastal water quality by reducing land-based sources of pollution and restoring natural habitats.[37] The guidance builds on the Department of Health coastal nonpoint pollution control program management plan (1996), implementation plan (2000), and action strategy (2004) to address land-based pollution threats.[38]

Added to this are efforts by local agencies that recognize the value of a collaborative and interdisciplinary approach to manage water resources. For example, the City and County of Honolulu Board of Water Supply is adopting a fresh approach to water management through the lens of *ahupua'a* practices in the twenty-first century. Moving consistently through their eight districts on O'ahu, they are developing watershed management plans by involving community members, other agencies, focus groups, and awareness of watershed history and current uses. Their goal is to determine a set of practices that can be exercised by entire populations in partnership with the primary mission of the Board of ensuring clean and abundant water for the island.[39] They have scheduled the development of individual watershed management plans through the year 2013 in a planning process that emphasizes the following:

- community participation and consultation
- holistic management of watershed resources
- alignment with important state and city policies and programs
- an action orientation: implementation of important watershed management programs
- *ahupua'a* management principles

A CONTROL PROGRAM

The July 2000 implementation plan published by the Department of Health Polluted Runoff Control Program recognizes that polluted runoff is a major statewide problem.[40] The plan presents an approach for handling key elements of nonpoint source pollution required by the EPA to qualify for federal funding. The plan identifies polluted water bodies (water quality limited segments) resulting from polluted runoff, many of which are coastal and estuarine environments.

Before 1994, Hawai'i did not have a storm water program. Consequently, storm drains and stream channels served as conduits for rapid transport, accelerating the accumulation of toxic materials in our waterways. But in 1994, the federal government initiated a change with the NonPoint Pollution Control Act.[41] With more than 400,000 residents calling the capital home, the City of Honolulu was the first in the state required by federal law to apply for a NonPoint Source (National Pollution Discharge Elimination System) permit. The permit obligated the city to institute a water-quality monitoring program. Amendments in federal regulations in 2004 call for all cities with more than 100,000 people to apply for nonpoint source permits.[42]

Yet even in places where existing regulations have changed or new legislation has been passed, nonpoint source pollution is not being reduced as much as expected or hoped. Critics say that within the state there is ambiguity and confusion over which agency should lead the way in coordinating the various water-quality programs and enforcing water-quality mandates. One agency continues to channelize streams while another calls for reductions in polluted runoff. One agency manages under an *ahupua'a* approach while another approaches the problem from the continental mechanism dictated by federal regulations. This lack of clarity among agencies means that the state has no integrated water-management plan for watersheds that are included in the Department of Health's Polluted Runoff Control Program of 2000, a part of the Clean Water Branch and Coastal Zone Management Program.[43]

The Polluted Runoff Control Program was initiated after a 1987 federal mandate, Section 319 of the Clean Water Act, that specifically addresses nonpoint source pollution. The program administers grants and projects that purify bodies of water impacted by nonpoint source pollution. Under this mandate, the federal government provides funds to state agencies for polluted runoff-mitigation programs and pushes for individual states to allocate matching funds in return. To date, Hawai'i has been somewhat lax about contributing money toward polluted runoff alleviation, which means that the state has trouble getting the federal dollars that require matching state funds. Not only do state budget decisions determine the number and types of projects funded but also the number of state employees dedicated to the issue of polluted runoff.

To quality for federal funding for its polluted runoff program, Hawai'i has successfully addressed key elements in its implementation plan for the polluted runoff program.[44] According to the EPA:

- The state program contains explicit short- and long-term goals, objectives, and strategies to protect surface and groundwater.
- The state strengthened its working partnership and linkages to appropriate state, regional, and local entities; private sector groups; citizen groups; and federal agencies.
- The state uses a balanced approach that emphasizes both statewide nonpoint source programs and on-the-ground management of individual watersheds where waters are impaired or threatened.
- The state program (1) abates known water-quality impairments from nonpoint source pollution and (2) prevents significant threats to water quality from current and future nonpoint source activities.
- The state reviews, upgrades, and implements all program components required by the Clean Water Act and has established flexible, targeted, and iterative approaches to achieve and maintain beneficial uses of water as expeditiously as practicable.
- The state identifies federal lands that are not managed consistently

with state nonpoint source program objectives. Where appropriate, the state seeks EPA assistance to help resolve issues.

- The state manages and implements its nonpoint source program efficiently and effectively, including financial management.
- The state periodically reviews and evaluates its nonpoint source management program using environmental and functional measures of success, and revises its nonpoint source assessment and management program at least every 5 years.

Fig. 8.10. Accelerated soil erosion is perpetuated by poor land management. In this West Maui case, agricultural lands lacking vegetative cover are allowed to lay open to the wind and rain. (Photo provided by U.S. Geological Survey)

In its September 2000 review of this plan, the EPA commended the state of Hawai'i for integrating pollution control measures under the federal Clean Water Act and the Coastal Zone Act.[45] The level of public outreach and participation during the development of this plan was noted, and the Hawai'i implementation plan was described as "providing the needed framework to guide the continued development and implementation of the Polluted Runoff Program in Hawai'i."

Whether directly or indirectly, current state and county regulations address all of the coastal nonpoint source pollution programs and management measures. However, a lack of funding and an insufficient number of staff impedes the success of these regulations. Even in places where the program has been implemented, the measures are not reducing the inputs from nonpoint source pollution as much as expected. It is critical for a state agency to assume responsibility for coordinating all programs and enforcing water-quality regulations, to ensure the health of our state waters.

PERSISTENT PESTICIDES

As large-scale agriculture took hold in Hawai'i during the early twentieth century, there was an escalating demand for fumigants, nutrients, pesticides, and fertilizers. Though groundwater sources typically rest hundreds of feet (m) below the leeward plains, runoff from heavily irrigated fields leached downward. Saltwater encroachment has long been the primary source of groundwater contamination, but the discovery of trace quantities of several toxic organic chemicals in well water has raised serious concerns.

Before the discovery of agrochemical contamination, it was thought that pesticide use posed no threat to human life. Because chemicals are highly volatile, their residues were expected to evaporate as they percolated down through the rock and soil before intercepting the groundwater table. However, because agriculture fields artificially accelerate recharge because of irrigation, the process of "leaching" chemical compounds out of the soil and into groundwater is sped up. Contaminated runoff percolates rapidly into groundwater, leaving less opportunity for evaporation and filtration. It is not surprising that the spatial pattern of contamination shows the highest concentrations of pesticides in the wettest agricultural fields, where percolation is highest.

Urban Honolulu streams contain high levels of the pesticides chlordane, dieldrin, and DDT.[46] Fumigants and insecticides such as DDT, aldrin, chlordane, heptachlor, and others used in termite control and agribusiness are present in Hawai'i's water. Studies by the U.S. Geological Survey have detected these "organochloride" pesticides in stream sediment and fish tissue in selected streams on O'ahu. Concentrations of these organochloride pesticides declined after bans on their use in the early 1970s and then appeared to level off in the 1980s. Even though they are no longer used, concentrations of these poisons continue to persist in O'ahu aquatic ecosystems. The sources of these poisons are agricultural and urban soils that erode and enter watersheds. Although little can be done to reduce the concentrations of these compounds in soil, controlling soil erosion (especially during land clearing and construction activities) could reduce levels of pollutants entering streams and ultimately estuaries and coastal marine ecosystems.

In central O'ahu, contamination by dibromochloropropane (DBCP), ethylene dibromide (EDB), and trichloropropane (TCP) forced the temporary closure of several drinking-water wells because of their suspected health risks. The public health hazards posed by agricultural contaminants are very real and potentially lethal. EDB and DBCP are now banned from use in the Islands. Other chemicals of concern are nitrates, trihalomethanes, atrazine, trichloroethylene, and carbon tetrachloride. Groundwater contamination maps showing chemical compounds found in Hawai'i's drinking water are posted annually on the Department of Health Web site.[47] These maps show locations of wells that are keyed into tables of contaminates and their concentrations as measured in Department of Health water samples.

Although the detection of contaminants in the 1970s and 1980s frightened residents with regard to the safety of their drinking water, they served as a warning for regulators and managers. Contamination of water supplies reaffirmed the basic importance of water for life and the concept of cause and effect within the hydrologic cycle. Thanks to these experiences, a more effective statewide system of monitoring has developed. Although we must now pay for our past mistakes, we are also learning how to avoid repeating them and how to mitigate future fumbles.

THE WATER WE DRINK: *WAI MĀNALO*

Frightening though some accounts may sound, the drinking water in Hawai'i remains some of the best in the world. The water we drink principally comes from groundwater, including the flow from artesian wells and springs, or from surface waters. The source of water plays a major role in determining our water quality. The headwaters of our streams generally have pristine chemical quality, but as streams travel through dense agricultural fields, residential developments, and urbanized centers, the flow entrains dissolved solids, nutrients, bacteria, sewage effluent, industrial wastes, and urban by-products. Because of its innate purity, groundwater is often preferred for municipal use. However, despite nature's capacity for filtering out unwanted substances, contaminants are finding their way into some subterranean aquifers.

Contamination, defined as any substance hazardous to health when present in sufficient quantities, is rare at the headwaters of streams or in the groundwater below natural forests. However, Hawai'i's highly permeable geology, though responsible for the abundance of groundwater, actually increases the mobility of contaminants, allowing them to flow quickly into aquifers. As water travels over the land or through the ground, it dissolves naturally occurring minerals and can collect substances resulting from the presence of animals or human activity.

According to both state and national law, citizens of Hawai'i have a right to clean drinking water.[48] Federal laws have become increasingly stringent; potable drinking water systems are prohibited from tolerating any contaminants that could compromise public health and must not contain pathogenic organisms, toxic material, or any other substances that may be deemed harmful. The national Safe Drinking Water Act of 1975 requires the EPA to establish clear drinking water standards for public water supplies and monitoring of municipal and privately owned water systems.[49] Enforceable drinking standards, called Maximum Contamination Levels, are required for every region.

The state legislature, in recognition of the need to support drinking water quality by protecting groundwater, authorized the Hawai'i Groundwater Protection Program[50] to monitor public drinking-water wells. The EPA and the Hawai'i State Department of Health also require the water departments on all islands to regularly test drinking water. Several localized contamination sites have been identified, though widespread contamination has never been detected. Though the state is confident that the lack of widespread contamination means that regular monitoring is sufficient to detect potential threats, federal agencies do not want a sense of complacency to develop. They recognize that although serious agricultural and industrial contaminants were not identified in initial surveys, they may still be present given the limited scope of early monitoring efforts.

In 1999, each household in the state began receiving Water Quality Reports in the mail.[51] The reports of drinking-water contamination are based on current Department of Health monitoring data for public drinking-water wells; however, each local water supply board is responsible for sampling on individual islands. The reports educate citizens about their drinking-water source, indicate any detected contaminants or elements in the water, and assure residents that drinking water meets all the safe drinking-water standards. The Department of Health is establishing sampling protocols for drinking-water wells including monitoring of microorganisms, total coliforms, fecal coliforms, *E. coli,* inorganic and organic chemicals, radionuclides, and turbidity. Each contaminant requires special tests for detection, yet critics of the program worry that not enough attention is being given to endocrine disrupters. This class of chemicals is thought to affect cells by interfering with the endocrine system, which controls reproduction, growth, and development in humans and many animals.

Only regulated compounds have established safety levels, known as "maximum contaminant levels" (MCLs). During routine sampling by the Department of Health across the state in 1998, trace amounts of organic chemicals were detected in separate water systems. Trichloroethylene, a common metal cleaning and dry cleaning fluid, was found in Honolulu Board of Water Supply pumps in Waipahu and Hale'iwa. The contaminants were at levels far below the federal MCL of 5 parts per billion (ppb). Isophorone, used in herbicides, paints, and adhesives, was detected in Punalu'u, but this contaminant remains unregulated to date, so no safe limits have yet been established in Hawai'i. The presence of ethylbenzene, a major component of gasoline, was confirmed in Volcanoes National Park but also at levels far below the MCL.[52]

WHAT ARE YOUR DRUGS DOING IN MY WATER?

In 2008 the Associated Press[53] conducted a study of pharmaceuticals found in drinking water across the nation. They particularly examined chemicals that are not regularly monitored by our agencies and found a vast array of dissolved compounds in drinking water, including antibiotics; anticonvulsants; mood stabilizers; steroids; medicines for pain, infection, high cholesterol, asthma, epilepsy, mental illness, heart problems, and sex hormones . . . that's right, sex hormones.

Although it is true that the concentration of these drugs is miniscule, measured in parts per billion or parts per trillion far below their original prescription level, it is also true that the impact of ingesting these compounds over long periods of time is unknown. The presence of so many prescription drugs in the drinking water of over 40 million Americans has health officials worried about the long-term consequences to human health. How these drugs interact with each other is not known. What their effect is over decades of use is not known. How they interact with the chlorine and fluorine that is often added to drinking water is not known. Worse, in most cases these drugs are not on the list of compounds regularly tested for by our water authorities.

How do these drugs enter the drinking water supply? People take pills. Their bodies absorb some of the medication, but the rest of it passes through and is flushed down the toilet. The wastewater is treated before it is discharged into reservoirs, rivers, or lakes on the mainland, and into the ocean and coastal lands here in Hawai'i. Then, some of the water is cleansed again at drinking water treatment plants and piped to consumers. But most treatments do not remove all drug residues. On the mainland, treated sewage has a high potential to find its way back into the drinking water supply, either through recharging aquifers or where surface water is used for drinking. A growing number of municipalities are intentionally recycling treated sewage back into the groundwater system, from whence it is eventually withdrawn for drinking—for example, in Los Angeles. Most of these contaminants are not removed by typical treatment; neither are they monitored by water boards.

The research team reviewed hundreds of scientific reports, analyzed federal drinking water databases, visited environmental study sites and treatment plants, and interviewed more than 230 officials, academics, and scientists. They surveyed the nation's 50 largest cities and a dozen other major water providers, as well as smaller community water providers in all 50 states. They discovered that the federal government does not require any testing and has not set safety limits for drugs in water. Of the 62 major water providers contacted, the drinking water for only 28 was tested. Among the 34 that do not test for drugs in drinking water are Houston, Chicago, Miami, Baltimore, Phoenix, Boston, and New York City's Department of Environmental Protection, which delivers water to 9 million people. "We know we are being exposed to other people's drugs through our drinking water, and that can't be good," says Dr. David Carpenter, who directs the Institute for Health and the Environment of the State University of New York at Albany.[54]

SWIMMING WITH BACTERIA

Frightened by a flurry of water-quality violations and media scares in the early 1990s, Hawai'i residents became increasingly concerned about the health risks of swimming at Hawai'i's beaches. Monitoring that is performed by the Department of Health and local agencies on a regular basis

of *Enterococcus,* a fecal bacterium, showed that waters off Kūhiō Beach and Keʻehi Lagoon are some of the most polluted in the state. When debris, including syringes, washed onto a few of Oʻahu's crowded tourist beaches in the early 1990s, at the same time that surfers and swimmers were reporting strange lesions on their bodies, a barrage of accusations started flying. The public demanded answers about the safety of Hawaiʻi's swimming waters.

Fears surfaced again after the 48-million-gallon (181-million-liter) sewage spill in the Ala Wai Canal during the heavy rains in the spring of 2006. The horrifying death of Oliver Johnson after he fell or was pushed into the Ala Wai shortly after the spill galvanized local fears of bacteria to the point that Honolulu and Waikīkī beaches and surf breaks were eerily empty for over 6 weeks after the spill.[55] Johnson died of massive organ failure brought on by septic shock caused by *Vibrio vulnificus,* a bacterium of the cholera family that multiplied in his body and eventually shut down his principal organs. An important aspect of Johnson's situation is that he did not go to a hospital until some time had passed after the first signs of infection, plus he had open wounds and suffered from chronic liver disease—both conditions that greatly elevate the risk from polluted waters.

The *V. vulnificus* that killed Johnson is from a family of bacteria that is "particularly happy" in seawater, say microbiologists. It does not usually attack people but can mutate into an invasive form capable of overwhelming the body's defenses, especially in patients with chronic liver disease. Most of Hawaiʻi's *Vibrio* cases are associated with wounds or blood infections and ear infections; the bacterium can cause vomiting, diarrhea, and abdominal pain. *Vibrio vulnificus* infections can be easily cured within the first day or two with antibiotics, but once it invades the bloodstream there is only a 50% chance of survival according to the Centers for Disease Control and Prevention.

In fact, deaths from waterborne bacteria encountered in our oceans and streams are relatively rare. There have been five deaths in Hawaiʻi from the family of *Vibrio* bacteria since 2001.[56] Also in Hawaiian waters are the potentially fatal bacterial infections leptospirosis, S*taphylococcus aureus,* and group A streptococci. Since 1974, a total of nine deaths in Hawaiʻi have been attributed to leptospirosis, a freshwater bacterium found in Hawaiʻi streams and splash pools.

Many swimmer infections that lead to severe complications are just a case of bad luck. One infectious disease specialist interviewed in 2006 said, "With all the people who get cuts and abrasions throughout the state of Hawai'i, only a small number come out with serious life-and-limb infections. It's an uncommon event, unless you're unlucky and pick up the wrong bug, the one that produces those chemical toxins" that can rapidly advance through the human body.[57]

So how do you protect yourself from waterborne bacteria? Health officials point out that an open wound or a weakened immune system put you at increased risk of contracting harmful bacteria in our waters. Certain conditions, such as warm seawater or freshwater streams and ponds likely to be contaminated with animal urine, boost the chances of exposure. The best protection is to avoid entering Hawai'i's streams and oceans when you have an open wound, especially if you have a weakened immune system.

A good cleansing with antibacterial soap and water to remove dirt and damaged tissue from wounds is still the favored treatment. Keeping cuts and scrapes clean and dry remains the best prescription for preventing skin infections that can worsen and invade the bloodstream, causing life-threatening complications. And with the speed at which these infections can spread, physicians urge people to seek medical care at the first sign of fever, nausea, and increased swelling, pain, and redness. If you have swum or walked through a stream or waterfall plunge pool and develop any flu symptoms, treat it as leptospirosis and immediately go to a doctor. If you have been in the ocean and develop an infection, do the same thing.

BEACH CLOSURES

Public reports in 2006 after the Ala Wai spill highlighted the fact that potentially harmful bacteria are present in the same waters where we swim and play—and have been all along. According to the Natural Resources Defense Council, the state has approximately 24 miles (38 km of shoreline considered safe, accessible, and generally suitable for swimming.[58] The process of establishing sampling protocols, indicator organisms, and beach-closing standards for these swimming locations is complex. Even though Hawai'i's bacterial standard is one of the strictest in the nation, with water samples having to be less than a geometric mean of 7 *Enterococcus*/100 ml compared with a nationwide value of 35/100 ml, there is still considerable controversy over implementation of state guidelines.

The state has proposed a two-part system of designating harmful levels of coastal bacteria. Two types of fecal indicator bacteria are used, *Enterococcus* and *Clostridium perfringens*. If the *Enterococcus* standard exceeds the permitted level, and *Clostridium* exceeds set limits, then the

Fig. 8.11. In March 2004, "Beach Closed" signs were posted at Waimānalo Beach on O'ahu due to sewage contamination. (Photo provided by Honolulu Star-Bulletin: http://archives.starbulletin.com/2005/07/29/news/story6.html [last accessed January 21, 2010])

state will close a beach. Yet, the federal government wants the state to close a beach even when just the *Enterococcus* levels are exceeded and will not accept the state's proposal until studies clearly quantify the health risks associated with a particular level of *Clostridium*. Getting such numbers will require extensive research in beach areas where there are a number of disease outbreaks, something Hawai'i has not yet experienced.

The standards of the EPA for *Enterococcus* were created from data collected in places like New York City and the Great Lakes, where pollution-related disease was frequent enough to conduct a large-scale study. But in Hawai'i *Enterococcus* may be found in uncontaminated soil because it grows naturally in our environment.

The bulk of the shoreline sewage problem is attributable to four sources:

1. Sewage spills from heavy rains that generate overflows in waste-water-treatment plants
2. Breaks in sewer lines carrying untreated sewage from homes and businesses

3. Electrical failures at pumping stations and treatment facilities that cause sewage spills

4. Polluted runoff

Periodically, a beach will close if fuel or oil from a boat threatens a swimming area. Wailua Beach on Kaua'i closed after approximately 16,000 gallons (60,500 liters) of diesel fuel spilled from a grounded fishing vessel in 1999.[59] It is interesting that 99% of beach closures in Hawai'i are due to spills of one type or another (fuel, sewage, or some other mechanical discharge), compared with 47% of postings on a national level. Three questions then arise: does Hawai'i truly have more spills than other parts of the country? Are we failing to detect contamination from other sources (such as polluted runoff)? Or do we have generally clean water that is only occasionally contaminated by spills? In answer, the EPA feels that many potential sources of bacterial pollution are not being adequately monitored or addressed in the state.

One of the major causes of beach closures in other parts of the country is polluted runoff, particularly in urban areas. In 1995, an extensive study of Santa Monica Bay verified the link between illnesses in swimmers and polluted runoff.[60] This study found that people who swam within 100 yards (91 m) of storm-drain outlets are 50% more likely to get colds, flu, sore throats, and diarrhea than those who swam farther away. The study concluded that as many as one in 10 of those individuals swimming near storm drains will experience symptoms similar to pathogenic exposure. Though this sort of exposure is not usually life threatening, the well-being of affected individuals can be compromised, and any viruses present in the water can be a serious health threat to children and the elderly.

Unfortunately, considerable amounts of misinformation about water quality continue to create anxiety about swimming in the ocean. One prevalent misconception is that all infections are from human-contaminated sources. Although swimmers are capable of transmitting infections to one another, such as *Staphylococcus aureus,* marine organisms can also carry communicable viruses and bacteria. The most common waterborne public health risk in Hawai'i is related to leptospirosis, contracted from the urine of mongooses and rats in freshwater. The Department of Health recommends that school-age children in particular should avoid all streams to reduce the health risks of leptospirosis.

In Hawai'i, only one beach, Lydgate Park on Kaua'i, closed in the late 1990s because of water-monitoring data.[61] Some people argue that this proves Hawai'i's excellent water quality, while others disagree, saying that if only one beach was closed due to monitoring data, then the water-monitoring program must be inadequate. Certainly in Hawai'i far more beaches are not being monitored than those that are.

Surfrider provides a $6 do-it-yourself water quality testing kit, approved by the EPA, to check for levels of coliform bacteria. Because the state is most likely to detect problems from known sewage spills, the threat of polluted runoff may well be overlooked under the current monitoring system. As Surfrider's program demonstrates, one way to protect the public from polluted water is to develop active citizen monitoring, where people claim responsibility for the waters they share.

Limited financial resources certainly play a role in the state's monitoring program, but a cost-benefit analysis shows that increased water monitoring would be a sound financial decision for Hawai'i. It would be sound planning for the state to protect the tourist industry by ensuring that our waters are safe. The state of California adopted a "right to know" bill that requires monitoring of all public beaches with more than 50,000 annual visitors and regular sampling near storm drains.[62] Critics of Hawai'i's program suggest that a protocol similar to California's should be applied to Hawai'i if the state is going to guarantee the safety of swimming water, especially against point source pollution.

The good news for those of us who wade, swim, fish, paddle, and surf is that Hawai'i's dirtiest monitored beaches are still clean by EPA standards. Based on counts of bacteria, viruses, and protozoan pathogens, one study identified O'ahu's dirtiest waters, such as Ka'elepulu Stream in Kailua Beach Park, as very poor, giving it a rating of 35 *Enterococcus* colony-forming units (CFUs) per 100 ml. Though poor by Hawai'i's standards, this is still within EPA safe-swimming guidelines.[63] Addressing the polluted runoff challenge may well be the best place to start to increase the quality of Hawai'i's waters.

MĀMALA BAY: THE FINAL WORD?

In 1990 the Sierra Club Legal Defense Fund and Hawai'i's Thousand Friends sued the City and County of Honolulu for violations of the Clean Water Act related to failing to upgrade sewage processing to the national standard of secondary treatment. The outcome of the suit was the creation of a study commission to establish the nature of ocean circulation in Māmala Bay (defined as the embayment between Diamond Head and Barbers Point) and the character of point and nonpoint pollutants and their potential impact on humans and the ecosystem.

Among the study's findings in 1995 was that sewage plumes from Honolulu outfalls were greatly diluted within the zone of the diffusers at the head of outfall pipes.[64] Point sources of pollution had comparatively minor effects on phytoplankton and benthic communities and there was conclusive evidence that nutrient enrichment in shoreline areas was closely related to nonpoint sources such as cesspool and groundwater drainage or to localized discharges such as from Fort Kamehameha outfall and the Ala Wai Canal. The study found that discharges from the Sand Island facility were able to reach most beaches in the bay, though only rarely, whereas those from Honouliuli were able to reach only western beaches. Consistent with this, the frequency of point source pollution on beaches was low, and the risk of contracting an infectious disease by bathing, swimming, surfing, or fishing in Māmala Bay waters was low. The Ala Wai Canal was determined to be a major source of contamination of Waikīkī Beach, and nonpoint sources were the primary cause of contamination of beaches in the eastern portion of the bay. Since the report was issued more than a decade ago, the Sand Island facility, in keeping with

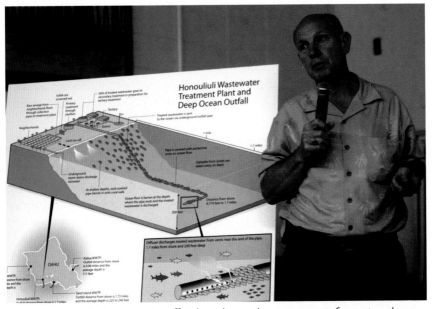

Fig. 8.12. At public hearings city officials and consultants present information about how Oʻahu's sewer collection and treatment system works. (Photo by F. L. Morris, provided by Honolulu Star-Bulletin: http://archives.starbulletin.com/2007/05/10/news/ story08.html [last accessed January 21, 2010])

the recommendations of the study commission, has moved to advanced primary treatment with the implementation of ultraviolet disinfection and increased removal of suspended solids.

WHO IS WATCHING OUR WATER?

Monitoring can be the most direct and defensible tool available for evaluating how water quality improves or does not, as a result of regulations and management actions. Hawai'i's Department of Health is the principal water-quality monitoring agency and is tasked with the challenge of implementing a successful program. Monitoring is expensive and, in itself, not necessarily justifiable unless the results or information can be used for specific purposes. The state's current monitoring plan incorporates four water-quality categories: core network, recreational bathing waters, watershed protection, and toxic contaminant screening. Federal agencies such as the National Oceanic and Atmospheric Administration (NOAA) and EPA give general guidelines for the state, but continental watershed models cannot necessarily be applied to Hawai'i's dramatically different environment.

The state of Hawai'i and the city of Honolulu have suffered thousands of violations under federal regulations and have been criticized over and over again for their water-quality management. Yet the greatest impediment to water resource management may be the governmental and regulatory environment in which it operates. Ironically, overlapping jurisdictions may actually impede cooperation between agencies, causing responsibilities to be shifted, ignored, and even resented. Compounding the problem is the recognition that agency programs may not be structured to solve complex watershed management problems. In the end one must ask, is there a demonstrable health problem?

The feasibility of establishing a comprehensive and successful water-monitoring and cleanup program is limited by the reality that watersheds, groundwater, and coastal waters are segmented under a multitude of private owners and county, state, and federal programs. For watershed management to be successful, all stakeholders must establish an overarching culture of positive support, common purpose, and mutual respect. This will require defining community goals, developing shared tools, and agreeing on common measures of success. How well we protect the remaining natural essence of our state's water supply and the cleanliness of natural coastal waters will foreshadow our success worldwide in preserving the ultimate island: Earth.

CHAPTER 9

Climate Change and Sea-Level Rise

Mālama o kekai, kekai o ke mālama
"Take care of the ocean and the ocean will take
care of you"

The Sun's energy heats the atmosphere and Earth's surface, thus driving the weather and climate. In turn, Earth reflects and reradiates a portion of this energy. Some goes back into space, and some is trapped by atmospheric greenhouse gases (water vapor, carbon dioxide, nitrous oxide,

Fig. 9.1. The atmosphere contains gases that trap heat. Natural levels of greenhouse gas warm the surface, making modern life possible. But burning fossil fuel releases more greenhouse gas, causing global warming. (Photo from NASA)

methane, and others). If greenhouse gases increase in abundance, the amount of stored heat also increases, warming the atmosphere. Without this natural effect, Earth's average surface temperature would be a chilly 0°F (-17.8°C) instead of its current 57°F (13.9°C),[1] and life as we know it would not be possible. This is known as the greenhouse effect.

Measurements indicate that the amount of carbon dioxide and other greenhouse gases in the atmosphere has increased over the past century.[2] Measurements also indicate that average global atmospheric temperature has risen by approximately 1.44 ± 0.32°F (0.8 ± 0.18°C) over the past century, and precipitation (expected to increase in a warmer atmosphere) rose an average 1.4 ± 0.5% per decade.[3] For example, 2009, 2007, 2006, 2003, and 2002 tied 1998 as the second warmest year on the instrumental record (extending back to 1880 or so).[4] This is remarkable considering that 2007 was a cool La Niña year and a time of low solar activity (the lowest in over 20 years), and 1998 was a warm year with high solar activity that has been dubbed the "El Niño of the century" (El Niño years tend to be warmer than others).[5] The year 2005 was the warmest on the instrumental record, and the period 2000–2009 the warmest decade. Long-term temperature records of Earth's surface (including the ocean) and atmosphere show a sustained warming trend over the past 50 years of approximately 0.36°F (0.2°C) per decade.[6] This trend, however, is characterized by highly variable temperatures from one year to the next: 2008 was cool and saw global average temperatures return to values not seen in 10 years; 2009 was warmer and achieved status (in a six-way tie) as the second-warmest year measured. This variability and the influence of shorter-term effects such as El Niño are superimposed on the steady long-term trend of warming.

Scientific researchers who study the greenhouse effect connect the unusual global warming witnessed now principally to gas emissions from our use of fossil fuels. These emissions have caused an increase in heat-trapping carbon dioxide, methane, and other greenhouse gases in the atmosphere. A number of researchers using mathematical models that reproduce the physics of heat movement in the atmosphere and ocean have shown that human carbon emissions are capable of causing (and are the most likely reason for) the observed warming of the late twentieth and early twenty-first centuries. This work is summarized in a series of reports by the Intergovernmental Panel on Climate Change (IPCC), which issued its first report in 1990.[7] Researchers, in addition, have collected and published, in peer-reviewed scientific literature, numerous observations of ocean warming, melting glaciers, rising sea level, reduced sea

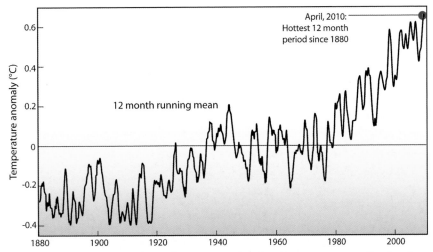

Fig. 9.2. Measurements indicate that average global surface temperature (land and ocean) has increased by about 1.4°F (0.8°C) in the past century. (SOEST Graphics, modified from NASA Earth Observatory, "2008 Global Temperature": http://earth observatory.nasa.gov/IOTD/view.php?id=36699 [last accessed January 5, 2010])

ice, changes in seasonality, shifts in the behavior of temperature-sensitive species, and other evidence supporting the scientific theory that Earth's surface environment is experiencing the effects of global warming as a result of human atmospheric pollution.

In 2006, the U.S. Congress asked the National Research Council (NRC) to study Earth's climate and report on the levels of warming in recent history. The NRC concluded that Earth's average surface temperature today is the highest of the past 400 years and probably the last 1000 years.[8] They state that Earth's surface is projected to warm by an additional (approximately) 3.6°–10.8°F (2°–6°C) during the twenty-first century. Global average temperature measurements by instruments indicate a near-level trend from 1856 to about 1910, a rise to 1945, a slight decline to about 1975, and a rise to the present. Global warming is also verified by several independent sources including the National Academy of Science and Engineering, the National Climatic Data Center, the National Aeronautics and Space Administration (NASA), the United Kingdom Meteorological Office, the Union of Concerned Scientists,[9] and more than 40 societies and academies of science including those of all the major industrialized nations.

Warming is documented by several types of sensors: weather balloon measurements have found that the global mean near-surface air temperature is warming by approximately 0.32°F (0.18°C) per decade;[10] satellite measurements of the lower atmosphere show warming of 0.4° to 0.47°F (0.22° to 0.26°C) per decade since 1978;[11] continental weather stations document warming of approximately 0.36°F (0.2°C) per decade;[12] and ocean measurements using various types of sensors show persistent heating since 1970.[13] NASA collects global climate data including land and ocean measurements and provides an annual report and periodic

Global and continental temperature change

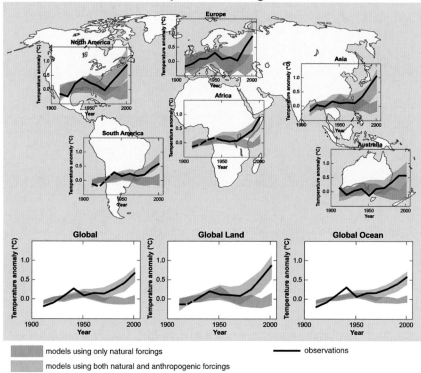

models using only natural forcings
models using both natural and anthropogenic forcings
observations

Fig. 9.3. Simulating surface temperature using computer models (blue = natural factors only; red = natural and human factors combined) and comparing the results to measured changes (observations of temperature change = black line) can provide insight into causes of major changes. The simulations represented by the blue band were produced with only natural factors such as solar variation and volcanic activity. Those shown in red were produced with human greenhouse gas production combined with natural factors. The red band shows that human factors combined with natural factors best account for observed temperature changes. (From IPCC 2007 Assessment Report [AR4]: http://www.ipcc.ch/pdf/assessment-report/ar4/syr/ar4_syr_spm.pdf [last accessed November 10, 2009])

updates.[14]Although the connection between human pollution and atmospheric warming cannot ever be indisputably proven, only inferred or correlated, the body of published research makes a compelling case that greenhouse-gas emissions are an important factor driving the rise in temperature. Critical to this case is that the physics of Earth's climate, when modeled on a global scale, requires the inclusion of human sources of greenhouse gases, of the type and quantity that have been measured, to accurately reproduce the observed changes in global climate over the past century. This does not exclude the possibility of a natural component to climate change, but the most likely explanation is one that includes human production of greenhouse gases. Indeed, a report issued by the U.S. Global Change Research Program stated in mid-2009, "Global warming is unequivocal and primarily human-induced."[15]

Are scientists in agreement about the human role in global warming? According to the vast majority of climate scientists, the planet is heating up and humans are the cause. In a survey of 3,146 Earth scientists, over 95% of the most highly qualified climatologists (those with more than 50% of their peer-reviewed publications in the past 5 years on the subject of climate change) agreed that "human activity is a significant contributing factor in changing mean global temperatures."[16] The survey found that as the level of active research and peer-reviewed publication in climate science increases, so does agreement that humans are significantly changing global temperatures. The divide between expert climate scientists (97.4%) and the general public (58%) on the subject is particularly striking and identified the need for more effective communication of scientific understanding to the general public. An earlier study published in 2004 came to a similar conclusion: there is strong scientific consensus on global warming and agreement that humans are the primary cause.[17] The study analyzed all peer-reviewed scientific papers (928 of them) between 1993 and 2003 using the keywords "climate change" and concluded that none of the papers disagreed with the consensus position.

Because it is the scientific consensus, in this chapter we assume that global warming is real, that it is driven primarily by human contributions to atmospheric greenhouse gases, and that the surface environment in many places, including Hawai'i, has begun to react to changes resulting from warming.

SOME OBSERVATIONS

To figure out whether Earth's surface is changing, scientists use temperature data from land weather stations, satellite measurements, atmospheric sensors, and ships.[18] These data are analyzed to determine Earth's average surface temperature for any year and trends over longer periods. For example, climatologists at NASA's Goddard Institute for Space Studies (GISS) in New York City provide an annual summary of the year's temperature trends,[19] and scientists at the National Climatic Data Center in Asheville, North Carolina, note that, on average, the linear trend of global warming for the past 50 years is nearly twice that of the past 100 years. Data summaries such as these are updated frequently and provide the public with access to current developments in the science of global warming. Analyses such as these document that the warming has not been globally uniform; in fact, some areas (including parts of the southeastern United States and parts of the North Atlantic) have cooled slightly over the last century. The recent warming trend has been greatest over North America and Eurasia between 40° and 70° N.

Since the industrial revolution, average global concentrations of carbon dioxide (CO_2) have increased 30%, exceeding the natural range over the past 15 million years;[20] methane (CH_4) concentrations have more than doubled, and nitrous oxide (N_2O) concentrations have risen by 15%.[21] These increases in greenhouse gases have enhanced the heat-trapping capability of Earth's atmosphere and caused important changes to air temperature.[22] Other research indicates that the amount of snow cover in the Northern Hemisphere and of floating ice in the Arctic Ocean has decreased.[23] Globally, sea level has risen 4 to 8 inches (10 to 20 cm) over the past century,[24] and the rate has accelerated so that now it is rising at approximately 1 foot (30 cm) per century.[25] Worldwide precipitation over land has increased and the frequency of extreme rainfall events has grown throughout much of the United States.[26] Notably, physical models predict that without human greenhouse gas production, solar and volcanic processes likely would have produced global cooling over the same period in which we have witnessed global warming.

The U.S. Global Change Research Program reports that climate change is apparent now across the nation.[27] Research findings published in peer-reviewed scientific literature reveal that trends observed in recent decades include rising temperatures, increasing heavy downpours, rising sea level, longer growing seasons, reductions in snow and ice, and changes in the amounts and timing of river flows. These trends are projected to continue, with larger changes resulting from higher amounts of heat-trapping gas emissions and smaller changes from lower amounts of these emissions. The observed changes in climate are already causing a wide range of impacts, and these impacts are expected to grow.

Researchers report that a number of ecological changes have already occurred in the United States over the past century paralleled by increases in average U.S. air temperature and changes in precipitation.[28] Warmer temperatures have resulted in longer growing seasons at the national level, altered freeze and thaw in the Alaskan tundra, and increased frequency of fires and other disturbances in U.S. forests.[29] Biologists report individual species such as Edith's checkerspot butterfly, brown and black bears, the red fox, and others have shifted north or to higher altitudes. Some species, including Mexican jays and tree swallows, have experienced changes in the timing of reproduction, as have plants such as forest phlox and butterfly weed. Research indicates that land-based ecosystems will need to move an average of 0.25 mile (420 m) per year to cooler areas over the next century if plants and animals are to remain in their comfort zones. Owing to topographic effects, the velocity of temperature change is lowest in mountainous ecosystems such as tropical and subtropical coniferous forests, temperate coniferous forest, and grasslands (0.05 mile [0.08 km] per year). Velocities are highest in flooded grasslands, mangroves, and deserts (0.8 mile [1.26 km] per year). High velocities suggest that the climates of only 8% of global ecosystems currently have residence times exceeding 100 years.[30] Forest growth is generally projected to increase in much of the east but decrease in much of the west as water becomes scarcer. Major shifts in species are expected, such as maple-beech-birch forests being replaced by oak-hickory in the northeast. Insect infestations and wildfires have increased as warming progresses. Although such changes illustrate efforts by species to adapt to a warming climate, these responses may alter competition and predator-prey relationships and have other unforeseen consequences.

Scientists hypothesize that changes in natural systems will continue and become even more apparent in the future, potentially resulting in degradation and loss of global biodiversity. With continued and more severe changes in the climate, the ability of wildlife to adapt through migration and physiological change will be increasingly limited. Furthermore, as species migrate with changing climate and adapt to new environments, animals will begin competing for habitat, potentially threatening the long-term survival of newly arrived species as well as long-time inhabitants. The challenge is even greater when considered along with the broad range of other environmental threats currently affecting wildlife, such as habitat loss, environmental contamination, and invasive species, as widely seen in Hawai'i.

HUMANS, FROM THE START?

Studies of climate change suggest that humans began their impact on climate with the domestication of animals and deforestation (due to early agriculture) over the past 8,000 years.[31] These activities occurred with the very dawn of our civilized communities. They produced sufficient greenhouse gas and destroyed enough natural carbon storage sites to prevent the natural decay of climate into the beginnings of a small ice age that we would be enduring today because of reduced exposure to solar radiation due to Earth's orbital configuration with the Sun. Comparison of the current climate with past episodes of similar climate preserved in ice cores[32] shows that concentrations of carbon dioxide started rising about 8,000 years ago, despite indications that natural trends in carbon dioxide concentrations should have been dropping. By about 5,000 years ago similar trends of rising methane concentrations were noted, correlating to the expansion of rice agriculture (requiring wetland production). Without these unexpected increases in carbon dioxide and methane (wetlands are copious producers of methane), current temperatures in northern portions of North America and Europe would be 5° to 7°F (2.8° to 3.9°C) cooler. Amazingly, a budding ice age, marked by the appearance of small ice caps in parts of northeastern Canada,[33] would probably have begun several thousand years ago.

Instead, Earth's climate has remained relatively warm and stable in recent millennia because of these agricultural influences (deforestation, wetland expansion, expansion of ruminant animals). When climate has temporarily cooled, such as the Little Ice Age, research has linked this to declines in human agriculture when population plummeted. The "black plague" in Europe, and the Spanish genocide in Central America, led to several episodes of dramatically decreased farming. Forests expanded into fallow farmlands and soaked up carbon dioxide in the atmosphere. This produced cooling events such as the Little Ice Age. These population-lowering events and their repeated correlation to global cooling episodes (recorded in ice cores) strengthen the link to land-use practices and the human connection to climate change.[34]

IS CLIMATE WARMING OR COOLING?

1998 was a record-setting year for warmth; 2005 was another. However, in 2008 global mean temperature dropped, returning to temperatures not seen since the mid-1990s (2008 was, nonetheless, the ninth warmest year on record). To the naked eye, a graph of annual temperatures from 1998 to 2008 looks as if global warming had stopped (when in fact average annual global temperature over the period still had a positive trend). This energized the political "climate denier" community; they seized the moment to influence national attitudes through a media eager to sell controversy. Suddenly "global cooling" was the sound bite of the day. This enticed blogs, Web sites, and other observers to conclude that global warming had come to an end. Thus it was an important paper that examined this question in the peer-reviewed journal *Geophysical Research Letters* in 2009.[35] David Easterling with the National Climatic Data Center and Michael Wehner with the Lawrence Berkeley National Laboratory examined the temperature record and applied computer models to reproduce and extend it.

Easterling and Wehner found that the observed record of average global surface temperature contains periods of no warming and even cooling (e.g., 1977–1985, 1981–1989) within the overall pattern of net warming that characterizes the end of the twentieth century and the beginning of the twenty-first century. They also found this pattern reproduced in climate simulations using computer models forced with increasing greenhouse gases. They showed that climate over the twenty-first century can and likely will continue to produce periods of a decade or two where the globally averaged surface air temperature shows no trend or even slight cooling (e.g., 2001–2010, 2016–2031) in the presence of longer-term warming. The authors ascribed this complex behavior to natural variability of the real climate system. To claim that global warming is not occurring on the basis of short multiyear time periods ignores this natural variability and is misleading.

No scientist expects global warming to be a smooth and consistent process from one year to the next. Nor will it get warmer everywhere at the same time. Like the rise in stock market value since the 1970s, climate is taking a bumpy ride of ups and downs as it undergoes a long-term increase in global temperature. Some of those who do not understand these aspects of global warming see every snowstorm and cool day as evidence that global warming is a hoax. But this is simply confusing "weather" for "climate." It has been said that climate is what you expect, weather is what you get. Among scientists, global warming is known as a "noisy" process, and researchers study the forest (long-term trends) as well as the trees (short-term events). In a blind test, the Associated Press gave prominent statisticians global climate data: they rejected global cooling and concluded that choosing a short interval (such as a decade) from among a long data set broke a cardinal rule of statistics, an error known as "cherry-picking the data."[36]

CLIMATE CHANGE IN HAWAI'I

Hawai'i's climate is changing in ways that are consistent with the influence of global warming. In Hawai'i:

- Air temperature has risen;
- Rainfall and stream flow have decreased;
- Rain intensity has increased;
- Sea level and sea surface temperatures have increased; and,
- The ocean is acidifying.

The U.S. Global Change Research Program studied the impacts of warming on U.S. islands in the Caribbean and Pacific. They released a report[37] in June 2009 that predicts climate impacts in Hawai'i will include the following:

- The availability of fresh water is likely to be reduced, with significant implications for island communities, economies, and resources. Indeed, as reported in chapter 6 scientists have already documented a long-term decrease in rainfall in Hawai'i, a decrease in groundwater recharge to streams,[38] and an increase in the intensity of the heaviest rain events.
- Island communities, infrastructure, and ecosystems are vulnerable to coastal inundation due to sea-level rise and coastal storms.
- Climate changes affecting coastal and marine ecosystems will have major implications for tourism and fisheries.

University of Hawai'i professor Tom Giambelluca and a team of researchers have documented a rise in surface temperature in Hawai'i.[39] They found a relatively rapid rise in the past 30 years (averaging 0.3°F [0.17°C]), with stronger warming at higher elevations (above 2600 ft [800 m]). The majority of the rise occurs as an increase in minimum daily temperature, with a trend about three times faster than that of the maximum daily temperature. In other words, the daily range of temperature in Hawai'i is shrinking, largely due to a rise in the minimum. Notably, over the approximately 85 years of records they analyzed, the temperature in Hawai'i has varied consistently with the Pacific Decadal Oscillation

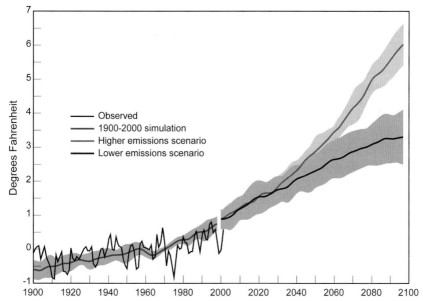

Fig. 9.4. Air temperatures have increased over the last 100 years in the Pacific island region. Larger increases are projected in the future, with higher emissions scenarios (computer simulations of increased heat-trapping gases) producing considerably greater increases. The shaded areas show the likely ranges and the lines show the central projections from a set of climate models. (From U.S. Global Change Research Program, 2009, "Global Climate Change Impacts in the United States": http://www.globalchange.gov/ (last accessed November 10, 2009)

(PDO) (see chapter 6). However, in recent decades the warming trend was dominant and, despite cooling associated with the PDO, surface temperature in Hawai'i has remained high.

There are six areas of general concern with regard to the impact of climate change on islands. To some extent these are interrelated: changes in the water cycle, public health and safety, ecosystems and biodiversity, ocean acidification and warming, impacts to food security, and sea-level variability.[40] These encompass several phenomena.

- Pacific islands will possibly be affected by:
  o changes in patterns of natural climate variability (PDO/El Niño–Southern Oscillation [ENSO], see chapter 6);[41]
  o changes in the frequency, intensity, and tracks of tropical cyclones;[42]
  o changes in ocean currents and winds.[43]

- Hawai'i has already experienced increases in extreme tides and greater frequency of marine inundation.[44]
- Sea-level rise, both long-term and episodic, is already an extremely important issue for many Pacific islands. Sea-level rise results in coastal erosion, inundation, and seawater intrusion into freshwater ecosystems and coastal agricultural zones.
- On low atoll islands where aquifer systems are near sea level, climate change and sea-level rise are adversely affecting water supplies through more frequent droughts, floods, and seawater intrusion into freshwater lenses.[45] On high islands, decreased long-term rainfall and increased storm intensity threaten water availability.
- Climate models do not agree on projections of future ENSO. However, more persistent El Niño–like conditions across the Pacific would lead to a reduction of freshwater resources where rainfall is tied to the ENSO.
- The number of intense storms (hurricanes, typhoons, and heavy rain events) is likely to increase.[46] Hurricane wind speeds and rainfall rates are likely to increase with continued warming. The peak speed of the most intense storms has been observed to increase already.[47] However, there is significant uncertainty about how increasing global temperatures will affect overall hurricane and typhoon frequency and tracks.
- It is possible that increases in the frequency or intensity of hurricanes would generally favor invasive species.
- Island biodiversity is threatened by invasive nonnative plant and animal species, as well as urban expansion, resulting in the highest extinction rates of all regions of the United States. Invasive species are often more resilient, and tolerant of drought and other environmental extremes. This gives them a survival edge over endemics.
- There is concern for increased extinction rates of mountain species that have limited opportunities for migration, and declines in forests due to floods, droughts, and increased incidence of pests, pathogens, and fire.
- The unique "cloud forests" located on some islands occupy a narrow geographical and climatological niche. A shift in temperature or precipitation patterns could cause this zone to shift upward enough to be eliminated.
- On islands, the majority of people, infrastructure, and economic activities are located near the coast, leading to dense areas of vulner-

ability. It is possible that the frequency of extreme events may increase over the next few decades to a century, thereby increasing the risk to public health and safety. Improving community resiliency to disasters is now a major goal of federal and local coastal management agencies.

- Coral bleaching associated with long-term chronic warming of surface waters has occurred in both the Pacific and the Caribbean since the 1990s. Hawai'i, in cooler subtropical waters, has not experienced severe bleaching.
- Ocean acidification threatens reefs and calcareous plankton in the oceans. This is discussed in greater detail in chapter 11.
- Tropical winds in the Pacific have weakened, a phenomenon that may have implications for changing rainfall.[48]

SEA-LEVEL RISE

The level of the sea is the product of several factors. For instance, when rain falls it may be absorbed into the ground or run off the land into the ocean. Groundwater does not contribute to the volume in the oceans, but changes in surface runoff may. In fact, by building dams on the world's major rivers we have changed runoff characteristics such that the equivalent of 1.2 inches (3 cm) of global sea-level rise is trapped in the world's reservoirs.[49] But with the world's major rivers now largely controlled by dams, further evading a bit of sea-level rise by building dams is no longer possible. However, this process is minor compared with the two primary drivers of sea-level change: warming the ocean and decreasing the global volume of ice.

In terms of heat content, it is the world ocean that dominates atmospheric climate. The oceans store more than 90% of the heat in Earth's climate system and act as a temporary buffer against the effects of climate change. For example, an average temperature increase of the entire world ocean by 0.01°F (0.01°C) may seem small, but in fact it represents a very large increase in heat content. If all the heat associated with this increase was instantaneously transferred to the entire global atmosphere it would raise the average temperature of the atmosphere by approximately 18°F (10°C).[50] Thus, a small change in the mean temperature of the ocean represents a massive change in the total heat content of the climate system. Of course, when the ocean gains heat, the water expands, and this represents a component of global sea-level rise.

Understanding how ocean warming and the resulting thermal expansion contribute to sea-level rise is critically important to un-

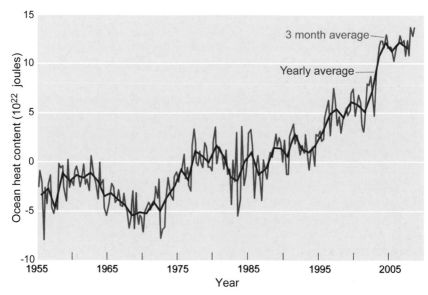

Fig. 9.5. The oceans store more than 90% of the heat in Earth's climate system and act as a temporary buffer against the effects of climate change. However, when the ocean gains heat, the water expands, and this represents a component of global sea-level rise. (SOEST Graphics, modified from S. Levitus, J. I. Antonov, T. P. Boyer, R. A. Locarnini, H. E. Garcia, and A. V. Mishonov, 2009, "Global Ocean Heat Content 1955–2008 in Light of Recently Revealed Instrumentation Problems," *Geophysical Research Letters* 36: L07608)

derstanding climate change and forecasting future impacts of the rising ocean. Researchers[51] found that from 1961 to 2003 ocean temperatures to a depth of about 2,300 feet (700 m) contributed to an average rise in sea level of about 0.02 inch (0.5 mm) per year. Although recent warming is greatest in the upper ocean, observations [52] also indicate that the deeper portions of the oceans are warming, estimated to cause sea-level rise of about 0.008 inch (0.2 mm) per year. Researchers have calculated that thermal expansion of ocean water is responsible for on average 0.2 inch (5 mm) per decade of sea-level rise during the twentieth century compared with 0.7 inch (17.8 mm) per decade in the first decade of the twenty-first century.[53]

Heat in the atmosphere also leads to melting of glaciers and sea ice, a decrease in the extent of snow cover, and shifts between snow-fall and rainfall. Glacier melting and snowmelt contribute to sea-level rise, whereas melting sea ice does not because it is already displacing its own mass in the oceans. However, as sea ice retreats, more open water is exposed to absorb heat from the atmosphere and from the Sun, further encouraging thermal expansion and contributing to heating of seawater in the Arctic and Antarctic. Using satellite imagery, researchers estimate that in 2007 the summer Arctic ice pack covered 1.6 million square miles (4.1 million km²), equal to just less than half the size of the United States. This figure is about 20% less than the previous all-time low record of 1.9 million square miles (4.9 million km²) set in September 2005.[54] Scientists describe the loss as "astounding," the most dramatic loss observed in the history of monitoring the Arctic ice pack. Most researchers had anticipated the complete disappearance of the Arctic ice pack during summer months after the year 2070, but by 2009 they speculated that losing summer ice cover by 2030, or even earlier, was not unreasonable.[55] Thus the stage is set for what scientists refer to as an "albedo flip." Albedo is a measure of the reflectivity of Earth's surface. That is, the former heat-reflecting sea-ice surface will become a heat-absorbing body of water.

Antarctica consists of three main geographic regions: the Antarctic Peninsula, West Antarctica, and East Antarctica. In West Antarctica, which has warmed 0.3°F (0.17°C) per decade at the same time that global warming was about 0.2°F (0.11°C) per decade, ice loss has increased by 59% in the early twenty-first century to more than 145 billion tons (135 billion metric tons) per year.[56] The yearly loss along the Antarctic Peninsula has increased by 140% to over 66 billion tons (60 billion metric tons). The East Antarctic ice sheet, by far the largest region of the continent, has an ice budget that is overall stable to slightly melting. It is experiencing melting along the coastal margin in warming seas and snow accumulation in the hinterlands. Overall the entire continent of Antarctica is experiencing net melting.[57] All three regions of Antarctica are warming, and overall ice loss in Antarctica increased by 75% in the past 10 years.

Until recently, the contribution of the Greenland continental glacier to sea-level rise has been unknown. Now, however, increased melting of the Greenland ice sheet has been observed, and it is known that the glacier is getting smaller. The balance between annual ice gained and lost is in deficit, and the deficiency tripled between 1996 and 2007.[58] In Greenland, the year 2007 marked a rise to record levels of the summertime melting trend over the highest altitudes of the ice sheet. Melting in areas above 6,560 feet (2,000 m) rose 150% above the long-term average, with melting occurring on 25 to 30 more days in 2007 than the average in

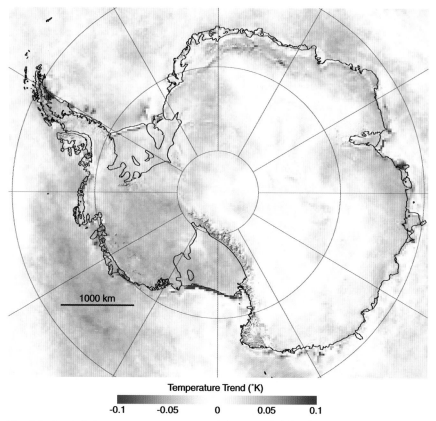

Temperature Trend (˚K)

-0.1   -0.05   0   0.05   0.1

Fig. 9.6. Overall, Antarctica is warming, and the annual deficit between ice accumulation and melting is accelerating. (From E. J. Steig, D. P. Schneider, D. R. Scott, M. E. Mann, C. C. Josefino, and D. T. Shindell, 2009, "Warming of the Antarctic Ice-sheet Surface since the 1957 International Geophysical Year," *Nature* 457:459–462)

the previous 19 years.[59] Scientists have found that glaciers in southern Greenland are flowing 30% to 210% faster than they were 10 years ago, and the overall amount of ice discharged into the sea has increased from 5 cubic miles (21 km³) in 1996 to 13 cubic miles (54 km³) in 2005, an increase of 250%.

Greenland's contribution to average sea-level rise increased from 0.09 inches (2.3 mm) per decade in 1996 to 0.2 inch (5.1 mm) per decade in 2005. This accounts for between 20% and 38% of the observed yearly global sea-level rise.[60] Two-thirds of Greenland's sea-level contribution is due to glacier dynamics (chunks of ice breaking off and melting), and one-third is from direct melting of ice. As glacier acceleration continues

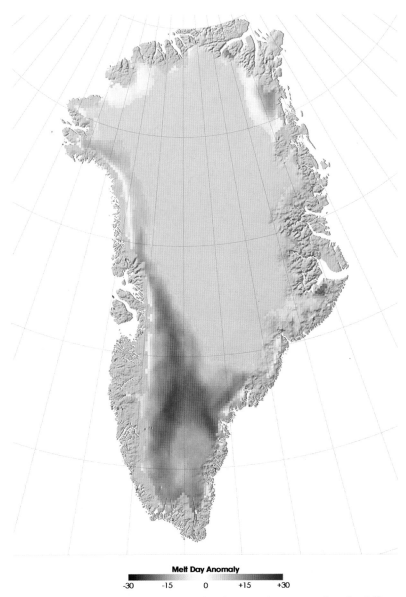

Melt Day Anomaly

-30 -15 0 +15 +30

Fig. 9.7. This map shows the Greenland melt anomaly, measured as the difference between the number of days on which melting occurred in 2007 compared with the average annual melting days from 1988 to 2006. The areas with the highest amounts of additional melt days appear in red, and areas with below-average melt days appear in blue. Although faint streaks of blue appear along the coastlines, namely in northwestern and southeastern Greenland, red and orange predominate, especially in the south. (NASA, Earth Observatory: http://earthobservatory.nasa.gov/Newsroom/NewImages/images.php3?img_id=17846 [last accessed January 21, 2010])

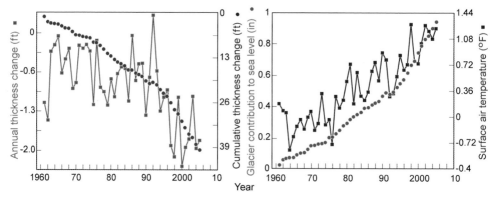

Fig. 9.8. (Left) Average annual and cumulative mountain glacier thinning (feet) since 1961. From 1961 to 2005, the thickness of mountain glaciers decreased approximately 40 feet (12 m). (Right) Cumulative contribution to sea-level rise from mountain glaciers (red) plotted with annual global surface air temperature changes (blue). (SOEST Graphics, courtesy of Mark Dyurgerov, Institute of Arctic and Alpine Research, University of Colorado, Boulder)

to spread northward from its current focus in southern Greenland, the global sea-level rise contribution from the world's largest island will continue to increase.

Retreating mountain glaciers are also contributing to sea-level rise.[61] In fact for the millions of people in communities that depend on seasonal ice melting as a source of fresh water, the retreat and eventual loss of these ice centers delivers a fundamental blow. Mountain glaciers are retreating and thinning in nearly every mountainous region of the planet. The cumulative mean thinning of the world's mountain glaciers has accelerated from about 6 to 13 feet (1.8 to 4 m) between 1965 and 1970 to about 40 to 46 feet (12 to 14 m) of change in the first decade of the twenty-first century.[62] Over the period 1961–2003, mountain glaciers contributed an estimated 0.02 inch (0.5 mm) per year to global sea-level rise, increasing to 0.03 inch (0.8 mm) per year for the period 1993–2003.[63]

Using the time it takes for radar to travel to Earth's surface and back, satellites using altimeters can measure the sea surface from space to better than 2 inches (5 cm).[64] The TOPEX/Poseidon mission (launched in 1992) and its successors Jason-1 (2001) and Jason-2 (2008) have mapped the sea surface approximately every 10 days for 17 years. These missions have led to major advances in physical oceanography and climate studies.[65] Altimeter measurements indicate that global mean sea level has risen about 1.7 inches (4.3 cm) from 1993 to 2008 at a rate of approximately

0.12 inch (3 mm) per year. However, this rise is not uniform across the oceans.

A map of altimeter measurements reveals the rate of sea-level change since 1993 on the world's oceans. Rates are contoured by color: light blue indicates regions where sea level has been relatively stable; green, yellow, and red show areas of sea-level rise; blue and purple indicate areas of sea-level fall. This complex surface pattern largely reflects wind-driven changes in the thickness of the upper layer of the ocean and to a lesser extent changes in upper ocean heat content driven by surface circulation.

Most noticeable on the map, sea-level rise in the western Pacific approaches 0.4 inch (10 mm) per year. This pool of rising water has the signature shape of La Niña conditions in the tropical Pacific. The sea-level buildup in the western Pacific coincides with the absence of strong El Niño events, with the last occurring during 1997–1998. As we described in chapter 6, the Pacific Decadal Oscillation is a basin-wide climate pattern consisting of two phases, each commonly lasting 10 to 30 years.[66] In a positive (or warm) phase of the PDO, surface waters in the western Pacific above 20° N latitude tend to be cool, and equatorial waters in the central and eastern Pacific tend to be warm. In a negative (or cool) phase, the opposite pattern develops. Hence, rapid sea-level rise in the western Pacific matches the current negative phase of the PDO. The degree to which this pattern contributes to the global mean rate of sea-level rise observed in satellite altimetry is not known.

Another important source of sea-level observations comes from the global network of tide gauges. Researchers[67] used this network in combination with satellite altimeter data to establish that global mean sea level rose about 7.6 inches (19 cm) between 1870 and 2004 at an average rate of about 0.05 inch (1.27 mm) per year (0.06 inch per year [1.5 mm] during the twentieth century). The sea-level rise trend increases over that period, with a notable acceleration around 1930. This is an important confirmation of climate change simulations predicting that sea-level rise will accelerate in response to global warming. If this acceleration remains constant then the amount of rise from 1990 to 2100 will range from 11 to 13 inches (28 to 33 cm). This is consistent with model results[68] in which global average sea level was projected to rise 7 to 23 inches (18 to 58 cm) by 2100. It was observed, however,[69] that the increase in sea-level rise at the end of the twentieth century is not statistically distinguishable from decadal variations in sea level that have occurred throughout the record.

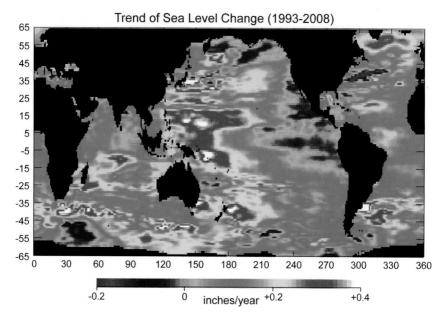

Fig. 9.9. Warming water and melting land ice have raised global mean sea level about 1.7 inches (4.3 cm) from 1993 to 2008 at a rate of about 0.12 inch (3 mm) per year. But the rise is not uniform. This image, created with sea-surface height data from the TOPEX/Poseidon and Jason-1 satellites, shows where sea level has changed during that time and how quickly those changes have occurred. The complex surface reflects the influence of warm and cool bodies of water, currents, and winds. (From NASA, Jet Propulsion Laboratory and CLS/Cnes/Legos)

However, further work with long tide-gauge records[70] revealed that sea-level rise acceleration may have started earlier, over 200 years ago. Researchers reconstructed global mean sea level since 1700 from tide-gauge records and concluded that sea-level rise acceleration up to the present has been about 0.0004 inch (0.01 mm) per year and began at the end of the eighteenth century. Sea level rose by 2.4 inches (6 cm) during the nineteenth century and 7.5 inches (19 cm) in the twentieth century. They also discovered quasi-periodic fluctuations with a period of about 60 years superimposed on the long-term history. On the basis of this analysis, they concluded that if the conditions that established the acceleration continue, then sea level will rise about 13 inches (33 cm) during the twenty-first century.

In their fourth assessment of global climate change, the IPCC[71] reported that sea-level rise since 1961 has averaged 0.07 (0.05 to 0.09) inch (1.3 to 2.3 mm) per year, and since 1993 it has averaged 0.12 (0.09 to 0.15) inch (1.3 to 3.8 mm) per year, with contributions from thermal expansion, melting glaciers and ice caps, and the polar ice sheets. However, they caution that it is unclear if the faster rate since 1993 reflects decadal variation or an increase in the longer-term trend. In other work, researchers[72] used tide-gauge trends to conclude that the recent sea-level rise acceleration reflects an increase that is distinct from a decadal variation and that increased rates in the tropical and southern oceans primarily account for the acceleration. The timing of the global acceleration corresponds to similar sea-level trend changes associated with upper ocean heat and ice melt. Whether the change is associated with the previously identified 60-year variations remains unclear.

In 2007 the IPCC estimated the contributions of ice loss on Green-

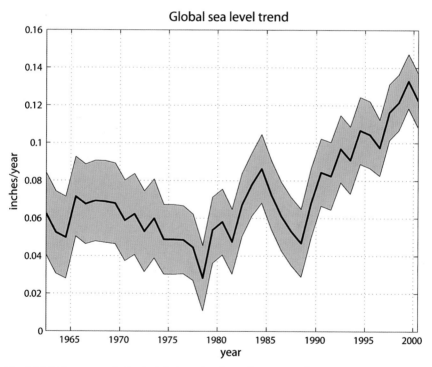

Fig. 9.10. Tide gauge–derived global sea-level trends estimated over 15-year segments and plotted versus the midyear of the segment. Note the rapid acceleration taking place after 1990. One standard error is indicated by the shaded region. (SOEST Graphics, from M. A. Merrifield, S. T. Merrifield, and G. T. Mitchum, 2009, "An Anomalous Recent Acceleration of Global Sea Level Rise," Journal of Climate 22:5772–5781)

land (about 0.008 inch [0.2 mm] per year) and Antarctica (about 0.008 inch [0.2 mm] per year) to global sea level. By summing these and other components of sea-level rise, researchers[73] found that from 1961 to 2003 the sea-level rise "budget" totaled about 0.06 ± 0.015 inch (1.5 ± 0.4 mm) per year. This agreed with their estimate of mean sea-level rise over the same period using observational tide-gauge data (0.06 ± 0.007 inch [1.5 ± 0.2 mm] per year). Tide-gauge data show that global average sea level rose between 4 and 8 inches (10 and 20 cm) during the twentieth century. But satellites that have been mapping the ocean surface for over 15 years have recorded accelerations in the rate of sea-level rise. Some research[74] shows an average increase in global mean sea level over 12 inches (30.5 cm) per century from 1993 to 2008 (approximately 0.12 inch [3 mm] per year). This rate is more than 50% greater than the average rate of the last 50 years.[75]

PAST SEA LEVELS

Sea level has changed throughout Earth history and the past 125,000 years or so have seen especially striking changes, principally because climate has moved between extreme states of global warming and cooling. This period is known as the last interglacial cycle, and it began with a warm period (probably warmer than today) that decayed into an ice age that, in turn, led to the current warm period. The reason for these changes is not fully understood, but it is thought that shifts in the amount of sunlight reaching continental lands in the Northern Hemisphere are at least partially responsible. Sea level changes as climate changes because heating and cooling the atmosphere affects the amount of ice on the planet and the amount of heat stored in the ocean.

Irregularities in Earth's orbit around the Sun, known as Milankovitch Cycles,[76] govern the amount of sunlight reaching the surface. Prolonged reductions produce cold spells including ice ages ("glacials") where sea level falls more than 400 feet (122 m) lower than currently. Prolonged increases produce warm periods ("interglacials") where sea levels are high, including the interglacial we live in now and have enjoyed for about the past 10,000 years.

Researchers hypothesize that the amount of sunlight at approximately 65° N latitude, where it is cold and continental land is dominant, can determine if snowfall melts completely during summer months. When summers are short and cool, snow builds up year after year, eventually forming continental glaciers of sufficient size that they exert their own influence on Earth's climate. Milankovitch Cycles are predictable. That

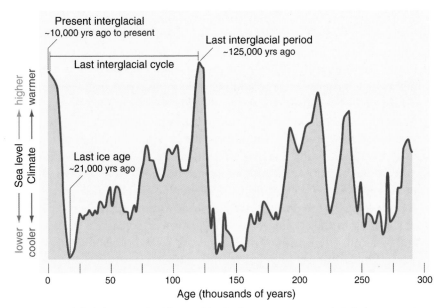

Fig. 9.11. Global climate is highly variable. Peaks represent interglacials (warm periods), and valleys represent glacials (cold periods). The last interglacial cycle is a period of about 125,000 years in which Earth has experienced two interglacials and an ice age. This pattern has many smaller warm and cold episodes. (SOEST Graphics)

is, the variations in exposure to sunlight are known to occur with frequencies of approximately 413,000, 100,000, 41,000, and 26,000 years. These frequencies explain some of the large-scale timing of glacial and interglacial climate. But climate is also highly variable, swinging between cooling and warming on time scales of centuries to thousands of years that are not fully explained by the longer-period Milankovitch Cycles. In general, scientists think of these variations as the result of complex, poorly understood environmental feedbacks involving heat and carbon dioxide storage in the ocean, sea ice advance and retreat, biosphere production and storage of greenhouse gases, the water cycle, changes in albedo, and other factors that can have global influences on climate.

It is somewhat mysterious why the current interglacial, known as the Holocene (meaning "recent time"; nominally defined as extending from 10,000 years ago to the present), has continued to be warm even though Milankovitch Cycles dictate that it should be cooling by now. Here again, climate variability is probably due to an environmental feedback, this time involving humans. The maximum solar radiation hitting the planet peaked 12,000 years ago and has since declined.[77] As mentioned earlier, one intriguing hypothesis holds that early human farming altered natural patterns of carbon dioxide and methane and produced an enhanced greenhouse effect beginning thousands of years ago.[78] That is the only seriously considered hypothesis for this problem, because (also as mentioned earlier) otherwise we would be in the midst of a small ice age with a growing ice sheet in northeastern Canada. For example, clearing and burning of forested land by Neolithic cultures resulted in excess amounts of carbon dioxide entering the atmosphere, potentially tipping climate toward a warmer world. Likewise, farming of rice (wetland creation) in Southeast Asia and domestication of farm animals released excess methane in the earliest days of civilization. Methane and carbon dioxide are two of the most powerful greenhouse gases and may have prevented Earth from sliding into the cooling phase foreshadowed by natural decreases in solar insolation. In support of this idea is the record of pandemics, which decrease global human population and allow cleared lands to reforest. The drawdown of atmospheric carbon dioxide this causes correlates to climatic coolings, such as the Little Ice Age, that characterize the past several thousand years. If this hypothesis is true, humans have been impacting climate much longer than we previously thought.

However, because of Milankovitch Cycles, today's warmth owing to humans has been seen before owing to naturally increased exposure to sunlight. In the last interglacial approximately 125,000 years ago,[79] which may have lasted between 9,000 and 18,000 years (interpreted from various types of geological records around the world such as fossil corals, ice cores, pollen, and stalactites), temperatures are thought to have been 1 or 2 degrees warmer than currently[80] and sea level as much as 10 to 26 feet (3 to 8 m) higher.[81] Notably, rates of sea-level rise may have reached 5 feet (1.5 m) per century, perhaps prophesying the flooding that is to come later this century.[82]

O'ahu contains a geologic record of the last interglacial sea level in the rock formation known among geologists as the "Waimanalo Limestone."[83] This limestone forms Popoi'a (Flat) Island in Kailua Bay, the rock shoreline of the 'Ewa Plain and the Wai'anae coast, and rings the ocean along the north shore of Mokulē'ia on the way to Ka'ena Point.[84] This limestone is a fossil reef that formed during the last interglacial and it contains fossilized coral heads, communities of mollusks and other invertebrates, and ancient algae. Detailed climate information buried in these same rocks, and similar ones on Bermuda, the Bahamas, the Mediterranean, Australia, and elsewhere, have fueled worldwide study of the last interglacial period.

Fig. 9.12. The rock shores of Oʻahu (Kaʻena Point) preserve a record of high sea levels and warmer seas from the last interglacial period. The yellow line highlights a large coral head that is about 125,000 years old. (Photo by M. Dyer)

What caused the high sea levels of the last interglacial period? For years paleoclimatologists thought that only the West Antarctic ice sheet was large and unstable enough to raise global sea levels by such alarming magnitudes in the time frame of only a few thousand years. However, when scientists reconstructed the evolution of the Greenland ice sheet by combining climate modeling with insights obtained from central Greenland ice cores, they discovered that it was considerably smaller and steeper during the last interglacial and plausibly contributed 13 to 18 feet (4 to 5.5 m) to global sea level.[85] They concluded that the high sea level during the last interglacial period most probably included a large contribution from Greenland meltwater and therefore should not be interpreted as evidence for a significant reduction of the West Antarctic ice sheet.

This finding suggests a major worry. Our modern industrialized society is threatened by sea-level rise this century and should be focused on Greenland as an important contributor to the volume of the oceans.[86] This is especially worrisome because research has revealed that the Greenland ice sheet responds rapidly to changes in atmospheric temperature in the North Atlantic and Arctic.[87]

The last ice age culminated between approximately 25,000 and 21,000 years ago. In the transition to the Holocene warm period, sea level rose rapidly between 15,000 and 6,000 years ago at an average rate of about 0.4 inch (10 mm) per year. Based on geological data among Pacific islands, by 5,000 years ago sea level reached and then passed its current position, rising a few feet (about 1 to 2 m) above the current level.[88] Fossil shorelines on islands throughout the Pacific indicate that these high oceans flooded coastal regions between 5,000 and about 2,000 years ago. By the time the first discoverers set foot on Hawaiian shores (ca. 1,500 years ago), sea level was falling below that height, and sandy coastal plains were widening onto the newly revealed seafloor. Although exact patterns are unclear, the period between approximately 2,000 and 500 years ago was characterized by a falling sea level among the islands of Oceania.[89] This time marked the exposure of many low-lying sandy coastal plains that became important Hawaiian community sites, including on O'ahu: Waimānalo, Kailua, Waikīkī, Kahana, Ka'ena, and elsewhere.

Sea level ceased falling a few centuries ago and has since been rising around the world.[90] Local evidence of this rise in sea level, and the coastal erosion it engendered, is found in the form of stranded fossil beach ridges of naturally cemented sandstone in the shallow waters on many fringing reefs around O'ahu (see chapter 10).

PROJECTED SEA LEVEL CHANGES TO 2100

The Intergovernmental Panel on Climate Change (AR4) has predicted future sea-level change to the year 2100.[91] They are not the first body to attempt this important exercise, nor are their latest results their first effort. The AR4 estimate in 2007 forecast a range of global sea-level rise from 7 to 23 inches (0.2 to 0.58 m) by the end of this century. However, these projections are based on thermal expansion and ice melting only. They do not include any component based on dynamic thinning (accelerated mechanical collapse of glaciers, especially coastal glaciers). Hence, it is widely understood that AR4 projections underestimate the potential for sea-level rise by the end of the century.[92]

Several studies published since suggest that future sea level will rise higher than AR4 projections. In one study, researchers compared climate model projections of future atmospheric warming and sea-level rise made in 1990 by the IPCC with observations of mean atmospheric temperature and sea-level position in 2006.[93] The results indicate that the climate system, in particular sea level, may be responding to global warming more quickly than models specify. Although the observed carbon dioxide concentrations follow the model projections almost exactly, the temperature changes are in the upper part of the range projected by the IPCC; notably, since 1990, the sea level has been rising faster than even the extreme scenarios projected by the models. The rate of rise for the past 20 years is 25% faster than the rate of rise in any 20-year period in the preceding 115 years. In their conclusion, the authors stated, "Overall, these observational data underscore the concerns about global climate change. Previous projections, as summarized by IPCC, have not exaggerated but may in some respects even have underestimated the change, in particular for sea level."[94]

In another paper published in late 2009, scientists estimated twenty-first century sea-level change on the historical relationship between twentieth-century temperature changes and sea-level changes.[95] This work concluded that a simple relationship linking global sea-level variations on time scales of decades to centuries to global mean temperature provides a basis for predicting sea level by the end of the century using future warming projected in AR4. The model projects a sea-level rise of 2.5 to 6.2 feet (0.75 to 1.90 m) for the period 1990–2100. The study authors noted that to limit the amount of global sea-level rise to a maximum of 3.3 feet (1 m) in the long run, reductions in greenhouse-gas emissions would likely have to be greater than those needed to limit global warming to 3.6°F (2°C), which is the policy goal now supported by many countries.

Another study[96] used a similar approach, developing an equation that estimates the relationship of climate to sea-level change over the past 2,000 years. The researchers found that sea-level rise by the end of the century will be roughly three times higher than predictions in AR4. They also concluded that even if temperature rise were stopped today sea level will still rise another 0.7 to 1.3 feet (0.2 to 0.4 m); actual cooling would be needed to stop the ongoing rise. The study estimated that sea level will rise 3 to 4.3 feet (0.9 to 1.3 m) by 2100.

Another projection, published in 2008, examined the contribution of dynamic glacier thinning to sea-level rise.[97] Workers filled in the component missing from AR4 projections, namely glacier calving and other types of mechanical ice collapse. The paper concluded that a most likely sea-level scenario by the end of the century will fall between 2.6 and 6.5 feet (0.8 to 2.0 m) with an emphasis on the lower portion of that range.

On the basis of these consistent findings by independent researchers using different methods, and the documented accelerations in melting of both the Greenland and Antarctic ice sheets and ocean warming, it is apparent that a sea level of approximately 3.3 feet (1 m) above the current level may be reached by the end of the twenty-first century.[98]

POTENTIAL FOR SEA-LEVEL RISE TO
EXCEED 3.3 FEET (1 M)

On the basis of the literature review in the preceding section, a 3.3-foot (1-m) sea-level rise by the end of the century is a reasonable planning target for local coastal zone managers that is well supported by the research community.

If greenhouse-gas concentrations were stabilized today, sea level would nonetheless continue to rise for hundreds of years.[99] After 500 years, sea-level rise from thermal expansion alone may have reached only half of its eventual level, which models suggest may lie within ranges of 20 to 80 inches (0.5 to 2 m). Glacier retreat will continue and the loss of a substantial fraction of Earth's total glacier mass is likely. Areas that are currently marginally glaciated are likely to become ice free. But it is unlikely that greenhouse gases will be stabilized soon, so we can probably count on additional atmospheric heating—and sea-level rise.

Researchers have pointed out[100] that observed sea-level rise has exceeded the best-case model projections thus far. Researchers have also suggested[101] that Earth's warming temperatures may be on track to melt the Greenland and Antarctic ice sheets sooner than previously thought and ultimately lead to a global sea-level rise of at least 20 feet (6 m). If the current warming trend continues, by the end of this century Earth will likely be at least 7°F (3.9°C) warmer than currently, with the Arctic at least as warm as it was nearly 125,000 years ago when the Greenland ice sheet was a mere fragment of its current size. Study leader Jonathan T. Overpeck of the University of Arizona in Tucson said, "The last time the Arctic was significantly warmer than present day, the Greenland ice sheet melted back the equivalent of about 6.5 to 10 feet (2 to 3 m) of sea level."[102] The research also suggests that the Antarctic ice sheet melted substantially, contributing another 6.5 to 10 feet (2 to 3 m) of sea-level rise. The ice sheets are melting already. The new research suggests that melting could accelerate, thereby raising sea level as fast as, or faster than, 3.3 feet (1 m) per century.

When it comes to whether or not global mean sea level will exceed 3.3 feet (1 m) by the end of the century, the state of California is taking no chances. In late 2008 Governor Arnold Schwarzenegger issued an executive order requiring all California resource agencies to assess the impact of future sea-level rise on their activities and to produce maps of the impact for public use.[103] The Pacific Institute was contracted by three state agencies to map sea-level rise throughout the state to a level of 4.6 feet (1.4 m). Among their conclusions, they estimated "that 480,000 people; a wide range of critical infrastructure; vast areas of wetlands and other natural ecosystems; and nearly $100 billion in property along the California coast are at increased risk from flooding from a 1.4-meter sea-level rise—if no adaptation actions are taken."[104] Because of the detailed nature of their assessment they were able to identify schools (140), police and fire stations (34), health-care facilities (55), EPA hazardous-waste sites (330), 3,500 miles (5,633 km) of roads and highways, 280 miles (450 km) of railroads, 30 power plants, 28 wastewater-treatment plants, and airports that fall within the zone of impact. The study stated the obvious, "the economic cost of sea-level rise suggests that while adapting to climate change will be expensive, so are the costs of doing nothing as substantial investments are already at risk and vulnerable. Because the economic costs of flooding are highly site-specific, regional analyses are critical for guiding land-use decisions and evaluating adaptive strategies."[105]

SCIENTIFIC RETICENCE AND SEA-LEVEL RISE

Global warming and its consequences present new challenges to modern society. Among these is a degree of uncertainty among scientists about how to communicate the dramatic environmental changes that may be in store. For example, in 2007 James Hansen of the NASA Goddard Institute for Space Studies published a paper about scientific reticence in *Environmental Research Letters,* an online science journal open to the public.[106] He proposed that normally quite articulate scientists were reluctant to communicate the full threat of potentially large and damaging future sea-level rise out of a misplaced sense of scientific conservatism.

Reticence, he argued, prevents researchers from expressing their full understanding of the implications of global warming on human communities. He expressed the opinion that professional reserve and cautious judgment may be hindering the ability of society to recognize and respond to the signs of catastrophic future sea-level rise. This pattern of reticence may lead to unnecessary future expense, damage, and even threats to public safety. In the words of Hansen, "We may rue reticence, if it serves to lock in future disasters." He described how a scientist who has made an important discovery may have concerns about the danger of "crying wolf" and that this danger is more immediate in his mind than concern about the danger of "fiddling while Rome burns." He cited studies of human behavior that show a preference for immediate over delayed rewards, which may contribute to irrational reticence even among rational scientists.[107]

As reviewed earlier, the basis for significant sea-level rise is compelling. Global sea level has accelerated, it is clearly responding to global warming of the atmosphere, and several studies converge on the likely reality that approximately 3.3 feet (1 m) of sea-level rise by the end of this century is imminent. Most major cities of the world are located along coastlines because they were founded during the era of ocean shipping. Low coastal lands often have average elevations of less than 20 feet (6 m), and with population growth communities have expanded into the lowest-lying exposed portions of those areas, in many cases within a few feet (1 m) of mean sea level. In Hawai'i, tens of thousands of people make their homes and conduct their businesses in this vulnerable area threatened by sea-level rise. Although scientific reticence may provide a safe alternative for the individual scientist, it threatens catastrophe for an uninformed public.

HIGHER SEA LEVELS IN HAWAI'I

In Hawai'i, sea-level rise is a particular concern. Riding on the rising water are high waves, hurricanes, tsunamis, and extreme tides that will be able to penetrate farther inland with every fraction of rise. The physical effects of sea-level rise fall into five categories. These are as follows:

- marine inundation of low-lying developed areas including coastal roads
- erosion of beaches and bluffs
- salt intrusion into coastal ecosystems (including taro farming)
- higher water tables
- increased flooding and storm damage when there is heavy rainfall

Currently, the potential range of impacts from any of these five has not been assessed in Hawai'i. This is because, until recently, estimates of future sea-level position continued to change on the international scientific stage and, it had been thought, until these centered on a widely agreed-upon value, efforts gained by an exhaustive study are vulnerable to going quickly out of date. Now, however, studies reviewed earlier identify that a rise of approximately 3.3 feet (1 m) is likely toward the end of this century. Formal planning is based on the "best available information." Clearly 3.3 feet (1 m) of sea-level rise by the end of the century fulfills this criterion, and it is time to make plans to adapt to this intrusion of the ocean in our communities and ecosystems. It must be recognized that the impacts of sea-level rise within our lifetime in Hawai'i are going to be significant and may lead to irreversible effects.

If a sea-level planning scenario were to be undertaken, a 3.3-foot (1-m) threshold would be appropriate and currently supportable from a scientific point of view. Whether this rise is reached in 50 years or 150 years is somewhat irrelevant. The scenario will be achieved in any case, and the potential impacts of that trend need to be examined sooner rather than later. As the ocean continues to rise, natural flooding occurs in low-lying regions during rains because storm sewers back up with saltwater, coastal erosion accelerates on our precious beaches, and critical highways shut down due to marine flooding. The Māpunapuna industrial district of Honolulu adjacent to the airport is a good example. If heavy rains fall during high tide, portions of the region flood waist-deep because storm drains are backed up with high ocean water. The undercarriages of trucks suffer a rust and corrosion problem because floodwaters become salty at high tide. Even when it does not rain, portions of the area flood with saltwater as it surges up the storm drains into the streets. Local

Fig. 9.13. The Māpunapuna region of Honolulu experiences flooding during heavy rains because it lies below sea level at high tide. The storm-drain system is typically inundated (to some degree) with saltwater, hence runoff cannot drain. This problem will spread throughout Hawai'i as sea level rises. (Photo by D. Oda)

workers have reported seeing baby hammerhead sharks in the 2-foot (0.6 m)-deep pools.[108]

Signs of sea-level rise impacts are widespread throughout the state. Not a year passes without some coastal roadways suffering temporary closure as high swells wash across them. These events prevent people from getting to work and to school and from bringing emergency vehicles to those in need. If it has not happened already, it will not be long before someone's life is threatened because an ambulance or fire truck fails to render timely assistance because the road has been closed due to high waves. Although the waves themselves are the apparent culprit, the operator behind the scene is more than a century of rising sea level that brings the oceans onto our coastal highways, as well as the vulnerable location of this infrastructure in the first place.

Another widespread sign of sea-level rise is the chronic erosion plaguing many sandy beaches in Hawai'i. A study[109] of every beach on Kaua'i revealed that 72% are eroding and only 28% are stable or accreting. Among eroding beaches the average rate is approximately 0.8 foot (0.24 m) per year, and on 21% the rate of erosion is accelerating. A beach is an accumulation of sand on the coast and any action that impacts sand availability has the potential to cause erosion. This can include engineering structures that impede sand movement, as well as seawalls that impound dune sand that would otherwise contribute to beach stability. A history of sand mining among the coastal dunes and beaches of Hawai'i contributes to the erosion problem, as does the overfishing of species that make sand on the reef. But as serious as these problems are, it is the rising level of the oceans that unifies and extends these different effects and ultimately leads to extensive beach loss throughout Hawai'i and the world (see chapter 10).

As unpredictable hazards such as hurricanes, heavy rains, high swells, and tsunamis continue to visit the Islands, it would be wise to incorporate the impacts and uncertainties of sea-level rise into the planning and development of our coastal zone. We should not ignore nature by hoping that these hazards will not reoccur and that sea-level rise will stop. The specifics of *how* we live with coastal hazards, and *if* we will reduce the likelihood of natural hazards becoming catastrophes, rest in our hands.

THE BLUE LINE

Using topographic data collected by the U.S. Army Corps of Engineers armed with a sensitive laser in the fuselage of an airplane, it is possible to map the contour line marking 3.3 feet (1 m) above current high tide in

O'ahu. This "blue line" identifies the portion of our communities that will fall below high tide when seas reach this milestone later in the century. This dramatic map has roughly 12 inches (30 cm) vertical accuracy, hence it is highly precise. Those lands that border the ocean are vulnerable to inundation by seawater during high waves, storms, tsunamis, and extreme water levels. Permeable hotel basements will be flooded, ground floors will be splashed by wave run-up, saltwater will flow into below-ground parking areas, and seawater will come out the storm drains on most of the streets in Waikīkī and along Ala Moana Boulevard. However, don't think that waves will be rolling down the streets and reaching the blue line. More likely, these lands lying below high tide in the future will be temporarily dry at low tide and during arid summers. But they will have high water tables, standing pools of rainwater, and backed-up storm drains when it rains and tides are high—much like Māpunapuna does today.

Where homes and roads lie next to eroding beaches, sea-level rise will undoubtedly lead to the construction of large seawalls lining most of our shores. Despite the wet conditions, most of the buildings will probably still be inhabited and residents will have to time their movement between the tides (just as they do today in Māpunapuna). In the McCully and Makiki areas residents will not see any seawater; instead they will see the wetland of the nineteenth century reemerging as the water table rises above ground level in some areas. If you want a glimpse of our future, visit Māpunapuna and talk to the local people.

How to Respond?

How should we confront sea-level rise before it overwhelms our ability to adapt?[110] The context of discussing the answer to this question lies in what we know and what we do not know. Global warming is under way. If greenhouse gas production continues without decrease, some very dangerous circumstances are likely to develop: severe drought, extreme sea-level rise, intense storm behavior, and various types of social problems including previously unseen disease characteristics and heat stress.[111] Environmental damage will likely be unprecedented, especially in Hawai'i where we have widespread cultural investment in coastal plain wetlands (taro), landownership along beaches and bluffs, fragile ecosystems on high mountain slopes being assailed by invasive species, and shifting water and temperature patterns. Unfortunately, despite vocal calls to curb greenhouse-gas production, anthropogenic carbon dioxide emissions have been growing several times faster since 2000 than during the previous decade.[112]

Fig. 9.14. The blue line is the contour of high tide when sea level is about 3.3 feet (1 m) above current level. Communities *makai* of the line are highly vulnerable to coastal hazards and flooding and should be the first targets to increase hazard resiliency and adapt to sea-level rise. (Matthew Barbee, Perspective Cartographic LLC)

However, assuming that some progress is made in shifting from carbon-based energy to renewable energy before 2020, and that significant reduction in greenhouse-gas emissions is under way before 2050, it is possible that the most severe impacts can be avoided by the end of the century. Compared with no reductions at all, emission reductions of about 70% are needed by 2100 to prevent roughly half the change in temperature and precipitation that would otherwise occur.[113] By 2100 the resulting global climate would allow for some preservation of seasonal Arctic sea ice, heat waves would be about 55% less intense, and sea-level rise from thermal expansion alone would be about 57% lower than if no reductions were achieved. This scenario corresponds to an atmospheric carbon dioxide concentration of about 450 parts per million and globally averaged surface air temperature of about 4°F (2.2°C) higher relative to 1980–1999. It is expected that thermal expansion and meltwater production would continue into the twenty-second century and for several centuries thereafter. The sea-level rise due to thermal expansion alone by 2100 would be approximately 1 foot (0.3 m), down from the more than 1.4 feet (0.4 m) (thermal only) expected if there were no mitigation.

An important aspect of global warming impacts in the Pacific region involves the storage of meltwater from Greenland (in the Atlantic Ocean) and Antarctica (in the Southern Ocean) before it makes its way into the Pacific Ocean. Models[114] indicate that after 50 years, most freshwater (melted ice) input from Greenland remains in the Atlantic Ocean. Ice melting around Antarctica also remains largely in the Southern Ocean. Hence, in the absence of catastrophic melting, Pacific islands may avoid major meltwater-driven sea-level rise until the second half of the twenty-first century and beyond. This is good news, of course, but given that sea level is already rising at over 3 feet (1 m) per century in the western Pacific as revealed by satellite altimetry (Figure 9.9), it is obvious that other processes are at work and there is still cause for careful planning and mitigation.

This raises the critical issue of local variability. A look at the local network of tide gauges reveals that the global acceleration in sea level is not yet detectable in Hawai'i. This is supported by the altimetry map, which shows Hawai'i in a narrow region of relatively slow sea-level rise. Yet as pointed out by researchers in mid-2009[115] there are significant departures from the global mean sea level that is modeled by the IPCC and others. The altimetry map makes the point more clearly than any words: sea level is highly variable from one spot to another. Of what use is the global average sea-level rise to a local planner when there are regions of sea-level rise that are dramatically faster and regions where it is actually falling? Micronesia is clearly suffering the effects of rapid sea-level rise, but here in Hawai'i rates are still at twentieth-century levels, and there are several places across the planet where local sea level has actually been falling over the period of satellite monitoring. Improving estimates of the spatial variability in future sea-level changes is an important research target for coming years. This local variability is one of several information gaps that need filling on the way to deciding what our adaptation response will be.

SEA-LEVEL RISK

An important information gap concerns mapping sea-level risk using the microtopography of the coastal zone. Following California, Hawai'i would benefit from having a set of coastal maps depicting areas vulnerable to sea-level rise at the highest possible resolution. The blue line was mapped using a topographic data set provided by the Army Corps of Engineers Joint Airborne Lidar Bathymetry Technical Center of Expertise (JALBTCX).[116] This mouthful of an acronym represents a marvelous

Fig. 9.15. The windward coastal plain near Kualoa, O'ahu; areas in red lie below 3.3 feet (1 m) above modern mean high tide. These low lands will be the first to experience drainage problems as sea level rises. (Matthew Barbee, Perspective Cartographic LLC)

group of federal cartographers that travel the world using laser altimetry (yes, like the satellite) from an airplane to map the coastal zone. With a lidar (LIght Detection And Ranging) data set it is possible to chart the 1-foot, 2-foot, and 3-foot (0.3-, 0.6-, and 0.9-m) contour lines in minute detail among our roads, buildings, and coastal plains in Hawai'i. Maps of Hawai'i sea-level rise vulnerability, much like the California maps, would fill the information gap of "where are the lowest-lying lands and what is on them?"

Another information gap concerns the water table. With rising sea level will come drainage problems. Land drainage is controlled by many factors, one of which is the location of the water table. Where the water table intersects the land surface it produces a wetland, an area of no drainage. Where low-lying lands and a high water table intersect, drainage problems will materialize. As sea level rises, areas of "no drainage" will expand, and new ones will materialize. Between the coast and the blue line scientists can map and model the water table exactly throughout the urban corridor and other vulnerable areas. The water table in the coastal zone lies on saltwater at approximately mean sea level. It rises and falls with the tides in the ocean and even with large sets of waves. It also rises and falls with the seasons of rainfall and individual rain events and storms. This behavior is as important to understand as the behavior of the ocean surface. The water table can be instrumented and monitored with gauges the same way we monitor sea level with tide gauges today. As we learn how the water table rises and falls with natural processes, and with patterns of urbanization, we will be able to more accurately map regions that are vulnerable to flooding from rising sea level in coming decades.

Flooding will occur as a combination of events, heavy rainfall at high tide for example, strong onshore winds and intense rains, and others. It will be (and is currently) the simultaneous occurrence of such events that produce the first impacts of sea-level rise. In some areas the water table will break the ground surface sooner than others, and these are the places where planners and landowners can focus an adaptation strategy. By mapping the behavior of the water table it will become apparent which lands are most, and first, vulnerable and which are next.

Researchers can also develop an improved ability to predict what lands will be overtopped by high waves. The system of wave buoys that surrounds the Islands provides us with an early warning of large waves arriving on our shores. Waves arrive in Hawai'i from all directions of the compass. North swells arrive in the winter, south swells arrive in the summer, and trade-wind waves and Kona storm–generated waves may arrive at any time of the year. Hurricanes are most likely in Hawai'i in late summer and through the fall. All coastlines in Hawai'i are vulnerable to wave overtopping. With a few days advance notice, it is possible for road crews to get outfitted, homeowners to shore up their houses, and communities to prepare for the high run-up of waves on eroded beaches and coastal roadways. By using combinations of wave run-up computer models, microtopography of coastal lands, and knowledge of arriving waves, the disruption to our lives caused by these events can be predicted, and prediction is the first step to mitigation.

An additional step that will be important for coastal communities is to identify which beaches we value to the point of conserving them into the end of the century. Most of our beaches are lined by development in the form of roadways, housing tracts, and even urban corridors and business districts. As sea level rises, these beaches will seek new sand sources, and eroding coastal lands that are sand-rich will be a major source. If, however, these lands are developed, history has taught us that, in Hawai'i at least, seawalls will materialize and impound potential new sand sources.[117] These will slowly wipe out the last beaches among our Islands over the course of the century. To preserve beaches, it will be important to identify coastal plains behind beaches that are sand-rich and prevent these lands from being developed so that they can be eroded by rising seas to provide the sand our beaches need. In the next chapter we discuss this problem and some potential tools to help manage it.

PLANNING STEPS

In addition to filling the information gaps that will allow the Hawai'i community to formulate a response to sea-level rise, there are planning tools to be considered. The question driving coastal planning is "How can we reduce the vulnerability of human communities and natural ecosystems to the negative impacts of sea-level rise?"

Step 1 is to acknowledge the reality of sea-level rise. This can be achieved by writing sea-level rise into our laws, public awareness efforts, and planning activities. Coastal planning in Hawai'i is a shared endeavor between federal, state, and county authorities. Planning is achieved through a system of permitting. If you want to build something in the coastal zone, say, a house, you need a permit. Depending on where you want to build it, various levels of government have the opportunity to comment, alter, approve, or disapprove your request. Currently, you are not required to consider the future threat of sea-level rise in where or how you build (or redevelop existing structures). This can change.

Fig. 9.16. A simulation of wave overtopping indicates that when sea level reaches about 2 feet (0.6 m) above current level relatively rare flooding of our low-lying coastal communities will become a more frequent event. (From S. Vitousek, C. H. Fletcher, and M. Barbee, 2008, "A Practical Approach to Mapping Extreme Wave Inundation:

Legend:
- Annual wave flooding
- 5-yr wave flooding
- 10-yr wave flooding
- 25-yr wave flooding
- Coastal erosion

Sea Level = +1 ft

Sea Level = +3.3 ft

Consequences of Sea-level Rise And Coastal Erosion," pp. 85–96, in Proceedings, Conference on Solutions to Coastal Disasters, 2008, Turtle Bay, Oʻahu, Hawaiʻi, April 13–16, American Society of Civil Engineers)

Step 2 is to require coastal projects to have elements that mitigate the negative impacts of sea-level rise. In the example of your house, it can be a requirement that it is designed and located in light of the risk. Obviously, if a few feet of sea-level rise are anticipated this century you would want to build your house with features to mitigate negative impacts. By shifting the planning process to a risk-based footing, guidelines could be implemented to improve the safety of your house and reduce negative impacts on the environment. Planning is already on a risk-based footing with regard to tsunami and storm surge, and there is a growing effort to plan for the risk of coastal erosion. But there are no planning requirements in Hawai'i with regard to sea-level rise.

Step 3 is to require that development plans contain an environmental assessment that appraises the risks associated with sea-level rise. This will not be particularly challenging. There are several professional tools in place for meeting such a requirement: engineering software for calculating wave overwash, estimates of sea-level rise during the course of this century can be made with reasonable authority, coastal erosion data are publically available or can be produced by consultants, and lidar data are publicly available that can be used by consultants working for applicants. Sea-level rise vulnerability assessments can be produced for permit applicants seeking permission to develop the coastal zone. Privately funded sea-level rise assessments can be created immediately, permit by permit, even as public data bases are also developed.

Step 4 is to redefine the special management area in light of sea-level rise impacts. The special management area is an official planning zone adjacent to the ocean that varies in width from place to place, typically less than 1,500 feet (457 m). However, given the rising water table and drainage problems related to sea-level rise, a simple distance from the shoreline may no longer be adequate. Low-lying lands many blocks from the coast are vulnerable to drainage problems related to sea-level rise (e.g., Māpunapuna). The special management area can be remapped on the basis of these realities.

As public information gaps are filled, areas that have high vulnerability can be considered for special management status. Step 5 is to designate no-build and no-rebuild zones. This would move the coastal community toward improved resiliency (the ability to recover from catastrophic events). But removing private land from the threat of development is difficult, expensive, and ideally requires an owner willing to form a partnership. The most-straightforward approach is to purchase the land or purchase restrictions on how the land is developed. That is, pay the landowner to not develop. This is called a conservation easement. Purchasing land is expensive—especially in Hawai'i. But tools exist, including increasing revenues to conservation land funds (already in existence for each county and the state), tax exemptions for conservation uses, gifting programs to transfer title, reverse mortgage purchases, transferable development rights, and others.

Step 6 is to employ new tools in protecting the coastal environment while at the same time building improved safety into new development. A common scenario is for a wealthy landowner to hire a battalion of experts to secure a permit for some activity, such as building a dream house along the shoreline. Permit authorities at times have trouble justifying restrictive steps in the face of so much assembled talent bent on proving that the planned activity is benign. One compromise would be to allow a development to proceed but require a deed covenant forbidding any future seawall construction (or any activity that could damage the environment or restrict public access), no redevelopment, and no rebuilding. Such a step would essentially declare the property a "no-rebuild zone" when it is damaged in the future by coastal processes. Along with this, design elements can be employed to mitigate impacts such as no slab-on-grade, nourishing the coastal dune with additional sand, and ensuring adequate public access in perpetuity. Planning for no-rebuild is a step toward recovering developed land and creating future beach preserves.

Last, we can begin "climate proofing" our communities. Allowing the continued development of accreted lands, such as still occurs on some of the last healthy beaches in Hawai'i, makes no sense. In an era of accelerated sea-level rise, this increases vulnerability. Climate proofing can involve steps such as raising roadbeds when they are due for maintenance, improving culverts and drainage features, adding one-way-flow gates to culverts to protect wetland agriculture on the coastal plain, reengineering ports, and planning for the future impacts of sea-level rise on our community infrastructure.

A QUIET THREAT

The insidious thing about sea-level rise is its slow pace and stealthy nature. Buried beneath the more obvious signals of daily tides, massive waves, storms and riptides, a couple of tenths of an inch (less than a centimeter) of sea-level rise in a year is not noticed. We see the ocean every day, and over our lifetimes it appears the same. But this casual acceptance belies the truth. Episodes of high sea level are already hitting the Hawai'i coastline with a hidden one-two punch that we ultimately feel in our bank accounts and in the loss of our favorite shores. Take the following case study that happened on the Maui coast in late summer 2003, for example.

As tides go, ours are not particularly large. They range 2 feet (0.6 m) or so between the daily high and low, and twice a month they reach up to 2.5 to 3 feet (0.8 to 0.9 m) as the Sun, Moon, and Earth align. In the heat of late summer they swell with the warm water and extend upward a few more inches. Bring in a summer swell event and the physics of the huge waves pushes seawater against the coast raising average sea-level position higher by an additional foot (0.3 m) or so. Combine the effects of swell energy and seasonal heating, throw in the monthly spring high tides, and before you know it, the fingers of high tide are reaching past the beach and onto our roadways and yards. Most beaches respond to this higher water by narrowing somewhat and experiencing some temporary erosion. As a seasonal event this usually falls within limits that residents of the south shores have learned to wait out until the calm of the fall and winter. In the winter, large swells on our north shores can do the same thing, again patiently waited out by abutting owners.

These were the conditions on Maui's west shore through early summer 2003. But by mid-July something was different. Beaches narrowed as usual and then wholly disappeared. Sharp cliffs of eroded land appeared on the water's edge. Palm trees, clumps of soil, and entire hedge rows on carefully manicured shorelines fell into the water. Beach paths, showers, and verandas were undermined by the lapping waves and subsided into muddy water. By mid-August it appeared that the erosion front would keep pushing landward into the front lawns and pool decks of hotels and time-share condominiums. The very foundations of 10-story visitor towers appeared vulnerable. Soon visitors and *kama'āina* alike were confronted with a scene of thousands of sandbags, teams of sweating workers, and noisy construction equipment shoring up the eroding coast.[118]

Desperate pleas cascaded on state and county authorities for permits to install legal armaments to save the land. The state promoted the temporary use of steel plates driven into the sand to stop the onslaught. County planners balked at the jarring image of pile drivers hammering in broad steel sheets on the delicate beaches. Scientists and engineers were recruited in the quest for answers and solutions. Angry phone calls between and among hotel operators, the mayor, county council members, state and county planners, Native Hawaiians denied access to sacred coastal sites, environmental groups, tourists, and time-share owners crackled across the phone lines throughout the weeks of August and

Fig. 9.17. Temporarily high sea level caused dramatic erosion on the Maui coast in 2003. (Photo by Z. Norcross-Nuʻu)

September. And then like magic, one day in October, the seas subsided and sands returned.

In the postmortem analysis that followed this expensive and chaotic episode, one group of inquisitive specialists turned for an explanation to the tide gauge at Maui's Kahului Harbor. And they found what they were looking for: high sea level.[119]

It turns out that during the spring and summer months of 2003 a series of extreme tides migrated through the Hawaiian Islands, culminating with the highest recorded sea levels in state history by late September and October. Considering that these extreme tides extended only 2 to 8 inches (5 to 20 cm) above the mean sea-level position and each lasted for a month or two, the drama of the Maui coast is eye-opening. Not waves, not storms, not currents, but short-lived and quiet, these high water levels plunged our system of coastal management into utter chaos, angry exchanges, and many hundreds of thousands of dollars of crisis spending. Is the state of Hawaiʻi ready to face sea-level rise? In a word, "no," we flunked the practice drill.

The source of these high-water events is an oceanographic phenomenon known as a "mesoscale eddy." Mesoscale eddies are broad mounds of high water two or three times larger than the Big Island of Hawai'i. With waters a few inches (5 to 20 cm) higher than the surrounding sea, these form in the tropical eastern Pacific and migrate toward Hawai'i with regularity.[120] Four eddies, each larger than the last, migrated through the Islands between February and October 2003. As ocean waters warmed and expanded into the late summer, and sets of moderately sized waves from the south arrived as usual throughout the season, the fragile sands of West Maui were stretched to the breaking point. The death blow arrived in the form of a strong eddy that began driving up sea level in late August, persisted throughout September, and culminated in the first week of October with an all-time high of Hawaiian sea level reaching 8 inches (20 cm) above the average water levels of the earlier winter months.

A COMING CALAMITY

As sea level continues to rise, smaller mesoscale eddies that previously passed unnoticed through the Islands play an increasing role in destabilizing our shorelines. In time, extreme tides related to eddies will grow in frequency and magnitude until they become a permanent condition—not because there will be more eddies or that they will be higher, but because minor eddies we never noticed before will have increasingly dramatic impacts due to sea-level rise. In 2003, we were given a glimpse of our sea-level future.

CHAPTER 10

Beach Erosion and Loss

*E ʻaihue ʻia mai kā kākou keiki, inā hoʻoheno ʻole kākou i
nā kahakai i kēia manawa*

"If we do not cherish our beaches now, then they are
stolen from our children"

Early in Hawaiʻi's history, government recognized the need to protect certain valuable resources: beaches were not among them. Watersheds were set aside to protect our source of water. Reefs were set aside to protect nearshore fisheries. Forests were protected for their ecological and

Fig. 10.1. As every new beach disappears, the quality of life for our children is diminished. (Photo by C. Fletcher)

climate functions. But beaches were never designated as a special environment worthy of protection. This was a mistake, because now beaches are being lost on every island due to seawall building and overdevelopment along our coastline, exacerbated by two centuries of sea-level rise. As a result, local communities and government agencies are engaged in a struggle to conserve beaches by building protections into the system of existing laws. This complex and controversial effort is highly taxing to the good will and resources of the public and their governmental partners.

Among Hawai'i's greatest treasures are her beaches. Their loss, whether measured in terms of disappearance, degradation, or access, is a loss to the ecosystem, the economy, and the culture. As we saw in the last chapter, sea level has been rising for over 200 years, it is rising now, and globally the average rate of rise has accelerated due to warming. Rising sea level reshapes beaches, causing them to migrate landward: precisely where we have built many of our communities. In addition to global

TABLE 10.1   SEA-LEVEL CHANGES THAT IMPACT BEACHES

Phenomenon	Time Scale	Magnitude (per year)	Predictability
Global warning	50 to >100 years	0.08–0.4 inches (2–10 mm)	Predictable
Interdecadal oscillation	10 to 30 years	4–16 inches (10–40 cm)	Unpredictable
El Niño	2–5 years	12–24 inches (30–60 cm)	6–9 months lead
Annual cycle	1 year	4–16 inches (10–40 cm)	Predictable
Intraseasonal oscillation	40–60 days	4–8 inches (10–20 cm)	15–30 days
Mesoscale eddy	20–60 days	2–6 inches (5–15 cm)	15–30 days lead
Fortnightly tide	14 days	2 inches (4–5 cm)	Predictable
Large-scale weather systems	3–7 days	4–12 inches (10–30 cm)	0–10 days lead
Tides	<1 day	2–3 feet (1 m)	Predictable
Storm surges	5 hours	6–25 feet (2–7 m)	0–2 hours lead
Wave group setup	20 minutes	3–4 feet (1 m)	10 minutes lead
Individual waves	15–25 seconds	3–50 feet (1–15 m)	Unpredictable

Source: R. Lukas, personal communication.

sea-level rise, land subsidence, tidal fluctuation, and mesoscale eddies, the level of the sea also varies on much shorter time scales. Waves and storm surges are the sea-level changes we can see altering the beaches daily. As sea level rises, either in the short term or the long term, the beach must shift upward and landward. This change in beach position is known as shoreline retreat, but typically it is described as beach erosion. It requires that a broad swath of the coastal zone be free of development so that the beach and community infrastructure do not collide.

Beach erosion is a term used to describe the actual loss of beach, as occurs temporarily on some of Hawai'i's shores during high wave months and is often seen permanently in front of seawalls where the coast has been chronically retreating. During beach erosion, beaches gradually narrow and may eventually disappear altogether because they are experiencing a sediment deficiency, a loss of sand.

Coastal erosion, another common term, describes the landward movement of the beach and loss of the abutting land in the process. Unlike beach erosion, sand is not necessarily lost during coastal erosion. This is because many of our low-lying coastal lands are composed of marine sands deposited there by persistent winds and ancient high seas from 3,000 years ago (the Kapapa Stand of the sea).[1] Coastal erosion releases this sand and it nourishes the transgressing beach, allowing it to stay wide and healthy even as the shoreline migrates landward.

MIGRATING UPWARD AND LANDWARD THROUGH SPACE AND TIME

"Migrating upward and landward through space and time" is a famous phrase among coastal geologists. It describes the behavior of a beach or other coastal environment such as a tidal wetland, a reef, or an estuary, when sea level is rising. It means that a coastal environment needs unrestricted access to the adjacent land or it will drown under the rising water. Rising sea level forces coastal environments to migrate up onto the land, replacing formerly terrestrial environments with tidal environments. Dunes become beaches, streams become estuaries, and fields (and neighborhoods) become wetlands. If this route of migration is blocked, for example by a seawall, the coastal environment will disappear.

In the past several centuries most beaches have had to adjust to rising sea level worldwide. A retreating beach simply changes its position in space as it migrates, maintaining the same general appearance as before and probably containing some of the same sand. If the abutting land is composed of sand, coastal erosion creates little problem for the beach because it uses this sand to sustain the environment. After all, sand is the lifeblood of beaches. However, migrating upward and landward through space and time does create a huge problem for buildings and roads built too close to the shoreline. And therein lies the rub—protecting the land with a seawall will lead to beach loss. Allowing the beach to continue migrating will damage roads and buildings that lie in its path. What should we protect—the beach or the buildings?

One obvious approach to this problem is to avoid it in the first place. Knowledge of where and at what rate beaches are eroding can be used to guide the way we build along our shoreline. Hawai'i planners know the detailed rate and location of long-term chronic erosion on Kaua'i, O'ahu, and Maui.[2] In fact, county authorities on Kaua'i and Maui use the rate of coastal erosion to determine where a building or road can be built. The rate of erosion is used to calculate a construction setback, which is the distance from the shoreline that determines where construction can take place. This planning tool attempts to avoid the coastal erosion problem by scaling the setback to the rate of erosion. On Maui, the construction setback distance is calculated as 50 times the annual rate of coastal erosion plus 20 feet (6.1 m) (a general buffer). On Kaua'i it is 70 times the annual rate of erosion plus 40 feet (20 feet for sea-level rise plus 20 feet [12.2 m] as a buffer). The idea behind these laws is to prevent the construction of a home or other building close to an eroding coast. Ideally erosion is avoided on Maui for 50 years and on Kaua'i for 70 years (the average life of a wood-frame house).

How much setback will be enough later this century as sea level approaches 3.3 feet (1 m) above the current level? This is a problem that has not yet been fully addressed by planning authorities. The Kaua'i setback does include a fixed distance to accommodate the effects of sea-level rise (20 feet [6.1 m]), but whether this will be adequate and for how long is unknown. Kudos to Kaua'i officials: they have made an inspiring attempt to deal with the problem.

The exact relationship between an increment of sea-level rise and an increment of shoreline retreat is not well understood. Hence, it is difficult to predict how much erosion will be caused by sea-level rise. But it is possible to make an estimate. The width, steepness, and geometry of a beach have a characteristic form that depends on the size of sand grains and the energy of waves. Coarse sand grains tend to build steep beaches, and fine sand grains tend to build beaches with gentle slopes. High waves will move these grains offshore and lower the slope, promoting gently dipping beaches that can dissipate wave energy farther offshore. These

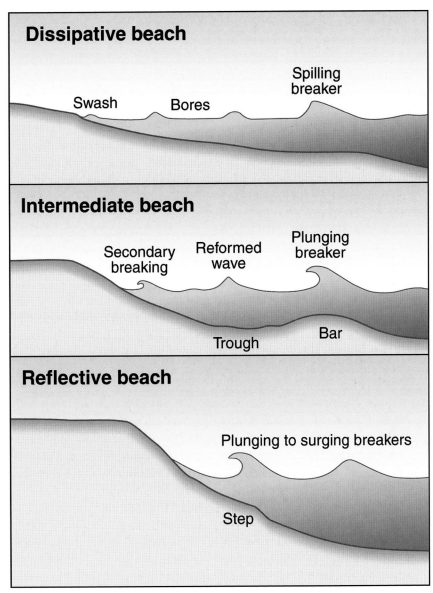

Fig. 10.2. A dissipative beach is one that has been temporarily eroded by high waves. The sand is transported offshore and it builds bars, rip channels, and a wide surf zone. As high waves decrease, fair-weather waves transport the sand back onshore as a shoreward-migrating bar and trough. Eventually, the bar "welds" onto the beach, the beach gains sand and grows steeper, the surf zone narrows, and a reflective condition develops. Many beaches in Hawai'i are reflective or in an intermediate state. (SOEST Graphics)

are referred to as dissipative beaches.[3] Beaches that are steep and tend to reflect wave energy back toward the ocean are called reflective beaches. Most Hawai'i beaches are reflective, or exist in a state of dynamic equilibrium among these factors, constantly shifting from reflective to an intermediate state and accumulating a range of sand grains as the tides, winds, and waves vary throughout the year.

Scientists use a beach profile, a topographic cross section of a beach, to describe these shifts in geometry. A beach profile is measured perpendicular to the shoreline and runs from the land to the sea. It includes the dune, the dry sand beach, the foreshore, and the offshore portion of a beach in the surf zone. The profile can be described by the distance (L) offshore where approaching waves first affect the sandy bottom and the depth at that place (D). Typically the ratio L/D varies between 10 and 100,[4] though it can range beyond these values depending on if the beach is narrow and reflective (closer to 10) or wide and dissipative (closer to 100).

As sea-level rises, a beach will try to stay in equilibrium with the new mean level of the ocean. To accomplish this, the profile shifts landward and upward to regain its equilibrium position; the L/D ratio of the native profile. It achieves this by eroding the shoreline, which on most natural beaches should consist of a sand-rich dune. If however, the shoreline consists of a lava outcrop, a clay bank, a seawall, or some other nonsandy barrier, the beach will experience a sand deficiency and will eventually disappear. The L/D value for Kailua Beach (for example) ranges between about 15 and 45. Hence, the change in D due to sea-level rise may translate into horizontal beach erosion one order of magnitude greater than D. With local sea-level rise currently approximately 0.06 inches (1.5 mm) per year,[5] this translates (using an L/D value of 25) to erosion of 1.5 inches (38 mm) per year. Clearly this is negligible. However, the average rate of coastal erosion that has been measured on the islands of Kaua'i, Maui, and O'ahu is slightly below 1 foot (0.3 m) per year.[6] Hence, if beach erosion were responding only to sea-level rise the average rate would be far less than we currently observe and would not have precipitated the need for the widespread coastal protection that characterizes our shores. Either our estimate of L/D is incorrect and should be closer to 200, or beaches are affected by more factors than just sea-level rise (which we certainly know to be true; for example, they can be affected by human impacts to sand availability), or some combination of these.

In truth, beaches are very dynamic and complex systems that respond to constantly changing ocean and wind conditions; reducing their behavior to a simple ratio of L/D is bound to be imprecise and is designed to give only a ballpark description. Nonetheless, if we stick with L/D ~10–100 as a guideline, the amount of shoreline retreat accompanying a 40-inch (1-m) rise in sea level by the end of the century may range from roughly 33 to over 330 feet (10 to 100 m). Considering that current average rates of erosion fit L/D ~200, planners should count on greater (rather than lesser) amounts of future shoreline change.

Do all beaches migrate upward and landward through space and time? No, beach accretion describes the process of a beach widening in the seaward direction. Such a beach is surely a rare treasure in this era of rising global sea level. Accreting beaches exist where there is a healthy and rich availability of sand. Two examples of accreting beaches include portions of Kailua Beach, Oʻahu, and Hanalei Beach, Kauaʻi. The source of sand to Kailua is thought to be a large sand field located offshore. Hanalei sand is probably supplied by currents (driven by trade-wind waves) that bring sand to the mouth of the bay from the Princeville and ʻAnini shoreline. Waves pick up this sand and deliver it to the beach. Of interest, both Kailua and Hanalei beaches have narrowed in the past half century despite their seaward advance, because aggressive vegetation such as morning glory and *naupaka* grow across the open sand faster than the beach is able to widen. Kailua Beach is actually narrower today than during World War II despite the fact that the water's edge is more than 30 feet (9 m) farther seaward.[7]

BEACH FLEXIBILITY

Found at the boundary between land and sea, the beach is one of Earth's most dynamic environments. It is said among coastal scientists that "the only reliable constant on the shoreline is the condition of perpetual change."[8] A beach evolves through time in a balance or dynamic equilibrium between several factors, including: quantity and type of beach sand, energy of the waves, sea-level change, and the shape and location of the larger coastline; when one factor changes, others adjust accordingly. The ability to change shape according to natural factors reveals it as an extremely durable system when left to its own resources. Evidence of this is the mere fact that we still have beaches despite more than 400 feet (122 m) of sea-level rise since the last ice age. Humans, however, have proven their ability to destroy this otherwise indestructible environment.

As beaches erode and eventually disappear due to some coastal management practices, the protection they offer as a buffer against storm waves also disappears. Homes along Hawaiʻi's shores will experience greater damage as sea level continues to rise and accelerate.[9] This is evident in the fact that one of our major insurers, State Farm, now no longer insures buildings within 1,500 feet (457 m) of the shoreline.[10] Other insurers are following suit. Meanwhile, for homeowners still able to get insurance, rates will continue to rise. It will likely become so bad that, as has now happened along the coastline of Florida, insurance for ocean flooding and storm damage can be obtained only from a centralized state-managed fund because there is no commercial profit to be made insuring buildings on the coast—they are too frequently damaged or destroyed, and homeowners cannot afford the appropriate premiums.[11]

DUNES: CANARY IN THE COAL MINE OF INTERAGENCY (NON)COOPERATION

One of the environments critical to beach survival is the coastal sand dune system. Unfortunately for Hawai'i, our coastal dunes are nearly extinct. Flattened and mined so that the view of the ocean from our pools and living rooms is unobstructed, the sand dunes that formerly lined our shores were nature's guarantee that beaches would have a savings account of sand during times of high waves, high sea levels, and storms. There are still some rare undeveloped shores where coastal dunes may be found and admired: Pāpōhaku Beach along the West Moloka'i shoreline; portions of Mokulē'ia on O'ahu's north shore; areas in beach parks where dunes have not been flattened such as Bellows Beach Park, Baldwin Beach Park, and the Waihe'e region on Maui; and others. Miraculously, Kailua Beach is trying to grow new dunes where it is still accreting in the central region. These are deserving of careful protection.

One reason dunes have not been protected in Hawai'i is because they often fall under county jurisdiction, and not all counties are aware of their importance. In Maui County, "filling" (changing the shape of) a dune is not permitted unless it is with beach-quality sand. Kaua'i County also has specific rules governing dune management. But in the case of the City and County of Honolulu, dunes are not protected other than that they may fall within the coastal construction setback—in the past they have been allowed to be landscaped out of existence. Many homeowners tend to flatten dunes without a permit because we have insufficient enforcement of coastal laws. This is a consequence of having planning and permitting personnel trained primarily in planning, rather than including a few coastal scientists among the team. After all, because a large aspect of the planning function is to preserve the environment, it would be rational to have some environmentally trained staff. Now, there are important

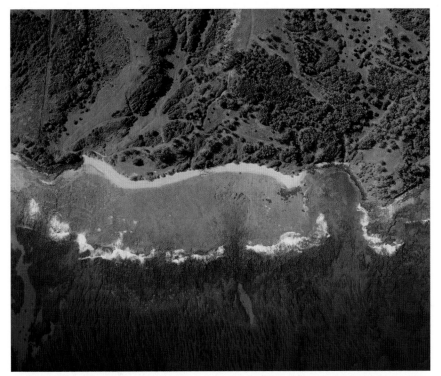

Fig. 10.3. Most beaches in Hawai'i lack abundant offshore sand sources (notice rocky shallow seafloor). The primary source of sand on a chronically eroding shoreline is the dune system located along its landward edge. But these have not historically been protected in the course of managing the Hawai'i shoreline. (Photo from Coastal Geology Group, University of Hawai'i)

exceptions to this generalization, but historically that was rarely the case. On paper, counties have the same goal of protecting the coastal environment as does the state. For instance, Chapter 23, Section 1.2 of the Revised Ordinances of Honolulu states that for the purpose of shoreline setbacks:

> It is a *primary policy of the city to protect and preserve the natural shoreline*, especially sandy beaches; to protect and preserve public pedestrian access laterally along the shoreline and to the sea; and to protect and preserve open space along the shoreline. It is also a secondary policy of the city to reduce hazards to property from coastal floods. (emphasis added)[12]

Unfortunately, the authors of this ordinance did not specify that preserving coastal dunes was a critical aspect of "protecting and preserving the natural shoreline," and management personnel apparently have not historically recognized that flattening dunes causes damage to the beach. On the other hand, why should they worry about beach loss when the beach falls under state jurisdiction, not theirs?

Along the majority of Hawai'i coasts, jurisdiction cuts straight down the center of the beach-dune system and awards half (the dunes) to the counties to manage and half (the beach) to the state to manage. Talk about a recipe for disaster—that is like asking one person to operate the brakes in your car while someone else operates the accelerator. If they don't talk to each other clearly and often and with a single goal, you can

Fig. 10.4. The state shoreline separates the sand-storing dunes from the sand-needing beaches—promoting the demise of each. (Photo from Coastal Geology Group, University of Hawai'i)

be sure that you are headed for a car wreck. Hence, because of multi-agency seawall building on eroding shores, we have lost one-fourth of the length of beaches around the island of Oʻahu.[13] Had city, state, and federal staff operated in an integrated fashion, focused on a single over-riding policy *"to protect and preserve the natural shoreline,"*[14] this level of environmental destruction ideally would not have happened.

Coastal agencies likely do not talk to each other clearly and often enough, and they may not have the same goal; the state Department of Land and Natural Resources wants to preserve environments, while some counties may not care as much about protecting the environment as they simply want shoreline uses to be legal—conservation versus compliance. How did simple compliance come to take precedence over preserving the shoreline? Many planning agencies are understaffed and overworked. These have evolved away from actually performing a conservation func-tion and into a daily routine of processing the torrent of permit applica-tions (to develop the shoreline) in a manner that achieves compliance with governing law. Actually achieving the purpose of preserving the shoreline environment as called for by county ordinance is too difficult with limited staffing levels.

A lack of adequate communication between agencies in different jurisdictions, and a focus on compliance and not conservation, is how one may get a permit to build a seawall on a retreating shoreline, or construct a swimming pool on what should be a dune, or bulldoze down a dune and plant grass to get an ocean view from the lānai. What should be a coordinated coastal zone management system with a unified goal (*to protect and preserve the natural shoreline*) is too often a chain of separate agencies that may resent the advice and comments offered by sister agen-cies and just as frequently ignore it. The thief of our beaches is the same system of management we created to protect it.

THE SHORELINE

The line of jurisdiction separating state and county lands is known as the certified shoreline, and it is defined by the upper reach of the waves on the beach annually, other than storm and tsunami waves.[15] It is the line from which the construction setback is measured. The construction setback is a distance from the shoreline that determines where you may build a house or engage in other land-changing activities. The certified shoreline is also the line seaward of which the public has access along the beach. The official Hawaiʻi "shoreline" tends to lie farther landward than the jurisdictional boundary of many other states and therefore gives the public greater access, but it has also spelled death to our system of sand-storing coastal dunes. Compared with the split-jurisdiction system we currently have, a single entity or a lead agency that manages dunes and beaches together would probably provide better conservation.

WHAT HAPPENS TO BEACHES WHEN SEA LEVELS RISE?

Although the relationship between shoreline position and sea-level rise is complex, sea-level rise ultimately forces a shoreline to migrate landward. This has been proved here in Hawai'i by the discovery of fossil, naturally cemented beaches several centuries to millennia old, located underwater offshore on our reefs and by the erosional response of beaches to extreme high tides that occur on occasion (see chapter 9). Though the migration of a beach under rising seas may not greatly alter its width if sediment continues to be available, its shape may change depending on the slope and topography of the land surface, changes in wave and current energy, and shifts in sand supply and movement.

As mentioned in chapter 9, the physical effects of sea-level rise can be grouped into five categories: marine inundation of low-lying developed

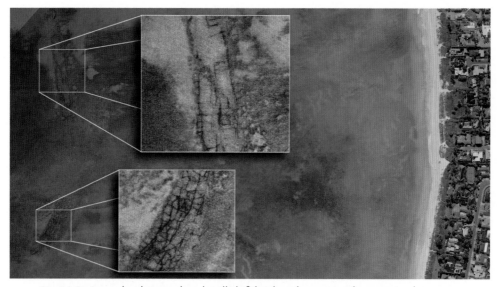

Fig. 10.5. As sea level rises a beach will shift landward to stay at the water's edge. Naturally cemented beach sand, marking the former position of the beach, can be stranded offshore by the migrating shoreline. These fossil beaches (example from north end of Kailua Bay on O'ahu) are proof that beaches migrate upward and landward through space and time. (Matthew Barbee, Perspective Cartographic LLC)

areas including coastal roads, erosion of beaches and bluffs, salt intrusion into coastal ecosystems, higher water tables, and increased flooding and storm damage when there is heavy rainfall. All of these effects have important impacts, but the first two have had and are continuing to have very dramatic impacts on coastal regions worldwide.

Coastal erosion is most likely a product of sea-level rise when no other obvious impact to sediment availability can be identified, although the exact relationship is not well understood. As mentioned earlier, a beach profile has a characteristic shape that depends on the size of sand grains and the energy of waves. As sea level rises, the profile shifts landward to regain this geometry and erodes the adjacent shoreline. Because this beach geometry is maintained, the rate of shoreline recession and the rate of sea-level rise have a unique relationship. This relationship indicates that by the end of the century, lands within approximately a few hundred feet (m) of today's shore are vulnerable to erosion. Next time you visit a beach, take a minute and assess the intense infrastructure, roads, houses, buried pipes, airports, and other community assets that lie within a football field's length of the shoreline—how will we ever deal with the coming calamity of sea-level rise? We need to frame and implement a statewide coastal adaptation plan (see chapter 9).

IS COASTAL EROSION A NATIONAL PROBLEM?

According to a study prepared for the Federal Emergency Management Agency (FEMA) by The Heinz Center for Science, Economics, and the Environment in 2000, approximately 25% of homes and other structures within 500 feet (152 m) of a U.S. shoreline (including the Great Lakes) will fall victim to the effects of coastal erosion within the next 60 years.[16] "This report, *Evaluation of Erosion Hazards,* provides for the first time a comprehensive assessment of coastal erosion and its impact on people and property along our nation's ocean and Great Lakes shorelines," said former FEMA Director James Lee Witt. "The findings are sobering. If coastal development continues unabated and if sea levels rise as some scientists are predicting, the impact will be even worse."[17]

The worst-hit areas include the Atlantic and Gulf of Mexico coastlines, which are predicted to account for 60% of nationwide losses.[18] Costs to U.S. homeowners will average more than a half billion dollars per year in the form of increased insurance premiums, replacement of damaged buildings, and engineering measures to counteract erosion. Additional development in high-erosion areas will lead to higher losses, according to the report. Highly protected areas of large east-coast cities will not be as adversely affected because they have already armored their shoreline, replacing the formerly natural environment.

As a result of this study, FEMA is recommending that coastal communities take steps to avoid erosion.[19] This avoidance policy is designed to circumvent the predicted future of high nationwide costs, human hardship, financial burdens to homeowners, and negative impacts to shoreline environments that suffer when erosion and human land use collide. How can you avoid erosion unless you know where it is happening? Proactively, the Hawai'i Department of Land and Natural Resources, the Hawai'i Sea Grant College, the counties of Kaua'i and Maui, and the City and County of Honolulu have been studying the erosion problem, and they offer several resources designed to inform developers and property owners if they have an erosion problem and how to manage it.[20]

WHERE DOES BEACH SAND COME FROM?

Sand supply remains an essential variable in sustaining Hawai'i's beaches. For example, a few beaches in Hawai'i have maintained their position over the last century, despite slowly rising seas, and Kailua, Kahana, Hanalei, and a handful of other beaches have actually grown seaward with time. As we mentioned earlier, this amazing behavior stems from a robust and uninterrupted supply of sand. In the case of Kaua'i's Hanalei Beach, the entire coastal plain of Hanalei underlying the north shore's picturesque village and taro fields has accreted seaward for nearly 4,000 years under a constant influx of sand. Interrupt this sand supply in the Princeville area and miles (km) away world-famous Hanalei Beach will suffer.[21]

Another example of the dominance of sand supply over sea-level rise is found at Kailua Beach, O'ahu. Sand from a deep channel cutting through Kailua's reef migrates onto the beach under certain wave conditions. Although local reefs originally created this sand, the majority of it now arrives on the beach through the offshore sand channel. Longtime residents of the beach can point to new sand dunes that have formed in the past decade, seaward of the previous dune ridge. But to preserve this gift, and to manage our beaches in accordance with FEMA's recommended avoidance policy, it is important that new development not be allowed to push its way farther seaward. As sea level continues to rise it is unclear how long beach accretion can sustain itself, and these days permission to build on accreting lands is a sure ticket to future seawalls and beach loss.

The typical image of a beach includes a smooth ribbon of white sand stretching into the ocean, but this does not always hold true because

Fig. 10.6. Much of the sand needed by Hawaiian beaches during periods of high waves and sea-level rise comes from dunes that store beach sand until it is needed by the beach to counteract erosion. When natural dunes are removed to make room for lawns and buildings this sand is no longer available to the beach, and it typically experiences chronic erosion. (Photos by D. MacGowan and Z. Norcross-Nuʻu)

beaches comprise different substances. In Hawai'i, beach sand originates from numerous sources including streams (very locally), eroding land, lava flows, biological erosion of submarine limestone, and the shoreward transport of skeletal remains of marine organisms. If you pick up a handful of white sand from your favorite Hawaiian beach, chances are it is hundreds to thousands of years old—the product of bioerosion (several types of critters chew through rocky reefs and produce sand in the process) on old fossil reefs located offshore with the addition of slow biological production of sand over great lengths of time.

Most of the sand particles you find on Hawai'i's white-sand beaches are actually pieces of coral, algae, sea urchin spines, and shells along with smaller mixtures of volcanic minerals and bits of rock.[22] *Koʻa* (coral) is a tiny animal that surrounds itself with its own hard home of calcium carbonate ($CaCO_3$). A coral reef is made by millions of *koʻa* all building new tiny homes on top of the old ones over centuries (see chapter 11). Animals such as the *uhu* (parrot fish) and the *'ina uli* (pastel sea urchin) create sand by scraping and boring into the reef for food and shelter. Sand is also full of tiny hard-shelled single-celled plankton organisms called Foraminifera, sea snails, and species of the algal genus *Halimeda* that harden and turn white after they die. Carbonate (white) sand in Hawai'i is a natural substance that actually "grows" on the reef. Look very closely at a handful of reef sand and you will find that many grains used to be alive.

On all islands, streams carry sand-sized volcanic fragments down from the mountains toward the shore, adding these to the reef detritus that collects on beaches. On the Big Island, volcanic clasts arrive in overwhelming quantity delivered by coastal currents or eroded from bluffs immediately behind beaches, creating some of its surreal black-sand beaches. Big Island black-sand beaches are formed from eroding volcanic sea cliffs, cinder deposits, and grains produced when lava chills rapidly to glass as it enters the cool sea. Green Sand Beach, probably the Big Island's most famed beach among sand and mineral collectors, owes its shade to the dominant green-tinted grains of olivine, a mineral of magnesium-iron silicate that is derived from eroded volcanic bedrock. On Maui, Red Sand Beach provides another example where beach sand is a direct product of erosion from adjacent reddish cinder cone deposits.

Although whitish carbonate sand is associated with reefs, the abundance of such organic fragments does not always mean that the beach sands' immediate source is from offshore. Much of the sand in eroding beaches comes from reworked dune sand and fossil beaches located landward of the modern beach. These older deposits collected over the past 4,000 years when sea level was a few feet higher (less than 2 m or so) than its current level. The higher Kapapa Stand of the sea retreated between 2,000 and 500 years ago and uncovered the coastal plains where we currently live, leaving behind stranded dunes and beach ridges: Waikīkī, Waimānalo, Kailua, Punaluʻu, Sunset Beach, Kapaʻa, the Mānā Plain, Kīhei, and many other flat, sandy coastal plains emerged as the seas withdrew. Before being developed, these places had topography like corduroy, made of undulating beach and dune ridges.

Now, because of a warming atmosphere and a warming—expanding—water column, sea level has reversed and is rising around the world, and on each Hawaiian island, and is reclaiming this land. With shoreline retreat occurring among so many Hawaiian beaches, the principal source of sand is through erosion of the adjoining land: if it is sand-rich with a healthy dune, the beach will benefit; if it is sand-starved with a degraded dune or a rocky backshore, the beach will suffer. A tenuous relationship exists between the slow production of sand by the reef and its persistent accumulation on a beach. An uncaring human hand can all too easily sever this fragile balance. Sand, like fresh water, is a resource to be understood, managed, and conserved. Historically, we have not done this well, as evidenced by the proliferation of seawalls and beach loss on every island.[23]

Fig 10.7. Waimea Beach, Oʻahu, 1949. Many Hawaiʻi beaches have been mined for sand in the past. These tracks in the sand are the work of bulldozers, back loaders, and dump trucks hauling sand away for various commercial uses. (Photo provided by Coastal Geology Group, University of Hawaiʻi)

WHAT HAPPENS TO A BEACH DURING A STORM?

A storm at sea is an unpredictable region of high winds that makes waves. Storm waves formed in the vicinity of Hawai'i are different from the seasonal large swells (formed by distant storms) that arrive every winter on the north shores and every summer on the south shores. Waves associated with local high winds are steeper and more erosive than swells of equivalent height, but both have an impact on a beach; when both hit a beach they tend to change the beach profile from a reflective to a dissipative state by removing sand from the shoreline and temporarily storing it offshore for the duration of the high-wave event.

When a storm encounters the Hawaiian Islands, bringing with it heavy winds and agitated waves, thick ribbons of beach sand get rearranged. As the first thumping waves rake sand into the ocean, it may pile offshore, forming a protective barrier for the beach—a sandbar. This barrier is characteristic of a dissipative beach profile, and it causes waves to break farther offshore. The now-flattened beach enables waves to expend their energy over a broadened zone, and by the time the wave hits the beach it has expended much of its force. Reefs also protect beaches by dispersing wave energy. In places like Waikīkī where the dune is gone and beaches are depleted of sand, water may surge onto adjacent terraces and into swimming pools during particularly energetic storms and swell events. In an undisturbed coastal environment, these waves would return to the ocean with eroded dune sand, nourishing the sediment budget of the beach in nature's example of effective sand management. But along much of our carefully landscaped coastline, these waves return instead with clumps of grass, dirt, patio furniture, and undercut pavement.

A lot of sand may also move sideways along the front of a beach due to longshore currents generated by high waves that approach the shore from an angle and are sometimes aided by flowing tide currents. For example, at Sunset Beach the winter wave energy is so high that even sandbars cannot form, and the majority of sand moves sideways to find shoreline segments offering some shelter from the wave energy. A particular beach may gain or lose sand in this process.

Once a large swell has passed, fair-weather waves may deliver sand back to the beach, and winds and high tides send it back into the dunes, restoring them, at least partially, to their original state.[24] However, if beach equilibrium is seriously disturbed by a major storm or high swell, the restoration process may take months, or even years, or never. In some

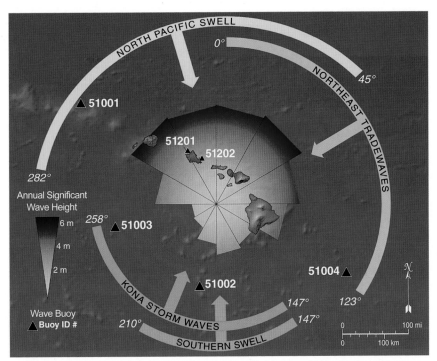

Fig. 10.8. Waves arrive on Hawaiian shores from all directions. This diagram plots the average height of the highest one-third of the waves (a statistical measure known as the "significant wave height") arriving from 30° segments of the compass (12 segments in all). The shade of red (dark to light) in each segment represents the deepwater significant wave height. Wave types include North Pacific swells, northeast trade waves, southern swells, and Kona storm waves. Locations of wave-monitoring buoys are also shown. (SOEST Graphics, from S. Vitousek and C. H. Fletcher, 2008, "Maximum Annually Recurring Wave Heights in Hawai'i," *Pacific Science* 62:541–553)

cases a beach may never fully recover its lost sand because downwelling currents moving offshore can carry sand beyond the edge of the reef into deep water. On other occasions, the recovery of natural beaches after high swell and storms can be complete, yet it may have found a new equilibrium landward of its original position, resulting in net shoreline retreat. If a storm or high swell causes permanent sand loss from the beach, shoreline retreat and the erosion of dune sand is one way to make up for it: maintaining a wide beach at the expense of a strip of abutting coastal land. These adjustments to high waves are one way that beaches reach new equilibrium with rising sea level. Sea level alone, rising a few tenths of an inch (1–3 mm) each year, does not have the energy to shift a beach. It is the waves that do the work for sea-level rise, repositioning the beach in the perpetual search for an equilibrium position.

Keep in mind that these descriptions are quite generalized. Every Hawaiian beach responds to high waves differently, just as each storm or high swell affects a beach in a different fashion. Portions of Waikīkī Beach get wider when there is high summer swell and erode when there are Kona storm waves. Sometimes the sand moves out to sea to an extent that it never returns to its original position. This is what scientists call the "depth of no return," and it varies from beach to beach. For example, off the east coast of the United States, the depth beyond which sand is less likely to return is usually considered to be 30 to 40 feet (9 to 12 m). Off Hawai'i, the depth of no return is likely to be more variable, because irregular offshore topography on the reef may trap some of the wave-derived sand. Aerial photos of Hawaiian shorelines indicate that some beach sand may move down well-developed channels on the seafloor and into deep water. At other locations the beaches are wide and accreting adjacent to reef channels, indicating that they are a source of sand moving toward the beach.

WHY IS THERE AN EROSION PROBLEM?

Fundamentally, the problem with erosion is linked to construction along the shoreline. Were there no homes, roads, or hotels on the beach, erosion would not be a problem. Erosion is a natural process where it is not the product of human actions. This being the case, the erosion problem is certain to increase in Hawai'i and elsewhere as Americans and others rush to the shore to build their homes and global warming drives sea-level rise (and perhaps increased storminess).

Hundreds of thousands of Hawai'i residents and millions of visitors every year use island beaches for recreation, sport, business, or sustenance. On any given beach, however, only a few dozen individuals at most have access to beachfront buildings, whether homes or businesses. Indeed, along most of our beaches many of the homes are rentals, owned by someone who does not live there and who spends the tourists' money in another economy. In conjunction with decades of poor sand management, this small group of shoreline dwellers has created the predicament of property vulnerable to coastal hazards. Likewise, these same few eventually create the beach erosion problem by building seawalls and other structures that protect their investment but destroy the beaches, which belong to the public. Indeed, the decision to build a seawall on a chronically eroding beach is a tacit statement by the owner and the permitting agencies that a private home, hotel, or roadway is more valuable than the beach. In many cases the decision has been made that the protection of a single home with a seawall takes precedence over the needs of the thousands of people who may use that beach and the natural ecosystem of which it is a part. So we ask you, which *is* more important: buildings or beaches? Property owners regard this question as absurd, but is it? Every year more seawalls are permitted, more houses are built on eroding shorelines, and Hawai'i continues to choose private land over the public trust.

WHAT CAUSES COASTAL EROSION?

Coastal erosion has many causes, not all of which are known or completely understood. Yet, there are three proven events that lead to shoreline retreat: (1) human interruption of natural sand movement and sand supply, (2) high waves and currents leading to natural deficits in sand supply, and (3) sea-level rise, which, with the energy of waves, drives the beach to reposition in a more landward location. Typically on Hawaiian coasts, all three of these operate together to varying degrees.

One of the principal causes of beach erosion is the obstruction of the natural sand supply to a beach. When sand supply is impounded by a seawall or other structure, it inevitably leads to deterioration of the beach. Once the beach has either significantly narrowed or completely vanished, longshore currents running parallel to the shore no longer have any sand to transport to neighboring beaches. These down-drift beaches then become targets of erosion, are starved of sand, and retreat landward at an increased rate. Thus shoreline armoring delivers a double whammy: (1) beach loss in front of the wall if the shoreline has been chronically retreating, and (2) sediment supply loss to the entire beach system.[25]

Kalama Park and Halama Street, Kīhei, Maui, is a classic example of beach management. In the early 1970s a massive 3,000-foot-long (914 m) stone revetment was installed at Kalama Park, and almost immediately its beach disappeared.[26] Much of the beach loss was due to placement loss, the simple loss of beach underneath the footprint of the structure because it was built on beach and dune sand. On the adjacent coast to the north, house after house was forced to build a seawall as the sand loss at Kalama infected the entire region. Dozens of walls went up, yard after yard (m) of beach disappeared over the next 30 years, and an embittered community fought with permitting authorities trying to stem the tide of illegal seawall construction. Regulatory battles were fought on each property. Meanwhile, that huge black wall still sits at Kalama holding its secret—if it were removed, the sands might return.[27]

Fig. 10.9. This stretch of Oʻahu's north shore illustrates many of the things we do wrong in managing our beaches. The sand dune was not protected; chronic erosion has narrowed the beach so that at high tide it is too narrow to enjoy, public access is difficult, and homes are vulnerable to tsunamis, storm surge, and high waves. In addition, homes that were once some distance from the ocean end up being damaged by sea spray and lose their value. Now, the priceless beach is gone to protect an inferior building. Sea-level rise will greatly worsen this situation, and the beach at the top will look like the beach on the bottom later this century if we do not make plans soon for an alternative future. (Photos by M. Dyer)

The development of our Islands over the last century has depended heavily on Hawai'i's beaches for a low-cost supply of sand. Used as an aggregate in concrete or spread on the sugarcane fields as fertilizer, sand has been collected for 150 years in Hawai'i. Even today, evidence of sand mining can be seen on beaches, and both private individuals and municipal agencies still conduct sand mining in some places. When you take your next chip shot out of the bunker, bend over and see if that sand looks familiar.

The Waimea Bay shoreline on O'ahu's north coast is now known for a cluster of offshore rocks that include the infamous "jump rock." Swimmers leap off a 30-foot (10-m) rock above the ocean and plunge into the shallow waves. It is astonishing that less than 100 years ago, "jump rock" was nearly buried beneath the sand, and the shoreline was located 200 feet (60 m) seaward of its current position.[28] Over the years, as the sugar companies demanded more and more sand for the production of sugarcane (calcium carbonate sand was used as fertilizer and for concrete production), sand was shoveled out by the truckload, until the shoreline eventually retreated past the excavated rocks. Luckily, Waimea Beach seems to have established a new position that is in equilibrium with the waves, and chronic erosion has ended. Continued sea-level rise, however, may change this situation at any time.

Because sea-level rise and beach and coastal erosion problems will probably only intensify over time, finding "tools" to manage these problems becomes crucial. This generation must strive to maintain the quality of the state's shorelines for the generations to come.

HOW FAST IS SHORELINE RETREAT TAKING PLACE?

No general answer can be given to the question of how fast the shoreline is retreating, because it varies across the globe. In Hawai'i, rates of retreat range from less than 0.5 foot to over 9 feet (0.15 to 2.7 m) per year. The average rate of eroding shorelines on Maui is about 1 foot (0.3 m) per year, and on Kaua'i it is about the same.[29] But every beach is unique, and each has its own pattern of responding to the multiple forces of sea-level rise, sand availability, and wave stress. To see erosion rates for specific coastal locations see the University of Hawai'i Coastal Erosion Web site: http://www.soest.hawaii.edu/asp/coasts/.

THE SHAPE OF THINGS TO COME

Hawai'i still has many wide and beautiful beaches, but there are too many places where the next generation is set to inherit a coastline ringed with decayed seawalls, narrow eroded beaches, and turbid polluted water. It is time for coastal communities to decide which beaches we will protect forever as a legacy for our children and grandchildren. If we do not act soon, we are guilty of shortsighted planning and dodging the politically difficult but critically important work of saving the Hawaiian shore.

Some of the Hawaiian shore is in great shape and has not experienced erosion or pollution. However, it is undeniable that over time the slow creep of erosion and pollution has expanded its presence on every island: 25% of the length of beaches on O'ahu has been lost or narrowed to shoreline hardening,[30] miles (km) of beaches have been lost on Maui,[31] over 70% of Kaua'i beaches are eroding and on one-fourth of these the erosion is accelerating.[32] Kona community groups are worried about how to recover lost beaches and protect the few that remain. In 2001 the Environmental Protection Agency issued a list of over 100 Hawaiian beaches, streams, and estuaries polluted with sediments, nutrients, bacteria, and trash.[33]

As these and many other sources document, major segments of the Hawaiian shore have been subjected to a combination of pollution and beach loss. This trashing of the coastline is only going to increase if we do nothing to slow, stop, and reverse the ill-advised pattern of building as close to the ocean as the law allows. For example, viewed from the sea, the Kāhala shoreline is a solid stone wall from Black Point to Wai'alae Beach Park. At high tide there is no beach to speak of, and public access is difficult and hardly worth the trouble. Lanikai beach continues to disappear. The beach there is less than one-fourth its original size.[34] Numerous access paths that in recent years led to sandy beaches now end in precipitous drops into turbid waves. Until recently the area immediately south of the remaining beach looked like a war zone of sand bags, loose boulders, and waves slapping the undersides of exposed housing foundations. Now it is simply a solid black wall in the water at high tide.

Kailua is one of the last long pristine beaches on windward O'ahu. But even Kailua is feeling the building boom. The current generation of homes is old and being replaced by large multistory cement manors only 40 feet (12 m) from the shoreline. A history of sand accretion is now being buried under poured concrete slabs as new homes are allowed to creep onto the dunes seaward of the existing house line. These homeowners will demand seawalls when erosion strikes—and it will strike as sea-level rises over coming decades—and those seawalls will kill Kailua Beach. Try walking the shoreline of 'Ewa Beach at high tide. You will have to run between wave crests along an endless avenue of seawalls. The beach is mostly a memory.

In fact, an evolutionary sequence has become apparent. Coastlines with modest development and a wide setback from the water are slowly changing into polluted urban strands. Waikīkī, Kāhala, Lanikai, Ka'a'awa, 'Ewa—each is a chapter in the urbanization of the Hawaiian shore. Will we let the last remaining beaches of Hawai'i fall prey to our questionable history of coastal planning? Contrary to recommended FEMA coastal construction guidelines, state setback rules do not recognize erosion patterns when siting a house on the beach.[35] They do not require any special coastal building code. They do not recognize lot orientation or road placement. And they require a standard setback from the shoreline regardless of how fast that shoreline is moving landward. Although most of the land *mauka* (inland) of the shoreline is managed by the counties, it is development on these lands that leads to beach impacts under state jurisdiction: another case of how coastal development and beach conservation are inappropriately managed by different jurisdictions.

As the miles (km) of damaged beaches throughout Hawai'i attest, this is a recipe for both environmental calamity and consumer fraud. Homeowners who benignly seek to live along a beach they love are being granted construction permits from authorities only to find themselves forced to destroy the beach to protect their homes. Establishing place-based beach preservation zoning districts where no coastal armoring will ever be allowed can save the remaining healthy beaches of Hawai'i from the historical pattern of degradation. On O'ahu, Waimānalo Beach, Kailua Beach, Mālaekahana, and Kahuku beaches are a precious heritage in need of protection. Sunset and Kawailoa beaches will disappear if armoring continues to be allowed where the shore is undergoing chronic erosion. Beaches at Mokulē'ia have already been damaged due to seawall construction. Leeward beaches are vulnerable to seawall construction to protect Farrington Highway. Agencies could commit to moving the road but never allowing damage to the beach. The Wai'anae coast has a rich system of long sandy beaches. For Wai'anae residents who want a glimpse of how highway protection is an effective destroyer of beaches, travel Kamehameha Highway from Kualoa to Ka'a'awa and see the future impact of sea-level rise.

STRATEGIC MANAGEMENT OF THE SHORE

On Maui, a new home must be set back a distance equal to 50 years of erosion plus 20 feet (6 m). On Kaua'i the setback accounts for the width of a lot, with wider lots requiring greater setbacks, and in general the setback accommodates 70 years of erosion plus a 40-foot (12-m) buffer zone. On O'ahu, agencies are considering place-based management plans that involve the community at all levels. Preserving future beaches will require creative tools such as these as well as an integrated coastal planning function that resolves the jurisdictional schism created by the certified shoreline.

Other tools will require purchasing land where we want future beaches to exist. A number of methods might be considered for this: reverse mortgages offered by the state and funded by general obligation bond issues wherein owners turn over homes at the end of receiving a fixed period of payments; a homeowner donation program; tax relief for businesses or families in exchange for later ownership transfer of coastal property to some beach authority; transferable development rights for businesses; county, state, and federal land conservation funds; land swaps of *mauka* lands in exchange for prime coastal lands; conservation easements that pay owners to manage their land for the environment; and others. A number of fees are generated in the coastal zone such as land lease fees, the real estate transfer tax, coastal property taxes, local business proceeds, and others. Some portion of these could be set aside to fuel a conservation fund for purchasing developed lands that are sand-rich along otherwise pristine beaches that are likely to need protection in a future of rising sea level. Of course it makes little sense to purchase rocky lands for beach conservation (unless it is to enhance public access, preserve a view plane, or some other conservation reason). These will yield no sand when eroded under high sea levels. Sandy lands are the primary target for conservation, and these could be pried free of the grasp of development so that they may respond naturally to coastal processes.

Also to be considered might be a moratorium on all seaward creep of existing development. This could include a mix of no longer allowing any building on accreted lands, amending state rules so that no new certified shorelines are allowed seaward of prior shorelines, disallowing the "string line" of neighboring properties to be broken by new building, and employing a combination of variable setbacks and disallowed new development on beaches with a combination of eroding and accreting lands. For beaches with a significant population of transient vacation rental properties, converting these to public lands would have less impact on the local community or local families because they are being operated as businesses and not as residences. Last, is there a legal basis for enacting a "freeze" on all existing coastal development? This question can be explored because the scientific certainty of global sea-level rise is already sufficient to support a statewide declaration of emergency planning for Hawai'i's beaches.

Hawai'i has largely escaped the dramatically accelerated beach erosion that will accompany sea-level rise rates in coming decades. Our history of erosion is mostly due to modest rates of historical sea-level rise, human management, and natural coastal processes. In other words, when accelerated sea-level rise does hit Hawai'i things are likely to get worse in a hurry.

WHY WORRY ABOUT BEACHES?

A beach's health serves as a barometer for measuring our success at living with the shore. The great irony of Hawai'i losing its beaches and dunes is that they represent the proverbial golden egg of the state's economy: beaches for tourism, beaches for coastal habitat, beaches as a critical point in the *ahupua'a* between land and sea, and beaches for the last line of defense against the impact of hurricanes, tsunamis, and high waves. And of course, beaches represent the quality of life in Hawai'i.

CHAPTER 11
Reefs and Overfishing

Hānau ka uku koʻakoʻa, hanu kana, he ʻākoʻakoʻa, puka
"Born was the coral polyp, born was the coral,
  came forth"

Aptly named the rain forest of the sea, reefs are rocky marine structures (made of a combination of coral and a hard form of algae)[1] that support one of Earth's most biologically diverse ecosystems. Although a large and interdependent life force relies on reefs for survival, they occupy a

Fig. 11.1. Reefs in Hawaiʻi are generally in good health. However areas under threat include locations where water quality is poor, excess sediment is eroding from watersheds, and overfishing has had negative impacts. There are growing concerns regarding ocean acidification due to carbon dioxide accumulation in seawater. Reefs continue to be deserving of our careful management. (Photo by Keoki Stender)

mere 0.2% of the world's oceans.[2] Not only are these underwater habitats crucial to supporting the ocean's biodiversity, they are home to 25% of all marine species.[3] But reefs also support the economy for those of us on terra firma. Within the United States, the ocean economy is generally proportional to the size of each state's economy, yet tiny Hawai'i defies this trend and has America's largest ocean economy.[4] Our ocean economy generated $5.4 billion in 2004, approximately 10.7% of the state gross domestic product.[5] Hawai'i's coral reefs alone are valued in total at $10 billion, and their direct economic contributions are rated at $364 million annually.[6] Moreover, coral reefs are the foundation for recreational and social activities statewide, and they are as critical today culturally and for subsistence as they were to ancient Native Hawaiians.[7] Hawai'i's isolation in the middle of the North Pacific has produced one of the highest rates of marine endemism in the world. Hawai'i is a global biodiversity hot spot. Home to 86 endemic reef fishes and over 1,250 unique marine species that are found nowhere else in the world, over 410,000 acres (165,921 ha) of coral reef habitat are found in the Hawaiian Islands, an area larger than the landmass of O'ahu.

In a scientific paper reviewing the status of the world's reefs,[8] a consortium of researchers concluded that the diversity, frequency, and scale of human impacts on coral reefs are increasing to the extent that reefs are threatened globally. Increases in carbon dioxide and global warming projected to occur over the next 50 years exceed the natural conditions under which coral reefs have flourished for the past half-million years. The authors of the report predicted that reefs will change rather than disappear as their living communities adapt to these stresses. Some species have already displayed greater tolerance to climate change and coral bleaching than others.

Locally, Hawaiian waters show a trend of increasing sea-surface temperature.[9] These rising water temperatures are a cause for vigilance to the threat of coral bleaching, which thus far has not been a major problem in Hawai'i. Coral bleaching occurs when corals expel their symbiotic algae during periods of stress induced by high water temperatures or other environmental stressors. Bleaching is a sign of poor health and may eventually lead to death of the coral. However to date there have been three documented coral bleaching events among Hawai'i reefs: 1996, 2002, and 2004. The 1996 event was associated with annual late summer heating, low winds, and high sea-surface temperatures that developed in waters around the Hawaiian archipelago. Areas of restricted circulation such as Kāne'ohe Bay got hit the hardest by inshore water temperatures that were elevated as much as 2–4°F (1.1–2.2°C) above normal levels. The other two events occurred among the reefs of the Northwestern Hawaiian Islands, and corals appear to have since recovered.[10] Bleaching has hit other areas of the Indo-Pacific and Caribbean regions with much greater intensity than here in Hawai'i, possibly because Hawai'i sits in the relatively cooler North Pacific waters of the subtropics rather than the tropics.

Among the Hawaiian Islands, managers of our reefs must deal with the negative impacts of recreational overuse, increases in ocean acidity and surface temperature, polluted runoff, coral disease, watershed development (and increased freshwater runoff), overfishing, alien species including noxious invasive algae, storm impacts, commercial trade of reef species, ship groundings, marine debris, and siltation. For instance, researchers have reported that in the Northwestern Hawaiian Islands there is 6.7 times more fish biomass on average than in comparable habitats in the main Hawaiian Islands. These data indicate that humans have reduced fish stocks in the main Hawaiian Islands to about 15% of what they once were.[11] Scientists estimate that more than 340 nonnative marine and

Fig. 11.2. Mud, in this case eroded from West Maui agricultural fields, can damage healthy reefs. A species of noxious alga, *Avrainvillea amadelpha* (known as "leather mudweed"), has taken over the surface of the former coral head. (Photo by Keoki Stender)

brackish-water species live in Hawaiian habitats, where they often out-compete Hawai'i's unique native species and disrupt native ecosystems. This challenge is immense, especially given that reef managers tend to have little authority in managing the same uplands and watersheds that pose the most intense threats to the reef community. The good news for Hawai'i, however, is that our reefs are generally in good shape[12] and that a widespread awareness of their importance and the need for their protection is foremost in the minds of most Hawai'i residents and visitors alike.[13]

OCEAN TRAVELERS

Although corals have a wide distribution in the world's oceans, the varieties that form reefs are typically restricted to relatively shallow, warm tropical and subtropical waters between latitudes 30° north of the equator and 30° south. In most cases clean, clear water is essential to coral health, although prolific reefs can be found growing where rivers empty into the sea and waters often turn turbid. For example, healthy coral communities grow near the mouth of the Hanalei River on Kaua'i and in Kāne'ohe Bay, both of which experience periodic muddy runoff during rainfall. However, landward proximities where corals are exposed to chronic mud buildup and/or freshwater influx during spawning season are unlikely to host healthy communities. Once coral larvae settle on a hard surface and become established, colonies can arise if conditions are suitable. Given enough time, coral colonies become thickets, and thickets build upward on the skeletal remains of older colonies, establishing a reef. Today, coral reefs are found in the low latitudes along continental coastlines, on the margins of volcanic islands, and as isolated coral atolls.

The marine waters of the United States and its territories play host to extensive reefs in the Atlantic Ocean, Gulf of Mexico, Caribbean Sea, and Pacific Ocean. The extended Hawaiian Island chain holds over 10% of the nation's shallow coral reef habitat, and much of that is included in the Papahānaumokuākea Marine National Monument,[14] the largest nature preserve in the United States.[15]

Like all Hawaiian life-forms, ancestors of the organisms that build our reef communities traveled across the vast Pacific to arrive on our shores.[16] Roughly half of native Hawaiian marine species are indigenous to the waters of Indonesia, the Philippines, and other islands of the Indo-West Pacific region. Another 10% to 15% are shared with the west coast of the Americas, about 13% are ubiquitous tropical marine species found across the oceans, and 20% to 25% are endemic to Hawai'i alone.[17]

These organisms were dispersed to Hawai'i as floaters, swimmers, and hitchhikers on the system of currents that circulates across the North Pacific. The Kuroshio Current from the Philippines and southern Japanese islands flows into the North Pacific Current that spans the midlatitude waters to our north. Both currents spin off periodic eddies that are probably responsible for the delivery of much marine biota to our island chain. Some attached to debris and grew to maturity on the journey. Others traveled in larval stage, gambling to hit shore before reaching adulthood.[18]

Although some species endured sustained journeys, most probably took advantage of shortcuts using fortuitous "stepping-stone" islands harboring abundant reef communities at intermediate positions across the ocean.[19] The trip is arduous and only the hardiest individuals of the most appropriately adapted species survived. Eastern Pacific reef communities (including Hawai'i) are notable for their low species count in comparison with those of the West Pacific.[20] The number of marine types decreases steadily along an eastwardly extending line. This biotic attenuation is a natural filter ensuring that those arriving and thriving in Hawaiian waters are survivors of a lottery in which there are not many winners.

THE BONEYARD

To a geologist, a reef is a mass of rock on the seafloor secreted by marine organisms. The planet's first reefs were built by photosynthesizing bacteria, among Earth's earliest life-forms, about 3.5 billion years ago. From fossils it is known that a variety of organisms have constructed reefs across geologic time. Clams, oysters, bryozoans, and sponges are all reef builders. The oldest corals date to about 500 million years ago, but these were solitary souls that did not build the vast colonies typical of modern reefs.[21] Corals similar to modern varieties have constructed reefs only during the past 60 million years, making the northernmost seamounts of the Hawaii-Emperor Chain perhaps one of the first places on Earth to host these special organisms.[22]

Hawaiian reefs comprise two components: biological and geological. The biological realm exists as a living veneer,[23] an organic seafloor community of coral, coralline algae, mollusks, fishes, echinoderms, bacteria, plankton, and others that draw nutrients and expel waste among the coastal waters of Hawai'i. The skeletal debris produced by this community is composed of calcium carbonate ($CaCO_3$). It accumulates as a geological product: whitish sediments or a rigid framework of rock made of the same secreted chemistry. While the organic community on the

Fig. 11.3. Coral reefs are living ecosystems built upon an accumulation of fossil skeletal debris composed of calcium carbonate (limestone, $CaCO_3$) that is secreted by coral, coralline algae, and other organisms. (Photo by Keoki Stender)

seafloor engages in the demands of metabolism, the sediments it produces are skeletal fragments collecting in voids and interstices of the rock framework beneath the seafloor. Like a boneyard of skeletal debris, these fragments consolidate into a mosaic of limestone composed of fossil coral and algae.

Over time, the upward accumulation of this skeletal pile kept pace with the sea-level rise that accompanied the end of the last ice age. As the reef matrix accreted upward, the living ecosystem built on the skeletal debris that collected beneath the living community. In shallow water, the living ecosystem must interact with the physical realm of waves and currents. Waves remove and redistribute sediments (carbonate sand and mud) produced by the reef community but also concussively impact the reef. This inhibits further upward growth, and the seafloor takes on a character adapted to withstand the high energy of shallow water. Instead of growing in relatively fragile branching forms, coral in shallow water grows as a thin layer that encrusts the seafloor. Hard algae known as coralline algae also accumulate that can withstand the hammering of waves. Shallow reef communities must withstand periodic inundation by flooding streams, intense sunlight, seasonal evaporation, and a mixture of land-

and marine-derived pollutants. Offshore, however, on the front of the reef below the waves, the action of accretion and reef building continues in deeper (30 to 100 feet [9 to 30 m]), cleaner, quieter waters.

BIRTH OF A REEF

Freckle-sized animals known as coral polyps, relations of the jellyfish, are the masons behind the intricate structure and chemistry of a coral reef. Corals begin their life as free-floating larvae, traveling ocean currents until they are able to adhere to a hard surface such as a cooled lava flow or ancient fossilized reef. Once grounded, larvae develop rapidly into polyps equipped with small tentacles and a tubelike body with a mouth at one end.

Coral polyps remove dissolved calcium from the surrounding sea-water and excrete it as a calyx, or shell, that armors the polyp when it is lying dormant or being threatened. To feed or expel wastes, the polyp extends its tentacles into the water column. One coral is bordered by several others, which in turn are bordered by dozens, a radiating condominium on the seafloor made of millions of individuals. Over centuries these continue to secrete their exoskeleton of calcium carbonate; some species form annual growth rings like a deciduous tree. Growing upward,

Fig. 11.4. Polyps of the coral *Montipora capitata*, a common species in Hawai'i. (Photo by Keoki Stender)

laterally, and even downward, corals grow within the governing authority of sunlight, wave energy, community competition, nourishment, and water quality.

PLANTS AND ANIMALS

Two organisms serve as principal architects of Hawaiian reefs: scleractinian (stony or hard) corals, and coralline and calcareous algae. Corals are animals. Most of them host microscopic single-celled plants, known as zooxanthellae, within their digestive system. This is a long-standing and successful partnership that gives corals their rainbow of colors and provides them with an additional food source through photosynthesis. These symbiotic algae are the reason most corals need sunlight, thus limiting stony coral development to shallow waters. In turn, the coral provides protection and access to light for the zooxanthellae.

There are over 50 species of coral found in the Hawaiian Islands, but only a few are common.[24] These grow in a range of forms designed to generally maximize the collection of sunlight and food, and minimize their exposure to stresses made by large waves. Stout branching, delicate branching, platy, encrusting, doming, mounding—these and other terms describe the numerous growth forms assumed by coral colonizers as they make the most of their environment. The more abundant Hawaiian genera include rice corals (*Montipora* species), lobe and finger corals (*Porites* species), cauliflower or moosehorn corals (*Pocillopora* species), and false brain corals (*Pavona* species).

Coralline algae and calcareous algae are members of a marine plant group on the reef that deposits calcium carbonate in its tissue. When the algae dies, it leaves a fossil skeleton behind that is hard, whitish, and essentially the same chemistry as the coral. These plants do not have real skeletons in the sense that animals do, but the limestone deposits they produce make it appear so. A few species of calcareous algae, such as members of the genus *Halimeda*, are not completely calcified. Their body segments alternate between calcified and noncalcified. The main segment, looking like a flat leaf, becomes firm with calcium carbonate, but the joints between these segments remain flexible. Hard plant debris builds up as piles of sediment in reef environments and is an important source of beach sand, making up over half the grains on many Hawaiian beaches. The coralline algae look like coral and grow in a binding and encrusting form on the reef, competing for space with corals. Most coralline algae are red, but there are some exceptions. A visit to any intertidal rocky coast in Hawai'i will reveal the encrusting coralline

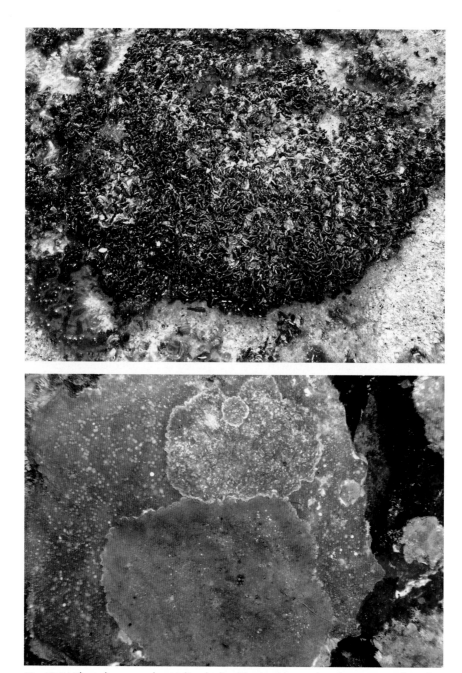

Fig. 11.5. The calcareous alga *Halimeda discoidea* (top) is an abundant source of sand to Hawaiian beaches. The coralline alga *Hydrolithon onkodes* (bottom) is an important component of Hawaiian reefs. (Photos by Yuko Stender)

community coloring the rocks a brilliant hue in between the rise and fall of the waves.

GEOLOGIC HISTORY

O'ahu has the best studied of the Hawaiian reef systems, but what has been learned about the geologic history of O'ahu reefs should only be extended to the neighbor islands with care—largely because of vastly differing environmental histories among the shores of the various islands. Ranging in age from decades to a few hundreds of thousands of years old, O'ahu's reefs remain relatively youthful compared with the million-year-long histories of the main shield volcanoes.

Contrary to popular belief, the seafloor around the main Hawaiian Islands is not replete with rich stands of fragile coral. Strap on a scuba tank and jump into our waters anywhere around O'ahu, Kaua'i, or Maui and chances are slim that you would encounter the idealized coral reef painted by the media and in tourism brochures. Instead, you would likely settle to the seafloor on a smooth limestone pavement, which at first glance you might take to be a submarine desert. This is the insular shelf, and it is a long-dead graveyard of fossilized coral and its algal partners.

Most of the insular shelf around O'ahu is a fossil reef that was last active 200,000 years ago.[25] Since then there have been two ice ages and with each this limestone was stranded high and dry a few hundred feet (50 to 100 m) above the ocean. In shallow water on the windward side of O'ahu north of Kāne'ohe Bay, the reef was last alive about 5,000 years ago, about the time researchers think that extraordinary large swells associated with strong El Niño years began hitting the Hawaiian Islands. Although these huge wave events occur only every few decades (such as in 1998 and 2009), they spell bone-jarring concussion and stress for the reef surface, and they shut down any hope of widespread coral growth on those exposed shores.[26] The same is likely true for all Hawaiian shores that experience north swells. South of Kāne'ohe Bay in Kailua and Waimānalo and in other protected locales such as West Maui, South Moloka'i, and East Lāna'i, the reef is largely protected from the most massive northerly swells. These sites play host to a diverse community of coral and algae in waters that are locally stressed by waves or land-based pollutants. In shallow waters, trade-wind waves, swells, heavy rainfall, runoff, and occasional storms generally limit widespread coral growth.

On the majority of the O'ahu shelf, a diver might find an occasional growth of coral sparsely distributed across the field of view, and there are plenty of nooks and crannies along the coast where coral can find

Fig. 11.6. At this location on Moloka'i, living coral consists of robust branching *Pocillopora meandrina*, doming *Porites lobata*, and encrusting *Porites lobata*. Most of these colonies are approximately the same size, and they are about the same age. The seafloor beneath them is a fossil reef over 5,000 years old. When a particularly large wave event comes along it will remove the living corals, exposing the 5,000-year-old seafloor. The age of these corals probably dates back to the last large wave event, 1998, when their predecessors were wiped out. These corals are likely to live only until the next large swell arrives. (Photo by C. Fletcher)

a safe home. But on the open shelf, most of the living community is a short, grassy-looking "turf algae"; rock-burrowing invertebrates such as various crabs, sea urchins, snails, and others, and scatterings of sand with stands of *Halimeda* and other algae. Clearly any coral growing on this ancient surface has as much chance of a long life as a mouse in a closet full of cats. Hurricanes, tsunamis, and heavy concussive swells generated by distant storms in the Southern Ocean and the North Pacific have all taken their turn at stopping any permanent reef accumulation for millennia. To understand this history, we need to understand the ultimate agent controlling reef growth: the position of sea level.

SEA-LEVEL HISTORY AND REEF ACCOMMODATION SPACE

In earlier chapters we learned that the blue skin of the sea is never still.[27] In addition to tides, waves, and storms, the ocean level is perpetually changing as global climate warms and cools,[28] glaciers wax and wane, and volcanic islands erupt and subside. Today the climate is turning warmer, and so the seas are rising. But in the past, Earth has been plunged into arctic conditions leading to the growth of 1- to 2-mile (1.6 to 3.2 km) -thick continental-scale glaciers that grew at the expense of the oceans. Water that evaporated from glacial-epoch seas fed expanding ice on most of the world's continents and thus the sea level fell.

The last ice age peaked about 20,000 years ago, and the natural manufacture of all that ice lowered global sea level more than 400 feet (122 m).[29] Ancient humans walked from Siberia to Alaska and from Asia to Indonesia across the exposed lands. Had the Hawaiian Islands been populated at that time, they might have walked from Maui to Lāna'i and over to Moloka'i. Reefs were abandoned hundreds of feet (50 to 100 m) in the air where their ecosystems died, partially dissolved, and recrystallized in lightly acidic groundwater and rainfall. Approaching the islands of Polynesia by boat, a voyager would have been greeted by forbidding gray cliffs on most shores, not welcoming soft sand.

But the last ice age was not the first time this had happened; detailed histories of global climate written in fossil plankton on the seafloor and buried layers of ice at the South Pole record that these ice ages have come and gone approximately every 100,000 years. In unpopulated Hawai'i, past ice ages were followed by a warm period with high seas, subtropical conditions, and renewed opportunity for reefs to grow along the shores— if there was room. Fossil reefs fill available accommodation space where a new reef would prefer to grow. If budding coral polyps are squeezed out between the ancient seafloor and modern wave scour, a new reef cannot form. Try as it might, a new reef community cannot get a foothold and build anew on the rooftop of its predecessor unless sea level goes unusually high and/or the regular occurrence of high waves, hurricanes, and tsunamis is somehow halted.

This does not mean that luxuriant coral growth is absent from our Islands. Witness the protected and isolated corners of the coastline such as Hanauma Bay, the south shore of Moloka'i, the south shore of Lāna'i, the west coast of Hawai'i, and other sheltered shores and embayments throughout the Islands. In these sites, protected from high waves, you will discover an undersea paradise of coral growth, reef accretion, and abundant marine diversity.

REEF ORGANIZATION

Scientists have discovered that corals grow from 0.4 to 4 inches (1 to 10 cm) per year.[30] This range depends on the level of disturbance experienced by a coral as well as the available light, food, and water temperature. Because of Hawai'i's location in the middle of the North Pacific, our reefs are pounded by violent swells arriving from the north and south and the buffeting effect of waves kicked up by the trade winds. This persistent wave environment influences the structure of coral reef outcroppings.

Reefs at greater depths receive less impact from nature's roughhousing than those in shallow waters, but they in turn are limited by the amount of sunlight they receive. Coral reef zones define the depth and types of species inhabiting each level. Every marine scientist has a favorite scheme for classifying reef zones, but a simple one includes the shore zone, inner reef or reef flat, reef crest, and outer reef or fore reef.

In the shore zone, the closest reef habitat to shore, hardly any coral develops due to breaking waves and large amounts of fresh water and silty

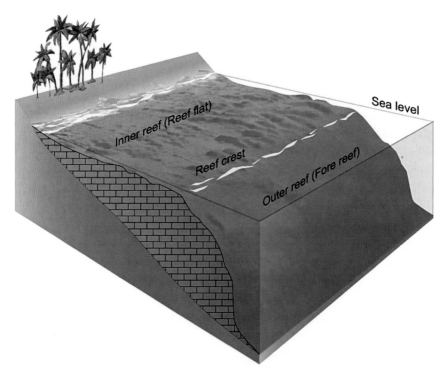

Fig. 11.7. The simple zones of a fringing reef include the shore zone, inner reef (or reef flat), the reef crest, and the outer reef (or fore reef). On some Hawaiian coasts a reef crest may be absent. (SOEST Graphics, modified from U.S. Geological Survey, Coastal and Marine Geology, "U.S. Coral Reefs—Imperiled National Treasures": geopubs.wr.usgs.gov/ fact-sheet/fs025-02/)

mud flowing into the ocean. The seafloor tends to be fossil reef limestone covered in a sedimentary blanket of both marine and terrestrial minerals. Still, a medley of fishes, seaweeds, and invertebrates can be found swimming in these shallow, warmer waters.

The inner reef is anywhere from 3 to 15 feet (0.9 to 4.6 m) below the ocean surface, depending on the locality and the level of tide. This zone is known for its accumulation of carbonate sand, the growth of coral heads usually of the genus *Porites,* and numerous invertebrates, fish, and algae species. Fishes here are primarily herbivores and nibble on algae growing on dead coral and seaweeds. As gill netting and other types of uncontrolled fishing decimate the herbivorous fish population, their role in limiting fleshy algae growth is diminished.[31] Overfishing results in coral death because alien algae can outcompete coral for space on the seafloor, often smothering coral beds under thick layers of fleshy tissue. Smaller *honu* (green sea turtle) are also commonly seen in the inner reef, nestled under hanging ledges and feeding on algae.

Beyond the inner reef, the reef crest marks the shallowest zone. Waves break directly on sturdy encrusting corals. Thick crusts of coralline algae that withstand exposure to the air during low tide may also characterize the crest. Some locations, such as Kailua Bay on windward Oʻahu, do not have a distinct reef crest; instead the reef flat grades directly into the fore reef, which drops off steeply to a depth of 80 to 100 feet (24 to 30 m) onto a sand-covered terrace.

The fore-reef zone is steep and most notable for its coverage of finger coral and other more delicate growth forms. Here may be found the highest density of coral coverage in the reef community. In the deep fore reef, most ocean waves have minimal effect on the slow-growing community. This protection enables the system of stony corals to accrete atop one another and lay down a calcium carbonate base in its most optimal conditions. Below about 80 to 100 feet (24 to 30 m) is a rubble zone defined primarily by coral blocks, rock, and sand that are swept by storm waves off the reef above. These provide a habitat for lobe and finger corals, an assortment of fishes, as well as sluggish sea cucumbers, lobsters, and the more adventurous open-ocean carnivores willing to wander into shallower waters.

HAWAIʻI'S REEFS

In the broadest sense, reefs grow as one of four types: barrier reefs, atolls, fringing reefs, and patch reefs. Hawaiʻi has no true barrier reefs, atolls populate the Northwestern Hawaiian Islands, fringing reefs are found attached to the shore among the eight main Hawaiian Islands, and patch reefs (smaller in scale than the other reef types) are well developed in Kāne'ohe Bay and occasionally elsewhere.

Thought by many to be a Hawaiian-style barrier reef, the ridge forming the seaward boundary of Kāne'ohe Bay is actually an old hardened sand dune formed during a period of lower sea levels approximately 80,000 to 110,000 years ago.[32] That was a time when the eastern shelf of O'ahu (and perhaps other islands) was covered in a broad field of forested sand dunes. Flooded today by high seas, opportunistic polyps and other reef denizens colonized the sandstone and have grown a coral-algal veneer upon its surface. Proof of this is found in the sandstone displaying dipping beds (former dune faces) composing Kapapa Island, Kekepa Island, and Lā'ie Peninsula and its system of small islets.

Seaward of the Nāpali coast on Kaua'i is a fossil coral plain 60 feet (18 m) deep that could be taken to be an old barrier reef. It ramps gently up from deeper water, forming the insular shelf. At its landward edge, it ends abruptly in a steep landward-facing wall that reveals its internal construction of fossil *Porites compressa* columns as long and thick as your forearm. This wall has retreated seaward under the relentless pounding from huge north swells; but with no offshore-facing reef front, this feature is no barrier reef. It is merely a former fringing reef being pounded out of existence by huge waves. The fringing reef grew between 8,000 and 5,000 years ago before the era of swells that mark more recent millennia.[33] Under incessant seasonal hammering, the seafloor is being lowered by the destruction of fragile fossilized corals to reach a depth that is more in equilibrium with the high energy.

The vast majority of Hawai'i's reefs are of the fringing variety. Fringing reefs attached to Island shores are forced to undergo the same evolutionary changes endured by the mass of the shield volcanoes. When an island is young and still under Pele's influence, lava entering the ocean prevents reef accretion. Although young coral colonies do grow, they are too often wiped out by lava flows and therefore are unable to have accumulated the mass of a true reef. But moving up the west side of the Big Island where the seafloor is not swept by high swells (because northerly waves are blocked by the other islands), the beginnings of fringing reefs attached to the land can be seen and beautiful coral gardens are found.[34]

North of the Big Island on the Kīhei coast of Maui and extending out from the North Maui shoreline are broad fringing reefs, demonstrating that reef development has a firm foothold on these stabilized volcanic coasts. Fringing reefs generally grow in size and become commonplace among the Islands north of the Big Island, but among the Northwestern Hawaiian Islands their shape fundamentally changes. Fringing reefs give way to submerged pinnacles, drowned platforms, and atolls as the volcanic shield structure subsides beneath the waves and the fringing reefs struggle to stay near the surface.

HUMAN IMPACTS

Because natural geological limitations to reefs (such as powerful waves from every direction and watersheds associated with prolific rainfall) are the rule on many Hawaiian shores, those locations where modern reef growth is successful become precious treasures worthy of special protection. Yet even the fossil limestone surfaces that mark most of our shallow seafloor and are largely devoid of stands of living coral provide important habitat for dozens of other species. The lack of high coral cover does not negate their value as sensitive environments in their own right and deserving of careful management.

The most significant human threats to Hawai'i's coral reefs are overfishing, land-based pollution, invasive algae, recreational overuse, and increases in ocean acidity and surface temperature associated with carbon dioxide buildup in the atmosphere.[35]

In the early 1900s, Hawai'i fishermen sold 3.5 million pounds (1.6 million kg) of reef fish a year, but by 1950 the annual catch had dropped to less than 1 million pounds (453,000 kg).[36] The yield of reef fish has been sinking ever since. One reason for the decline is that humans tend to target the biggest fish. The largest fish are also the most fecund members of a community. As the most successful individuals are depleted, the community becomes skewed to small, immature, and less prodigious reproducers. With the selective removal of adults that are proven successful breeders, the productivity of the entire population has declined.

Another problem is land-based stressors such as polluted runoff, soil eroded from open agricultural fields and construction sites, and the increased freshwater runoff related to our urbanized watersheds.[37] Runoff in our rainfall-intense climate can introduce silt into coastal waters within hours, and throughout the rainy months of winter reefs that are unlucky enough to be located near fallow agricultural fields may be physically buried by the red mud. Well known to interisland travelers are the long sections off Moloka'i, Maui, and Lāna'i coasts that turn from aqua blue to muddy brown during heavy rains.

Other potential sources of coastal pollution include cesspools, leaks and breaks in sewage delivery pipes buried below ground, sewage injec-

Fig. 11.8. Polluted runoff, including eroded soil, is carried through our system of channelized drainage onto the reefs. (Photo provided by Coastal Geology Group, University of Hawai'i)

tion wells in the coastal zone, and accidental discharges of untreated sewage from wastewater plants during heavy rainfall events.[38] Each of these presents some cause for concern because their potential impacts are poorly understood. Hawai'i has an estimated 100,000 cesspools, more than any other state, but new cesspool construction is now banned on O'ahu and Kaua'i, and heavily restricted on Maui, Moloka'i, and Hawai'i. Leaky sewage delivery pipes are a focus of intense repair and reconstruction programs. Although this is an expensive and time-consuming effort, it is high on the list of "to-do" projects among politicians and environmentalists.

The impact of sewage-injection wells on reefs is also not well understood. A 2007 study by the U.S. Geological Survey[39] on a single well indicates that a plume of sewage does move toward the coast and may discharge to the ocean fairly close to shore and in water less than 100 feet (30 m) deep. However, although this wastewater is not stripped of all nutrients and pathogens, it is treated to the secondary level and is presumably filtered somewhat in the groundwater environment. According to an e-mail broadcast by the Maui Surfrider Foundation on November 6, 2008, "Although the county incorporated biological nutrient removal systems that reduced nitrogen discharges by 60%, sewage wastewater continues to contribute to the harmful algal and bacteria blooms that smother our coral reefs, adversely affecting marine life." The basis for this statement is unknown. At sewage-treatment plants, accidental discharges continue to be worrisome, especially during periods of intense rainfall. But upgrading in-ground pipes and expansion of key treatment plants should offer some mitigation of this problem. As in many of our public sewage debates, when it comes to the impact of pollution on Hawai'i reefs, arriving at a hard-and-fast conclusion that everyone can agree on is proving elusive.

Seven major ocean outfalls discharge treated sewage into Hawai'i's coastal waters. Five of these, Sand Island, Honouliuli, Wai'anae, Kailua, and Hilo, expel fluids into waters that lie deeper than 130 feet (39.6 m) and thus are thought to have little impact among our reefs.[40] Two other outfalls, associated with East Honolulu and Fort Kamehameha treatment plants, discharge in shallower waters, but the waste is treated to the secondary level, and at East Honolulu studies show that there is no impact to coral communities.

Just offshore of the Waikīkī Aquarium (among other places) grows the invasive alga *Gracilaria salicornia*, nicknamed "gorilla ogo." This aggressor has pushed aside native *limu* and altered the ecology of the coral reef in the Waikīkī Marine Life Conservation District.[41] Although a dozen volunteer cleanups have hauled out tons of the plant, marine botanists predict that stronger methods are needed to control the species. It was only a few decades ago that this reef had 65 to 80 species of native algae. Now the rampant ogo blankets the environment from the reef crest to the shoreline. Similar invasions of algae have occurred along other Hawaiian shores, and the question of whether human sources of nutrients are feeding this problem remains unanswered. One study of a small region in West Maui found that over 90% of the nitrogen delivery came from human sources. However, since that study (1997), nearly all sugarcane and pineapple farming in the region has stopped, and current levels of nutrient loading to coastal waters remain unmeasured.[42]

As locals and visitors alike increase their desire to experience firsthand the beauty of our reefs, they threaten to love it to death. All the attention paid to corals has the potential to damage them. Oils from human hands can be harmful, and heavy footsteps, fin kicks, anchor damage, and even accidental brushes and bumps can all add up to a significant impact given the thousands of enthusiasts that seek out the seafloor.

However, despite these sources of human stress, as concluded in

The State of Coral Reef Ecosystems of the Main Hawaiian Islands (2008), "many of Hawai'i's coral reefs, particularly in remote areas, are still in fair to good condition."[43]

INVASION OF THE ALIENS

There are six most important targets in the battle to suppress noxious seaweed that is smothering certain sections of Hawai'i's reefs: *Acanthophora spicifera* (spiny seaweed), *Avrainvillea amadelpha* (leather mudweed), *Gracilaria salicornia* (gorilla ogo), *Hypnea musciformis* (hookweed), *Kappaphycus alvarezii,* and *Eucheuma denticulatum* (smothering seaweed). This squad of hitmen has the ability to completely smother a reef in a matter of years, especially when aided by overfishing, muddy runoff, and coral disease. Eradicating these persistent invaders is "incredibly difficult" according to marine managers,[44] but there have been successes. The famous supersucker, a large-mouthed vacuum hose aimed by a scuba diver at clumps of seaweed that sucks them up and dumps them on a waiting barge overhead, has successfully cleared sections of reef in Kāne'ohe Bay. The process cannot remove every last sprig of algae, but once the supersucker puts a major dent in the population, a healthy community of

Fig. 11.9. Volunteers remove *Gracilaria salicornia* from the reefs of Waikīkī, O'ahu. (Photo by Bruce Casler)

reef fish and protection as a special marine conservation district is all that is needed to keep the invaders at bay.

This problem is more than an esoteric issue among reef lovers. Noxious algae that has proliferated along the Maui shoreline washes up on beaches by the ton. The stinking mess drives away tourists, locals, and anyone else who is looking for a day on the beach. Wafting over the first few blocks of coastal communities is the stench of rotting seaweed, so bad that hotel occupancy and property values have nose-dived as a result. In Kīhei, the problem drove the community to buy a bulldozer-*cum*-Zamboni to clean the beaches. This machine skims the sand every day and clears off the algae, building massive piles of the stuff that has to be trucked away.

But battling the growing tide of seaweed with reactive measures like the supersucker and the beach Zamboni is only temporarily effective. What is needed is improved understanding of the root causes of algae growth. Managers are asking "What are the primary factors that promote invasive algae growth?" The typical answers (which we have discussed already) include excess nutrients in coastal waters, overfishing, suppression of healthy coral communities by polluted runoff and other stressors, and the introduction of a new species outfitted with the right survival skills to exploit an open niche in the ecosystem. With so many factors confusing the problem and the likelihood that different factors dominate in different parts of the Hawai'i coastal ecosystem, it is no wonder that managers are constantly engaged in an uphill battle. As anyone who owns a boat knows, working on the water is a logistically challenging and expensive operation. Adding to the problem is the fact that reef managers have no authority over the lands where much of the problem originates— the watershed. Reef managers are faced with a major challenge getting this problem under control.

STATUS OF HAWAI'I'S REEFS

As a general rule, Hawai'i's reefs are in good shape. Widespread coral bleaching (when a stressed coral polyp expels its zooxanthellae symbionts) related to high water temperatures that has been seen in the Caribbean and Indian Ocean has not been observed in Hawai'i. Human impacts to reefs operate largely at the localized level, in embayments and restricted segments nestled in the nooks and crannies of our shoreline. Given that land-derived stressors are aimed like a gun at the backs of our localized reefs and the offshore open ocean shelf often lacks accommodation space for reef growth due to wave stress, our reefs are squeezed from two sides and need all the help they can get from management authorities to resist the impacts.[45]

Unfortunately, most of the manageable problems threatening Hawai'i's reefs are either a direct or an indirect result of human enterprise. Reefs have survived millions of years of Earth's catastrophes and are indeed highly resilient systems, but much of that resilience is in the form of an ability to recover from trauma, not resist it in the first place. Also, many natural traumas are short-lived events such as storms, whereas human stressors tend to be persistent and unrelenting over time. Once human abuse is lifted, the reef stands an excellent chance of recovering, but allowing damage on the rationale that future recovery is likely is hardly the basis for an appropriate management system. In any case, scientific understanding of reef ecology is not sufficient to say if total recovery has happened in the past or if a carpet of recovered living coral masks the permanent loss of other, less-visible but still valuable members of the ecosystem. Not to be forgotten, as human stressors accumulate on our reefs, the very environment they occupy is shifting as well. Ocean acidity and surface-water temperature are both on the rise due to the buildup of carbon dioxide in the atmosphere, an alarming trend given the already heavy burden of human impacts laid on the reef systems of the world.[46] Also worrisome is the increased shoreline erosion accompanying accelerated sea-level rise that will release mud and other land-bound pollutants into the reef environment. On one hand, rising sea level should lift wave base, the depth to which wave scour limits coral development. This will open new ground for reef accretion. On the other hand, sea-level rise will increase erosion patterns along the shore, potentially releasing damaging sediment and other buried pollutants.

Until overfishing is effectively controlled and land-based stressors to reef communities such as siltation (mud) and polluted runoff are mitigated, Hawai'i reefs will retain their status as fragile environments worthy of special management efforts. Especially troubling is the fact that agencies tasked with reef management have no jurisdictional authority among the watersheds that pose such great risks—yet another sign, like the loss of our beaches, that environmental managers have yet to achieve a system of conservation that integrates enough government agencies to be effective.

DISEASE

Coral disease has only recently received study. Research cruises among the eight main Hawaiian Islands have revealed the presence of eight coral diseases afflicting the three major coral genera: *Porites, Montipora,* and *Pocillopora.* Generally, disease is widespread but still at a low level, with *Porites* trematodiasis and *Porites* growth anomaly being the most often observed (seen at between 60% and 70% of surveyed sites). The cause and carrier of these diseases is not known. Perhaps it is a virus that afflicts the coral polyp, and all researchers can do at this stage is observe disease occurrence and attempt to understand the patterns of behavior.

Diseases have also been seen among other reef dwellers. Plants among a patch of *Halimeda kanaloana* were observed to turn yellow and shed their segments off West Maui in the summer of 2006. The green sea turtles of Hawai'i have famously suffered from fibropapillomatosis, a disease that results in the growth of external and internal tumors, perhaps brought on by a version of the herpes virus. Surveys on Moloka'i suggest that the prevalence of this disease has been declining for the past 5–8 years. Skin tumors, parasitic protozoans, and other irregularities have been spotted on local reef fish such as endemic butterfly fishes, goatfishes, and snappers.[47]

OCEAN ACIDITY AND BLEACHING

From the reef's standpoint, increasing the amount of carbon dioxide in the atmosphere is not a good thing. Two problems result: ocean water acidifies[48] (actually, it turns less basic) and reduces the rate that a calcium carbonate skeleton can be secreted; and the ocean surface heats up, disturbing the symbiotic algae that corals depend on for food and energy.[49]

Since the beginning of the industrial revolution the release of carbon dioxide (CO_2) from human fossil fuel burning has resulted in atmospheric carbon dioxide concentrations that have increased from approximately 280 to 390 parts per million (ppm).[50] The atmospheric concentration of carbon dioxide is now higher than experienced on Earth for at least the last 800,000 years and probably over 15 million years and is expected to continue to rise at an increasing rate, leading to significant temperature increases in the atmosphere and oceans in coming decades.[51] The oceans have absorbed approximately 525 billion tons (476 billion metric tons) of carbon dioxide from the atmosphere, or about one-third of what we have released by burning fossil fuels. In doing this the ocean has provided a benefit by reducing greenhouse gas in the atmosphere and minimized some of the impacts of global warming. However, absorbing carbon dioxide is having negative effects on the chemistry and biology of the oceans.

Modeling studies and global seawater measurements have revealed

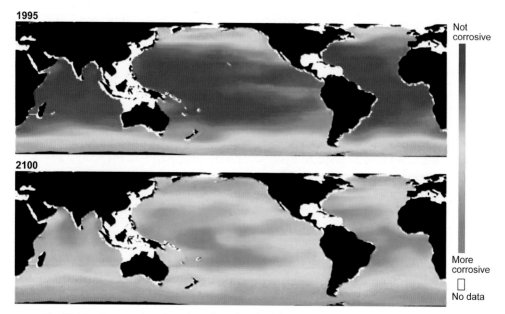

Fig. 11.10. Oceans absorb carbon dioxide, which lowers pH, thereby increasing acidity. Scientists have already measured a 30% increase in the acidity of some surface waters, and they predict a 100% to 150% increase by the end of the century. As dissolved carbon dioxide increases, fewer carbonate ions are left in seawater for organisms to use to build their shells and other hard parts. (SOEST Graphics, map from National Geographic: http://ngm.nationalgeographic.com/2007/11/marine-miniatures/acid-threat-text)

that the absorption of carbon dioxide is lowering seawater pH (making it more acidic).[52] The pH of ocean surface waters has already decreased by about 0.1 units from an average of about 8.21 to 8.10 since the beginning of the Industrial Revolution. Computer models estimate that by the middle of this century atmospheric carbon dioxide levels could reach more than 500 ppm, and near the end of the century they could be over 800 ppm. This would result in an additional surface-water pH decrease of approximately 0.3 pH units by 2100.

When carbon dioxide reacts with seawater it reduces the availability of dissolved carbonate. Carbonate (CO_3) is vital to shell and skeleton formation in corals, marine plankton, some algae, and shellfish. This phenomenon, ocean acidification, could have profound impacts on some of the most fundamental biological and geochemical processes of the sea in coming decades. Plankton is a critical food source that supports the entire marine food chain. Declining coral reefs will impact coastal communities, tourism, fisheries, and overall marine biodiversity. Abundance of commercially important shellfish species may decline, and negative impacts on finfish may occur. This rapidly emerging scientific issue and its potential ecological impacts have raised serious concerns across the scientific and fisheries communities.

By watching corals under conditions of increased acidity in a specially designed aquarium, researchers observed[53] a 50% decrease in skeletal growth at the same time that the photosynthesis of the symbiotic algae within the coral increased, probably because plants absorb, and feed on, carbon dioxide. But the excess growth of zooxanthellae does not help the host coral. The usually mutually beneficial relationship between guest algae and host coral breaks down. A similar breakdown has been widely reported when corals are exposed to elevated nutrient concentrations: the plant benefits and the animal coral suffers. Competition for carbon between the algae and the coral may be the explanation, but this phenomenon is not fully understood.

Some scientists believe that Hawai'i's reefs are more vulnerable to the impacts of ocean acidification because we are located north of the tropics in cooler and more isolated waters. Cool water absorbs carbon dioxide in greater quantities than warm water, potentially causing acidification levels to rise faster here. Hawai'i corals may be exposed to dropping levels of the dissolved carbonate they need to build their exoskeletons. In July 2008, scientists at the International Coral Reef Symposium in Florida declared acidification the largest and most significant threat faced by oceans today. In the worst-case scenario, global estimates show that coral reefs could be gone by the end of the century.[54]

Bleaching is what happens when zooxanthellae leave the coral, which tends to happen when water temperatures rise past about 90°F (32.2°C). Corals tolerate a narrow temperature range between approximately 77° and 84°F (25° and 29°C) depending on location. Corals bleach in response to prolonged temperature change and not due to rapidly fluctuating temperatures. They will also bleach in response to other stress events and environmental changes including disease, excess shade, increased levels of ultraviolet radiation, sedimentation, pollution, and salinity changes.

When corals bleach they commonly lose 60%–90% of their zooxanthellae and each zooxanthella may lose 50%–80% of its photosynthetic pigments.[55] The pale appearance of bleached corals is due to the calcareous skeleton showing through the translucent tissues of the polyp. If the stress-causing bleaching is not too severe and if it decreases in time, the affected corals usually regain their symbiotic algae within several weeks or

Fig. 11.11. These three images from Australia's Great Barrier Reef illustrate a sequence of degraded states that may result from the combined impact of ocean acidification and increasing sea-surface temperatures. In the lower left, each image shows the global atmospheric carbon dioxide concentration and the resulting sea-surface temperature increase leading to the degradation. (From Hoegh-Guldberg et al., 2007, "Coral Reefs under Rapid Climate Change and Ocean Acidification," Science 318:1737–1742)

a few months, and the coral recovers. If zooxanthellae loss is prolonged, such as if the stress continues and depleted zooxanthellae populations do not recover, the coral host eventually dies.

Scientists predict global surface temperatures to increase as much as 10°F (5.5°C) by the year 2100.[56] Although the effects of climate change on global coral-reef ecosystems will vary from one region to another, many species of coral are already in water that is near their thermal threshold. Any additional warming will reduce their growth and affect their health. The combined effect of warming and acidification, in an era of already significant human impacts, threatens reefs and thus constitutes a global problem.

LOCAL REALITIES

Coral reefs provided early Hawaiians with their primary source of protein, as well as endless hours of recreation. It is not surprising that the coral polyp was entwined with their earliest and most basic philosophy of existence. Within the *Kumulipo*, the Hawaiian creation chant, coral polyps were brought forth directly following man and woman.[57] To protect this precious resource, a system of *kapu* similar to the ones imposed on fresh water ensured the reefs' vitality. Stewardship of reef systems has been and continues to be vital for Hawaiian culture, yet over the last century the trend of balanced resource management system has shifted.

Currently, the Islands are managed at federal, county, and state levels. Resources, including marine and freshwater resources, ocean water and navigable streams, parks, and coastal lands, are all managed separately under a mosaic of government jurisdictions and agencies. Land is managed separately from the ocean, ocean water is managed separately from its inhabitants, and the ocean floor is managed under a set of rules different from all of these. Similarly in piecemeal fashion, agencies often have overlapping jurisdictions but no governing mandate to integrate their activities. This vision of management once again deviates dramatically from the original, integrated management system of the place-based *ahupua'a*. As natural resources continue to be impacted by human activity, human population and its wastes to climb, and climate to change, it is critical for resource assessment and monitoring to occur continuously and under an integrated panoply of agencies.

If we, as a community, wish to minimize our impact on the natural world, we must deepen our comprehension of what is needed to preserve coral reefs. Both resource assessment and monitoring are appropriate tools for highlighting the most effective measures of action, though monitoring methods used for one aspect of an ecosystem do not necessarily apply to another. Keeping track of fishing pressure requires a different set of tools than, say, monitoring the discharge of polluted waters from fallow agricultural lands. Therefore, a variety of techniques need to be available and applied, with continued reviews and updates under an integrated set of rules.

Sadly, over the last few years, dramatic modifications in budget priorities at both state and federal levels have left marine community management undermined. Yet, without continual funding enabling scientists to identify threats to the seafloor, outline mitigation actions, fully understand reef natural history, and assist in creating new management efforts where outdated ones have failed, the value and natural beauty of Hawai'i's reefs will continue to be vulnerable on populated shorelines. The loss of seafloor communities implies serious environmental, economic, and cultural implications for not only Hawai'i but also the rest of the world.

HAWAI'I REEF PROGRAMS

In 1994, the International Coral Reef Initiative (ICRI) was created at the First Conference of the Parties for the Convention on Biological

Diversity.[58] This international agreement emerged out of the recognition that the world's coral reefs are facing alarming degradation around the world, primarily due to human-related pressures. According to the *Status of Coral Reefs of the World: 2004*, approximately 20% of the world's coral reefs are beyond repair and 70% of the world's reefs are threatened or have already been destroyed, up from 59% in 2000.[59] In the 2008 *Status of Coral Reefs of the World* the expert opinion of 372 coral-reef scientists and managers from 96 countries and states was that the world has "effectively lost" 19% of the original area of coral reefs since 1950, 15% of coral reefs are in a "critical" state with loss possible within the next 10 to 20 years, 20% are seriously "threatened" with loss predicted in 20 to 40 years, and 46% of the world's reefs are regarded as healthy and not under any immediate threat of destruction except for "currently unpredictable" global climate threats.[60]

If the health of the planet's coral reefs is not fostered, this decline could lead to the loss of much of the world's reef resources within the current century. Some of the ICRI objectives call for countries to include the protection, restoration, and sustainable use of coral reefs in existing local, regional, and national development plans. ICRI also calls for increased capacity for the creation and execution of policies, management, research, and monitoring of coral reefs in member countries.

This international initiative spawned the Hawai'i Coral Reef Initiative Research Program (HCRI-RP) in 1998.[61] This initiative is managed in conjunction with the Hawai'i Department of Land and Natural Resources/Division of Aquatic Resources and the University of Hawai'i. By supporting scientific research and monitoring, results provide resource managers with information that helps them prevent and possibly undo the damage that has already been inflicted upon seafloor communities. Also, Hawai'i's Living Reef Program, a public education campaign to increase awareness about the necessity of healthy reefs, was launched in 2004.[62]

According to *The State of the Coral Reef Ecosystems of the Main Hawaiian Islands* (2008), coral ecosystems in Hawai'i range from fair to excellent condition but are "threatened by continued population growth, over-fishing, urbanization, runoff, and development."[63] The report also stated that sources of pollution, coastal developments, and aquatic alien species threaten the health of Hawai'i's reefs. To maintain Hawai'i's high quality of life and dynamic economy, continual, sufficient funding is imperative to hone coral reef management and disseminate educational material that promotes reef conservation.

OVERFISHING

In 2007, a paper published in the journal *Marine Ecology Progress Series* reported that total fish biomass in Hanauma Bay and 10 other protected areas under state management was 2.7 times greater than the biomass in comparable unprotected areas.[64] The study also reported that in the Northwestern Hawaiian Islands there is 6.7 times more fish biomass on average than in comparable habitats in the main Hawaiian Islands. These data indicate that humans have reduced fish stocks in the main Hawaiian Islands to about 15% of what they once were. The study author, Alan Friedlander of the National Oceanographic and Atmospheric Administration, concluded that the 11 state marine protected areas are too few, protecting only 0.3% of the Hawai'i coastline, stating, "If you want to rebuild fish stocks, you need to stop fishing in at least 20% of Hawai'i's waters and regulate fishing in the rest."[65]

Within the sea a complex network of predator-prey relationships exists, so to maintain Hawaiian fisheries it is necessary to preserve the biodiversity of the community. For example, with traditional fishing methods the spawning and growth to adulthood of species were attentively guarded and regulated as they related to phases of the Moon and seasons of the year.[66] Only a certain amount of fish were harvested at any given time, with certain times of year, type of fish, gender of fish, and size of fish being carefully restricted. All of these management techniques displayed a deep comprehension of the underwater landscape and its geobiological function.

The late 1800s witnessed a breakdown of Hawai'i's traditional resource-management system. By the turn of the century, traditional fisheries-management practices were virtually eradicated. As the new cash economy multiplied across the Islands, so too did the commercial landing of fish. Over the twentieth century, tourism, resident population, and shoreline recreation increased dramatically. The face of Hawai'i's fisheries was forever changed. What once served as the source of a subsistence lifestyle became the carrot for recreation and a commercial career. Fishing techniques underwent dramatic changes. The areas that were fished increased outward from the Islands to support the amount of catch occurring annually.

In 1900, the population of the Hawaiian Islands was 150,000, with a total fish catch (oceanic and inshore) of 6.2 million pounds (2.8 million kg), valued at $1.1 million.[67] Today the population is almost 10 times that amount at 1.2 million, with an annual fish catch of 23.4 million pounds

(10.6 million kg) valued at $59 million dollars. The catch has grown approximately 400%, but the value has increased exponentially, depicting one of the primary incentives behind the trend of fishery growth. It is certainly easier today to pull hundreds of pounds (kg) of commercial fish out of the ocean than it used to be with an old-school hand net; where once a rod or net was used, motorized vessels now restlessly pace ocean waters scooping up the ocean's creatures.

A report from the U.N. Food and Agriculture Organization on the state of marine fisheries worldwide says that 52% of the oceans' wild fish stocks are fully exploited.[68] Of the rest, 23% are lightly or moderately exploited and still offer some scope for further fisheries expansion, 16% are overexploited, 7% are depleted, and 1% are recovering from depletion, meaning that they have no room for further expansion. These assessments are found in the most recent *Review of the State of World Marine Fishery Resources.*[69]

At the same time, however, not all the catch statistics of Hawai'i's fisheries can be considered reliable due to the fact that a large amount of resident catch from recreation and a subsistence lifestyle goes unreported. In addition, commercial fishers have a habit of underreporting their catch. According to the Pacific Fisheries Coalition, the nearshore recreational catch is likely equal to or greater than the nearshore commercial fisheries catch, and recreational fishers take more species using a wider range of fishing gear.[70] Unlike most other coastal states, Hawai'i doesn't require recreational fishers to have a license. Also, a large percentage of the fish collected for the world's home aquariums is taken from Hawaiian waters, with no regulations limiting the size, number, and season for most of the desired species.[71]

Bigger fish are in no less trouble. Shark, tuna, marlin, and other top undersea predator populations inhabiting the middle of the ocean, thousands of miles (km) from the nearest land, are being replaced by smaller, less-desirable rays and other fish, according to a study by the Pew Institute for Ocean Science.[72] Studying an area of more than 6,000 square miles (15,539 km²) in the middle of the equatorial Pacific Ocean south of Hawai'i and north of the Fiji Islands, researchers Peter Ward and Ransom Myers from Dalhousie University in Nova Scotia found that sharks, tuna, and other top-of-the-food-web fish are half the size and their populations 80% smaller in numbers than they were 50 years ago.[73]

There are several potential reasons for fish (both pelagic and near-shore) decline: wasteful fishing practices, habitat destruction, lack of fisheries knowledge, inadequate enforcement, increased human population and market demand, too few larger fish spawning, improved fishing technology, invasive species, lack of detailed scientific data in support of new management efforts, and others. This complex network of cause and effect makes managing the problem difficult, and getting any number of stakeholders to agree on a management strategy becomes an exercise in compromising one's view of the world.

HAWAI'I'S FISHERY TYPES

Hawai'i's marine fisheries[74] are divided into three geographical areas: the eight main Hawaiian Islands, including their surrounding reefs; the Northwestern Hawaiian Islands (NWHI), a 1,200-mile (1,900-km) string of mostly uninhabited reefs, shoals, and atolls stretching northwest from the main Hawaiian Islands (now closed to fishing unless it is for subsistence by Native Hawaiians and the catch is consumed within the monument boundaries); and the mid-North Pacific Ocean, ranging from latitude 40° N to the equator and from longitude 145° W to longitude 175° E.

Hawai'i's fishing boats can also be divided into three types:

Large-scale commercial fishing vessels: Although smaller than vessels in most U.S. and foreign fleets, almost all of Hawai'i's large-scale commercial fishing vessels are less than 100 feet (30 m) in overall length. These include the older *aku* boats, which use a pole-and-line fishing technique, tuna longline wooden-hulled fishing craft, modern tuna and swordfish longline vessels, distant-water albacore trollers, and multipurpose vessels that catch bottom fish and spiny and slipper lobsters. These vessels are allowed to fish up to 1,000 nautical miles (1,600 km) off Hawai'i's shores, throughout the mid-North Pacific, and some span the South Pacific. However, most of these vessels function within 200 miles (321 km) of the main or Northwestern Hawaiian Islands and head out for around 2 to 3 weeks at a time.

Small-scale commercial fishing vessels: The boats in this category include tailored and moored boats between 12 and 45 feet (3.6 and 13.7 m) in length. Most of these vessels use trolling and hand-line techniques. The target species include tuna, billfishes, mahimahi, ono, and bottom fishes for the trollers and hand-liners. These vessels typically stay within 10 miles (16 km) (single-day trips) of the main Hawaiian Island shores.

Small-scale recreational, part-time commercial, and subsistence fishing: This category uses the same types of boats as in the previous category. Charter fishing boats are also included in this category. The target species for this segment of the fishery include a variety of reef species, as well as the more familiar tuna, billfishes, mahimahi and ono, bottom fishes, and crustaceans.

In Hawai'i, the list of fishing techniques is long and creative. They include longline; *aku* pole-and-line; deep bottom, inshore, and tuna hand line; trolling; diving; net; trap; rod and reel; handpick; and a few other miscellaneous techniques.

Longline fishing is carried out by using gear that consists of one main line that extends over a nautical mile (1.6 km) in length, and attached to it are a number of secondary lines, hanging with baited hooks. The mainline bobs just below the ocean surface; it is unlawful to engage in longline fishing within state waters or to sell or offer for sale any marine life taken with longline fishing gear within state waters. Longline fisheries target blue marlin, striped marlin and other billfishes, mahimahi, ono, and opah.

The *aku* pole-and-line fishery (also called the bait-boat fishery) in Hawai'i targets mostly skipjack and juvenile yellowfin tuna. This technique uses live bait to attract tuna to bite on barbless hooks with feathered skirts. Currently, most Hawai'i catch is sold on the fresh-fish market.

Two types of hand-line fishing exist. The first, ika shibi, targets squid and small 'ahi. The hand-line fishery is done at night, and most fishermen use a parachute anchor to slow the vessel's drift while fishing. The other method, called palu-ahi (chum tuna), deploys a weighted bag stuffed with chum to attract tuna to baited hooks. The Hawai'i offshore hand-line fishing grounds are primarily at seamounts and weather buoys 30–200 miles (48 to 321 km) from shore. Bottom fishes typically include seven deep-water species: onaga, ehu, *kalekale*, *'ōpakapaka*, *'ūkīkiki* or gindai, *hāpu'u*, and *lehe*. Laws governing bottom fishing generally allow use of traditional hand-line gear only, to limit the volume of fish caught.

Trolling is conducted by towing lures or baited hooks from a moving vessel using rods and reels or hydraulic haulers and outriggers. Some will also use a hand line, especially to target ono and tuna. Commercial troll fisheries target ono, mahimahi, and large yellowfin tuna. Trollers fish where water masses converge, where underwater features create sharp depth changes, at fish-aggregation devices, and near flotsam and foraging seabirds. Both commercial and recreational fishers can sell catch to cover their expenses. Trolling is the most popular pelagic fishing method in Hawai'i.

Bottom-trawling fishing boats drag nets along the ocean bottom in extremely deep water near underwater seamounts. This technique destroys

Fig. 11.12. Studies show that uhu, or parrot fish, biomass at heavily fished sites around O'ahu is only about 3% of that in remote parts of the Islands. Parrot fish are herbivores that control algae and maintain the health of reefs. (Photo by Keoki Stender)

deep-sea habitats containing deepwater corals, sponges, and other fragile organisms in their paths.

SUSTAINABLE FISHERIES: ARE THEY POSSIBLE?

There are few sustainable fisheries.[75] In unexploited populations of top predators such as *ulua*, *'ōpakapaka*, onaga, and ehu, the much greater fecundity of larger individuals determines that a relatively few large individuals can provide a substantial portion of the genetic input to the next generation. However, fishermen tend to target large fish, and after awhile only small individuals remain. In heavily fished populations, nearly all the genetic input is from smaller fish, which in turn produce small fish that reproduce early. Once reproduction starts, fish put their energy into producing eggs and sperm and their growth rate slows down.

With continued targeting by fishers, eventually large individuals disappear from the population, and the biomass of the entire fishery is reduced. Because of this, fisheries management is evolving into an approach that sets a maximum and a minimum size. The minimum size is that which permits at least one spawning to occur before harvesting is allowed. The maximum size is that which allows the largest females with the most eggs and that are genetically most robust to survive and reproduce season after season. Fishing-gear restrictions protect the prereproductive fish; no-take nursery areas protect the largest females. Quotas and bag limits can be used if the intermediate size classes need some protection.

Owing to the poor state of Hawai'i's coastal fisheries, Hawai'i managers have undertaken a number of steps to improve the status of fisheries.[76]

Resource-management tools that are available include the following:

- Catch reporting and licenses: these involve the commercial fishermen in keeping a log of what they catch each month.
- Artificial reefs: these are meant to increase fish populations by increasing habitat.
- Bag limits: these should ensure that enough fish are left to reproduce and that each fisherman gets a fair number of fish.
- Closed seasons: these ensure that fish get a break from fishing pressure during their spawning seasons.
- Education: knowledge of past and current fish abundance and fishing practices, rules, and regulations keep fishers knowledgeable of natural fish abundance and tools used to make and keep fish populations healthy.
- Community-based management: this involves local knowledge, observations, expertise, and work to manage fish resources.
- Fisheries management areas: these are areas closed to certain types of fishing gear or closed for a certain period of time to allow fish to recover.
- Gear restrictions: these limit too-efficient, nonselective, or habitat-destructive equipment such as unregulated lay-nets.
- Fish replenishment areas: these are areas permanently closed to fishing to ensure healthy stocks of fishes of different sizes and species and keep intact whole ecosystems.
- Minimum size limits: these allow members of a species to reproduce and replace themselves.
- Stock enhancement: this activity takes wild fish and breeds them in captivity and then releases them into the wild.

Size limits have limited effectiveness in bottom-fish areas because fish often die when they are caught in deep water and are then brought to the surface. Mortality also results from handling, hooking, and the general trauma of being hauled in. When fishing levels are high, accidental mortality of small fish that are thrown back is enough to dissipate any benefits of a size limit.

Limited-entry schemes, which require a database on users and numbers of fish removed, reduce or restrict the number of participants. Limited entry has not been previously utilized as a management method in bottom fisheries because the government has no statutory authority to limit users and because bottom fish are harvested in Hawai'i in small quantities by a large number of local fishers, most of whom are fishing as recreational fishermen. However, in 2009 a limited-entry closure was discussed in the media for the Hawai'i bottom fishery.

Hatchery programs and stock-enhancement programs release young fish into the wild to build up local populations. Hatcheries must incorporate genetic material from wild stocks on a regular basis to avoid problems from interbreeding. Although these programs sound like promising methods for rebuilding fish populations, they are expensive and have not generally proven effective for increasing overall abundance.[77] Although stock enhancement is not realistic for bottom fish, 'ōpakapaka and ehu may have some aquaculture potential. Hawai'i has had some success with seeding certain coastal areas with hatchery-reared moi and mullet.

Artificial reefs provide cover from predators and substrate for food and, if properly placed, may increase total biomass. Some critics are concerned that these reefs may be acting like fish-aggregation devices drawing both fish and fishers to the same environment. Before placing any artificial reef structures, it is important to examine a prospective area; what may seem like a barren wasteland may in fact be a productive bottom-fish nursery. Derelict vessels and concrete fish habitats have been placed at selected locations in Wai'anae, Maunalua Bay, and Waikīkī reefs on O'ahu.

A number of community-based programs have been developed in recent years. These focus on revitalization of local traditions and resource knowledge, with emphasis on strengthening local accountability for the sustainability of marine resources. The state has been encouraging community-based management since 1994, and a number of community-based efforts have been established.

Perhaps the most important steps yet to be taken are to implement a fishing license program and vastly increase the coverage of marine life conservation districts in Hawai'i. A licensing program would allow improved collection of scientific data, offer important restrictions on fishing pressure in nearshore waters, and provide an opportunity to educate the fisher (especially visitors to Hawai'i). Banning gill nets would also be a strong move because these "walls of death" can be extremely destructive. Gill nets are currently heavily regulated with net openings, restrictive hours of deployment, and the requirement of constant monitoring providing important improvements to the practice of netting. But enforcement is a perpetual problem in Hawai'i, and if they were made illegal it would be clear to all persons that when they see a gill net in use, it is an illegal use.

MARINE PROTECTED AREAS

One of the most widely used tools to control fishing impacts is the Marine Life Conservation District (MLCD).[78] These are implemented by government authorities to conserve and replenish marine resources. MLCDs typically allow only limited fishing (or disallow it altogether)

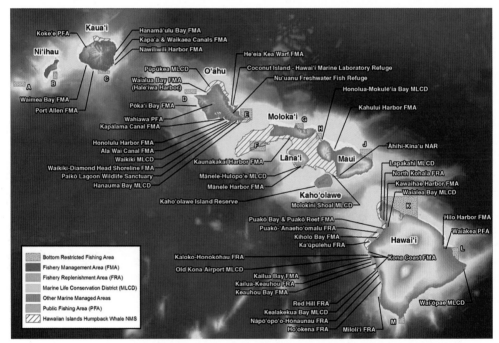

Fig. 11.13. The first marine life conservation district in Hawai'i was Hanauma Bay, O'ahu, established in the fall of 1967. Fish populations increased at a phenomenal rate, and as a result the bay has become world famous as a premier dive and snorkel site. Since then Hawai'i has developed an array of marine protected areas, but scientists call for more. (Hawai'i Department of Land and Natural Resources, Division of Aquatic Resources: http://hawaii.gov/dlnr/dar/regulated_areas.html)

and ban or stringently limit other uses that consume reef resources. They provide fish and other aquatic life with a protected area in which to safely grow and reproduce, and are home to a great variety of species. MLCDs are most popular as sites for snorkeling, diving, and underwater photography. Hawai'i now has 11 MLCDs and 21 Fishery Management Areas (FMAs).[79]

MLCDs are established by the State of Hawai'i, Department of Land and Natural Resources, following suggestions for additions to the system that come from the state legislature or the public. The Department of Land and Natural Resources' Division of Aquatic Resources conducts surveys of marine ecosystems on a statewide basis and also recommends areas that hold promise of successful restoration for MLCD status.

To commit an area for conservation status, it is evaluated by the Department of Land and Natural Resources with regard to public accessibility, quality of marine life and future potential value, safety from a public usage standpoint, compatibility with adjoining area usage, and minimal environmental or ecological changes from an undisturbed natural state. The area should have recognizable boundaries so that users can easily identify it and it can be readily enforced by conservation officers employed by the Department of Land and Natural Resources. It is important that the district be of sufficient size that daily and seasonal fish movement patterns fall reasonably within the protected area, ensuring that a population of fish remains protected throughout their normal activities. Movement patterns of many reef fish are not well understood, but research continues to improve our understanding of this aspect of MLCDs.

If an area being considered for protection meets the criteria of a MLCD, state scientists employed by the Division of Aquatic Resources conduct fish surveys, topographic analysis, and establish a baseline of ecologic health for the area. This ensures that future research can determine the performance of the district. Input from public, governmental, and private agencies is considered in making a final determination. Final approval is obtained from the Board of Land and Natural Resources and the governor.

The most important fishery management objective is to protect the long-term health of fish stocks to ensure that marine ecosystems are not damaged. Many populations of exploited fish are declining in numbers and size despite the best efforts of fishery managers. Sustainable fisheries have become an unreachable goal under current management approaches. Marine protected areas (MPAs) offer a way out of this downward spiral. If some of the larger, more fecund, and genetically more robust fish are fully protected from harvesting, those fish will provide a dependable quantity and quality of offspring.

The actual level of protection within different types of MPAs in Hawai'i can vary. The most effective MPAs protect ecosystem structure and function by including a core of no-take reserves in which any extraction of living organisms is prohibited. Because ocean currents transport eggs and larvae over large distances, networks of no-take reserves, or Kapu Zones, are needed to achieve the stock-rebuilding objective.

Fishery Management Areas (FMAs) are areas that are closed to certain fishing gear or activities while remaining open to others, or areas that are closed for a length of time and later reopened to allow fish populations to recover and grow to harvestable size. The Hawai'i bottom-fish plan designates 20% of important bottom-fish habitat as no-fishing zones for bottom fish around the main Islands. Many wonder when the same level of protection will be levied upon the nearshore, reef-based fishery, as called for by researchers.[80]

MLCDs may permit some extractive activities, including certain kinds of recreational fishing such as pole-and-line, spearfishing without scuba, and certain types of nets. Commercial fishing is generally forbidden. Over 1,300 acres (527 ha) of coastal waters have been designated as MLCDs. There are MLCDs at Hanauma Bay, Pūpūkea, and Waikīkī on O'ahu; Lapakahi, Kealakekua Bay, Waialea Bay, Wai'ōpae tide pools, and the Old Kona Airport on the island of Hawai'i; Molokini Shoal and Honolua-Mokulē'ia Bay on Maui; and Mānele-Hulopo'e on Lāna'i. Only three, at Hanauma Bay, Waikīkī, and Wai'ōpae tide pools prohibit all harvesting.[81]

Natural Area Reserves, wildlife sanctuaries, and other reserves and refuges are closed to all extractive types of fishing and gathering, except perhaps Native Hawaiian harvesting. They include 'Āhihi-Kīna'u Natural Area Reserve on Maui, Kaho'olawe Island Reserve, and Coconut Island— Hawai'i Marine Laboratory Refuge on O'ahu. Marine sanctuaries such as the Hawaiian Islands Humpback Whale National Marine Sanctuary usually allow commercial and recreational fishing, although some parts of a sanctuary may be set aside as no-take reserves.

THE SPILL-OVER EFFECT

Reserves serve as natural hatcheries, replenishing fish populations regionally through egg and larval spillover beyond reserve boundaries. The dispersal of eggs and larvae from no-take marine reserves to surrounding areas can maintain and improve fishing in adjacent areas because large individuals in the reserve escape capture and their total egg production is much higher than smaller members of the same species. The size and abundance of exploited species also increase in areas adjacent to reserves. Fishers excluded from marine reserves generally experience significant benefits because fishing in neighboring areas is vastly improved.

Marine reserves also create economic opportunities that contribute as much or more to Hawaiʻi's economy than commercial fishing. This is because marine reserves are excellent sites for ecotourism and ocean wilderness tours, scientific research, marine education, recreational snorkeling and diving, underwater photography, and cultural activities. However, even nonextractive uses can alter and damage reserve ecosystems.

Because it is usually easier to prevent environmental damage than to repair it later, caution dictates that in the absence of sufficient information on which to base safe and reliable estimates of the effect of an activity,

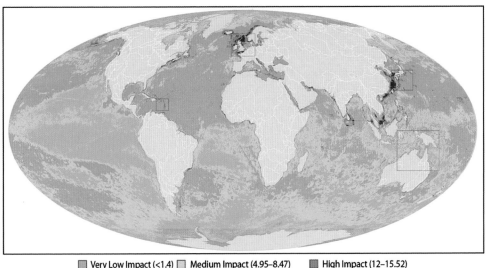

☐ Very Low Impact (<1.4) ☐ Medium Impact (4.95–8.47) ☐ High Impact (12–15.52)
☐ Low Impact (1.4–4.95) ☐ Medium High Impact (8.47–12) ■ Very High Impact (>15.52)

Fig. 11.14. Global map of human impacts to marine ecosystems. Impacts include effects of primitive fishing methods in less-developed countries, destructive commercial fishing in the open seas, pollution from industrial chemicals, invasive species, nutrient pollution in coastal areas, ocean acidification, oil rigs and other platforms, ocean-based pollution, commercial shipping, population pressure, climate change to sea-surface temperature, and increase in ultraviolet radiation. (National Center for Ecological Analysis and Synthesis, "A Global Map of Human Impacts to Marine Ecosystems": http://www.nceas.ucsb.edu/GlobalMarine)

the burden of proof shifts to those proposing activities that may have a negative effect on the ecosystem.

DEMONSTRATED BENEFITS

Marine reserves can produce long-lasting and rapid increases in abundance, diversity, and productivity of fish populations.[82] Fish size and reproductive output are known to increase within reserves. Decreased mortality, decreased habitat destruction, decreased extinction, and balanced, healthy ecosystems also result. Marine reserves provide sites for collecting valuable fishery-independent data. Larger reserve sizes result in increased benefits, but even small reserves have positive effects if they are part of a network. Size and abundance of harvested species increase in areas adjacent to reserves.

Networks of reserves buffer against localized changes in the environment[83] that threaten fish populations and provide significantly greater protection for marine communities than a single reserve. Reserve networks that span large geographic distances and encompass substantial areas protect against catastrophic events and provide stable platforms for dispersal of sustainable marine communities in the long term.[84]

CHAPTER 12

A Responsibility to Nurture the Land

Ua mau ke ea o ka ʻāina i ka pono
"The life of the land is perpetuated in righteousness"

In a traditional Hawaiian approach to conservation, humans play a central role in managing resources that are taken from the natural world. People are not viewed as a problem; rather, they are viewed as part of the living universe. Communities have a responsibility to nurture the land in a reciprocal and sustainable manner. The land is typically used, not left alone, unless for reasons related to the use. The Hawaiian attitude boils down to establishing a relationship between people and their lands. For Native Hawaiians, these islands are the *one hānau,* the "birth sands," and so it is a responsibility to know the lands and care for them.[1] Practiced ideologically, a relationship grows out of long-term, intimate knowledge of species and ecological systems. This relationship is not about preserv-

Fig. 12.1. The Great Seal of the State of Hawaiʻi includes the state motto adopted by the Kingdom of Hawaiʻi in 1843: Ua mau ke ea o ka ʻāina i ka pono.

ing lands in isolation; it is about conserving them as a resource for sustainable use. In this sense "conservation" and "use" are deemed compatible. Hence, it may differ from the traditional Western environmentalist approach of a "hands-off" management style. Rather, sustainability in the traditional Hawaiian sense is about "hands-on" in a sustainable manner. The trick, sometimes achieved and sometimes not, is to reach that ideal level of sustainability.

Achieving this requires an intimate knowledge of the land and ocean, its resources, and its natural biogeochemical processes. Today, only a few people remain intimately connected to the land; hence our knowledge of these processes has decreased, and true understanding is often replaced by subjective attitudes and vocal opinions. Certain pervasive issues aggravate this situation:

- A thus-far successful reliance on imported resources leads many to lower Island sustainability as a priority.
- The critical work of monitoring the environment goes unfunded in shrinking budgets. For instance, the number of stream gauges in Hawai'i has decreased 300% from a high of nearly 200 in the late 1970s to fewer than 50 today.
- People tend to react when they see change. But slow changes can accumulate without obvious visual cues over decades. The net deterioration can be significant.
- Construction continues on accreting shorelines because of property rights, despite knowledge that sea-level rise will eventually lead to accelerated coastal erosion and marine inundation.
- Attempts by agencies and community groups to improve protection of reefs, coastal water quality, and beaches are made difficult by jurisdictional barriers.
- Personal and commercial interests prevent taking clearly needed steps such as expanding marine life conservation districts, creating new rules on safe building in the coastal zone, preparing for climate change, and improving freshwater management.
- Hawai'i climate is warming,[2] sea level is rising[3] and coastal erosion is chronic,[4] rainfall and stream flow are decreasing and storm intensity is increasing,[5] sea-surface temperatures are rising[6] and the ocean is acidifying[7]—these are all measured, factual phenomena in Hawai'i—yet state and county governments

refuse to mobilize a serious response to threats to Hawai'i communities from global warming.

Although there are many natural hazards and stewardship issues in Hawai'i, there are also overarching themes that point to pathways for managing these. A unified governmental vision, a more holistic view of natural systems that is place-based, and a reintroduction of *ahupua'a* principles of respect and responsibility are a few of the shared means to address the various issues discussed in this book. Overall, perhaps the most important key that remains unturned in Hawai'i's system of natural management and hazard mitigation is the creation of integrated tools and principles that vary with the place. Beaches, reefs, water resources, coastal communities, and natural hazards are all highly unique from one island place to another—ideally each place would have its own specific rules of management that incorporate integrated communities and government guidance under one arrangement.

Modern governance relies on legislators and other elected officials to set priorities and engage in cost-benefit analysis to allocate funds. Unfortunately, in a struggling economy and a state with a multitude of needs, long-term answers are often expensive and lack urgency. But there is compelling necessity for robust solutions and to inspire individuals to demand the change we need. Throughout this book, problems are identified as well as potential new attitudes and management tools with which to address them. This chapter serves to condense that information and reiterate the pressing issues. Resting in our hands is the future possibility that natural hazards do not become catastrophes and natural resources do not continue to disappear.

THE PROBLEMS WE FACE

The failure to fully protect our shorelines occurs where federal, state, and county authorities neglect to work for a common vision, where good laws are weakly enforced, and where rules and regulations do not fit the personality of the environment. Perhaps predictably, the more federal, state, and county governmental agencies that get involved to help protect the environment, the less protected the environment is—due to fractured enforcement, a lack of coordination, and strengthening of jurisdictional divisions. Land is managed separately from the ocean, ocean water is managed separately from its inhabitants, and the ocean floor is managed under a different set of rules from all of these. Any advancement of useful policies happens by overcoming these barriers—not because this regulatory framework provides a convenient venue.

The lack of a common vision or a coordinated environmental plan between federal, state, and county affects several aspects of environmental stewardship in Hawai‘i. For example, some say that there is ambiguity and confusion over which agency should lead the way in coordinating various water-quality programs and enforcing water-quality mandates. This lack of clarity among agencies means that Hawai‘i has no integrated water-management plan for watersheds. What should be a coordinated coastal-zone management system with unified goals is too often a chain of separate agencies that may resent the advice and comments offered by sister agencies and just as frequently ignore them. And too often the goal of overworked and understaffed authorities is to simply comply with statutes, not take steps that are consistent with the statute to conserve the resource.

In some cases, regulations enacted to protect our environment, in practice, provide guidelines for how to legally damage them. Permitting happens along the shoreline in piecemeal, parcel-by-parcel fashion, without guidance from an overarching vision for the future of the resource. The minimum 40-foot (12-m) setback for buildings along the coast, for example, does not provide nearly enough room for the preservation of sand dunes and other natural bulwarks against erosion, tsunamis, and hurricanes. Yet a half century of development has blossomed on our beaches under this rule. Where did the magic number of "40 feet" come from? The unsubstantiated story goes that it was a compromise distance that no one liked but everyone could live with when the state coastal zone management law was being voted on decades ago. There are still no rules requiring appropriate building practices or parcel layout, and public access to beaches is severely limited and growing more scarce with each new building. At no place in Hawai‘i statutes will you find guidance for dealing with sea-level rise or climate change. Acting entirely within their legal rights, developers flatten dunes to give tourists a better view of the beach they unwittingly damage. The beach is part of a connected system between the land and sea and should be regulated that way. But it is not; half—the dunes—is given to the counties to manage, and half—the open sand—is given to the state to manage.

Many land-use practices are perpetuated by a desire to strengthen our economy and yet, in the long term, serve to deteriorate ecosystems and Hawai‘i's natural beauty—the very attraction that sustains our visitor economy and provides most of us with jobs. Often the profits generated

Fig. 12.2. Beach management is a tricky issue that requires balancing the sometimes competing demands of local, state, and federal agencies; landowners; public groups with specific points of view; and the visitor industry with appropriate treatment of the coastal ecosystem and coastal hazards. (Photo by C. Fletcher)

by these actions end up going offshore anyway. A short-sighted, profit-only mentality of land use will ultimately cost us money and, more important, our island way of life. Take, for example, our coral reefs. Hawai'i is America's largest ocean economy, with 10.7% of the state gross domestic income coming from ocean-related commerce.[8] The most significant human threats to Hawai'i's coral reefs are overfishing, land-based pollution, invasive algae, recreational overuse, and increases in ocean acidity and sea-surface temperature associated with carbon dioxide buildup in the atmosphere. There are local solutions to some of these dangerous influences; we need only the political will to implement them.

We can recognize and account for the fact that water quality, potable water availability, and watershed and coastal ecosystems are diminishing in proportion to the pace in which land is converted under pavement. It is remarkable that best management practices do not dictate the use of permeable concrete. The very first use of permeable concrete in Hawai'i was made at the University of Hawai'i Mānoa campus only in 2008—and its use is not required anywhere else in the state. Unlike natural landscapes where rainwater pools in gullies and filters slowly into the ground, urbanized streets, parking lots, driveways, rooftops, and gutters divert polluted runoff into channels, culverts, and storm drains headed at high velocity into our coastal waters. This artificial maze of conduits alters the natural course of the runoff, accelerating its pace, damaging streamside vegetation, widening channels, and depriving groundwater resources of their recharge while carrying the polluted discharge into the ocean. The potential for polluted runoff to affect coastal and marine environments is immense in Hawai'i because most of the population and its wastes reside within a few miles (km) of the shoreline. The Islands' water quality is closely linked to the choices that will be made in coming years about how the vast agricultural fields will be used. Flattening dunes and building seawalls, channelizing streambeds, and blanketing our land with impermeable surfaces is a cocktail mix for wetland loss, beach narrowing, ecosystem damage, and lower water quality and abundance around the state.

Water issues include restoring aquatic ecosystems and watersheds in places that have been damaged by community development and historic patterns of water withdrawal. Place-based watershed-management plans, such as are being developed by the City and County of Honolulu Board of Water Supply, are based on public education, input, and access to all aspects of the *ahupua'a* that are integral to water conservation. Watershed-management plans are holistic, meaning that they consider people and their relationship with the land, natural processes of the water cycle, aspects of land use, and methods of restoring watersheds. By formulating an environmentally holistic, community-based, and economically viable watershed-management plan, governing agencies may provide a balance between the preservation and management of watersheds, and sustainable groundwater and surface water use and development to serve current users and future generations. Watershed-management plans, as conceived by the Board of Water Supply, have five objectives: (1) promote sustainable watersheds; (2) protect and enhance water quality and quantity; (3) protect Native Hawaiian rights and traditional customary practices; (4) facilitate public participation, education, and project implementation; and (5) meet future water demands at reasonable costs.[9]

There are also important waste issues that constantly draw on our attention and finances. Raw sewage spills, leaks in the pipelines, treatment-plant problems, sewer-line blockages, structural failures from corrosion, sewage infiltration to groundwater, groundwater infiltration to overwhelmed treatment facilities, limits in sewer-line capacity, and poor maintenance programs, together with Hawai'i's moist environment, have created a persistent and pervasive water-quality hazard on O'ahu that rises anew with each rainstorm.

Consider the case of natural hazards. The state has two oil refineries (Tesoro and Chevron), both located only 10 feet (3 m) above sea level in Campbell Industrial Park. A direct hit by a hurricane, a large earthquake, or a moderate tsunami could render these inoperable. Without gasoline to refuel and return to the mainland, what airline or barge operator will bring us the food and goods we need? Earthquakes, hurricanes, and tsunamis all may cause significant damage and injury, and we will be in direct need of outside assistance. With little to point to in the way of Island sustainability, Hawai'i is vulnerable to escalating problems associated with the next hazard strike.

POTENTIAL TOOLS

Both individual residents and governmental officers have a hand to play in the future of our disaster preparedness and environmental well-being. None of us can stop a natural hazard from coming, but all of us can prepare and utilize mitigation techniques. Experts have gathered substantial information about hurricanes, earthquakes, and tsunamis to guide us in reducing our chance of catastrophe. It is incumbent on citizens of Hawai'i to avail ourselves of these materials and start making wise choices that enhance our own safety as well as the sustainability of our environment.[10]

For example, one can mitigate hurricane damage by building away from the coast, ridgelines, and exposed hillsides; building homes with continuous load paths from the foundation to the uppermost components of the roof; and minimizing external features such as eaves, balconies, and other structural components that are easily ripped off by high winds. We can improve resistance to earthquakes by establishing a continuous load path from the roof to the foundation, by using flexible materials in the original design, and by making sure the structure is firmly attached to its foundation.

We can avoid mass wasting by avoiding construction on or at the base of steep slopes; avoiding areas at the mouths of gullies, gulches, and narrow stream valleys; keeping away from regions where scars of previous events can be identified; and identifying areas where other indicators of downslope movement exist, such as leaning trees or structures and hummocky topography. If a building is already located on an unstable slope, only a few options exist to reinforce the soil. One can revegetate the hillside with native plants, grade slopes to a more stable angle, install drain systems to dewater slopes, anchor one's house to bedrock, and avoid increasing the load on the slope. One could always consider relocation if these options are not effective.

And, of course, we can engage in good old-fashioned environmental protection by reducing, reusing, and recycling when we can; not building where future seawalls will be needed; being mindful of the chemicals we dump on our streets; using permeable materials when building driveways and other projects; reducing our carbon footprint; and being an advocate in the halls of our legislature for responsible stewardship.

By their nature, the state and counties shoulder the greatest responsibility for both our environmental well-being and disaster preparedness. It is in them we trust—for better or worse. To make our regulatory system work, it must be adjusted to view natural resources as more than a patchwork of parcels of sky, land, and ocean.

STEPPING BACK TO MOVE FORWARD

Our leaders are tasked with preparing us for future threats as well. How we confront sea-level rise may make the difference between prosperity and bankruptcy (both economically and environmentally). Researchers can map the water table throughout areas where urban development will experience drainage problems. The water table can be monitored with gauges the same way we monitor the sea level with tide gauges today. As we learn how the water table rises and falls with the tides, large waves, and rainfall, we will be able to more accurately map regions that are vulnerable to flooding from rising sea level. Some areas will be inundated sooner than others, and these are the places where we can focus an adaptation strategy.

Researchers can develop an improved ability to predict what lands will be overtopped by high waves. The system of buoys that surrounds the Islands provides us with an early warning of large waves arriving on our shores. With a day of advance notice, it is possible for road crews to get outfitted, homeowners to shore up their homes, and communities to prepare for the high run-up of waves on eroded beaches and low roadways. By using combinations of wave run-up models, detailed topography of coastal lands, and knowledge of arriving waves, the disruption to our lives caused by these events can be estimated, and estimates are the first step to mitigation. As seas rise the frequency of wave inundation will increase. Eventually, when sea levels rise to approximately 2 feet (0.6 m) above current level, community inundation may become an annual event. We can either meet this with awareness and improved resiliency or be victims and leave our communities to suffer. Something as simple as avoiding poured cement home foundations and using post-and-pier construction instead (so that waves will roll beneath homes instead of into them) may buy a community the additional time to raise an entire generation.

Sea level is rising; it has accelerated since the twentieth century. The atmosphere continues to set new records for warmth, causing ocean waters to expand, and ice around the world is melting and the water is flowing into, and raising the level of, the oceans. Establishing a strategic coastal policy, and actually seeing the positive end product of it, are two events likely to be separated in time by many decades. What does it mean to adapt to climate change? How will such a policy fit into the governing laws of the land? What actual tools would be used to implement adaptation? And how can adaptation be performed fairly for the thousands of beachfront property owners? These and other questions need a discussion that has yet to transpire in Hawai'i.

NEARSHORE RESOURCES

To maintain Hawai'i's high quality of life and dynamic economy, continual, sufficient funding is imperative to restore and maintain healthy coral-reef ecosystems. It is important to enhance collaboration among stakeholder groups through improved information exchange and communication. The Hawai'i State Department of Land and Natural Resources, Division of Aquatic Resources, has undertaken a number of measures to improve the management of marine resources. A few of these measures include changes in minimum size limits for certain resource species, the initiation of marine recreational fisheries surveys, and changes to the rules governing marine protected areas.

These are important steps, but many call for three additional critical moves: ban all gill netting, put into management status at least 20% of the state shoreline, and implement a fishing license program. Today there are rules that govern the use of gill nets in Hawai'i. Unfortunately it is impossible to enforce these on all shores all the time. In 2008 and 2009 gill nets killed monk seals, turtles, sharks, and other unintended bycatch. Banning gill nets will rouse vocal opposition. Some (not many) will claim that gill nets are their prime source of food and livelihood. They will claim native gathering rights. But these people are choosing to make netting their source of food and their livelihood—they do have options. Unfortunately, while we worry over the plight of the few, damage to coastal fisheries by nets continues, and that is one community that does not have options.

Researchers have called for placing at least 20% of the shore under protected status. Scientists familiar with the problem of overfishing in Hawai'i report that total fish biomass in Hanauma Bay and 10 other protected areas under state management is 2.7 times greater than the biomass in comparable unprotected areas. They report that in the Northwestern Hawaiian Islands there is 6.7 times more fish biomass on average than in comparable habitats in the main Islands. These data indicate that humans have reduced fish stocks in the main Hawaiian Islands to about 15% of what they once were. Currently the 11 state marine life conservation districts in Hawai'i offer protection to only 0.3% of the Hawai'i coastline. If we are truly serious about restoring our coastal fishery we can stop fishing in at least 20% of Hawai'i's waters and regulate it in the rest.

A number of communities throughout the state are currently strengthening local influence and accountability for the health and long-term sustainability of their marine resources through revitalization of local

Fig. 12.3. Controlling the spread of alien algae will require improving our understanding of the relative roles of polluted runoff, sewage disposal, overfishing, and ecosystem dynamics. Ocean acidification will play an increasing role in this issue as calcifying organisms struggle in the changing chemistry of ocean water. (Photo by Keoki Stender)

traditions and resource knowledge. Marine reserves can produce long-lasting and rapid increases in abundance, diversity, and productivity of fish populations. Fish size and reproductive output is known to increase within reserves. Decreased mortality, decreased habitat destruction, decreased extinction, and balanced, healthy ecosystems also result.

A fishing license program allows several things to happen at once. Applicants for a license will learn about the problem of overfishing, about proper fishing techniques, about species that are endangered, and about areas to avoid. This public education is targeted and specific and can be achieved at this level of specificity through a licensing program. Licensing also provides assistance to enforcement by providing a quick and easy criterion for enforcement officers ("May I see your fishing license please?") and by educating a segment of the community that may not be reached by officers. Also, a licensing program allows for scientific data to be collected through monitoring of catch and fishers. The casual visitor or local who grabs a three prong and heads for the beach will then be an educated fisher and is likely to think twice before aiming at and killing that fish that used to be anonymous but now looks like the picture of the endangered species in the education pamphlet.

NO MANDATED EXPERTISE

The state currently struggles with an information gap. Geologic analysis can be part of every development and an early factor in siting everything from homes and roads to pathways and parks. Yet the state of Hawai'i has no lead science agency. In fact, there is literally nowhere for citizens to turn for geologic information because we are the only state in the union without a state geological survey. Neither is there a collated body of literature on rock and mineral resources, sand and gravel reserves, soil erosion, mass wasting, flash flooding, impacts of climate change, earthquakes, or any number of geological issues of critical importance to public safety and resource management.

Scientists at the University of Hawai'i, the U.S. Geological Survey, the Pacific Tsunami Warning Center, and other locations are happy to lend their know-how either by contract or out of good will, but public safety should not pivot on this happenstance arrangement. A state science office would provide scientific information to the public and to managers of public safety and welfare. Indeed, some of the current problems in resource management may be rooted to some extent in this lack of knowledge. Although it sounds like a call for bigger government, in truth reorganizing and unifying existing expertise under the needed mission would initially fill the void. Such an office would offer the advice that sister agencies need for decision making. Civil Defense is meant for managing people and mitigating hazards; a geological survey would be responsible for managing scientific information. To the detriment of public health and safety, we lack the science component.

WHAT A WASTE

Although many legislators may be tired of hearing about it, there remain serious waste issues in Hawai'i. It is sound planning for the state to protect the tourist industry by ensuring that our waters are safe. The state of California recently adopted a "right to know" bill that requires monitoring of all public beaches with more than 50,000 annual visitors and regular sampling near storm drains.[11] A protocol similar to California's could be applied to Hawai'i if the state is going to guarantee the safety of swimming water, especially against polluted runoff.

Shifting sentiments are beginning to redefine the way Hawai'i manages water, and a suite of solutions is being discussed. New sewage-treatment plants can be located close to agricultural production to reduce distribution costs. For many of the megahotel complexes the reuse and treatment facilities are integrated from the beginning. Wastewater-treatment plants can be designed with reuse as part of a single system, supplying all the needed water to irrigate their golf courses, lawns, and open space. Merging water-supply offices with wastewater managers under one roof promotes more urgent examination of the long-range costs that effluent reuse can avoid.

Monitoring cesspools and septic systems has been a difficult and costly dilemma for the state. Ideally, Hawai'i would monitor the safety and condition of each cesspool and septic field, but rarely are funds allocated specifically for that purpose. Monitoring is made even more difficult by the fact that there are no records of how many private systems are in place or of how much wastewater they generate. Hawai'i will need to dedicate a worthy portion of its financial resources to the waste problem because it is likely affecting stream and coastal water, and our public health and safety.

CLIMATE CHANGE

Hawai'i's climate is changing in ways that are consistent with the influence of global warming. In Hawai'i:

- Air temperature has risen;
- Rainfall and stream flow have decreased;

- Rain intensity has increased;
- Sea level and sea surface temperatures have increased; and,
- The ocean is acidifying.

Because these trends are likely to continue, scientists anticipate growing impacts to Hawai'i's water resources and forests, coastal communities, and marine ecology. There is a significant need for sustained and
enhanced climate monitoring and assessment activities in Hawai'i; and
a compelling requirement for focused research to produce models of
future climate changes and impacts in Hawai'i. The challenge posed by
climate change in Hawai'i is enormous, and unfortunately there is no
state-level authority to which we can turn for information. Here again
is an important role for a state science office. But for now, unless you
follow the scientific literature, we remain in the dark regarding potential
impacts.[12] Global climate models that provide Earth-scale understanding
of climate change effects do not even recognize features on the small scale
of our Islands. We will learn more when our computing power improves
and regional assessments provide consistent results.[13] But meanwhile, to
take climate change seriously, managers need some prediction of how
rising atmospheric and ocean temperatures will impact the system of
moisture delivery, winds, and ocean currents that govern our climate and
its products.

Climate change threatens water resources in Hawai'i. Researchers
have already detected a decrease in rainfall and long-term stream flow.
Rainstorms are projected to be stronger but less common such that we
will continue to see a decline in overall precipitation but an increase in
storm and flood intensity. Given our high dependency on external food
sources, how will decreased rainfall affect food production in Hawai'i?
How will falling water tables be replenished if climate change results in
decreased recharge?

Look to the mountains and see the nearly perpetual band of clouds
on their slopes. Scientists infer that rising temperatures in Hawai'i could
result in a shallower cloud zone because of a possible rise in the lifting
condensation level, which controls the bottom of the cloud, and a decline
in the height of the trade wind inversion, which controls the top of the
cloud. Where clouds intercept the land (forming fog) they deposit water
droplets directly on the vegetation and soil. This process is a significant
source of water to the mountain ecosystems of Hawai'i, especially at
windward exposures. With a smaller cloud zone, less cloud water would
be available to these important forests.

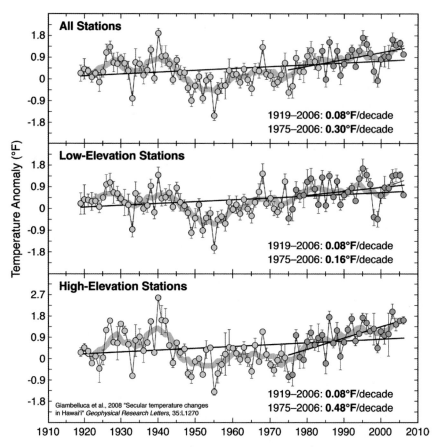

Fig. 12.4. Hawai'i is getting warmer. Data show a rapid rise in air temperature in the past 30 years, with stronger warming at high elevations. The rate of temperature rise at low elevations, 0.16°F (0.087°C) per decade, is less than the global rate; however, the rate of warming at high elevations in Hawai'i, 0.48°F (0.268°C) per decade, is faster than the global rate. Most of the warming is related to a larger increase in minimum temperatures compared to the maximum-a net warming about 3 times as large- causing a reduction of the daily temperature range. (T. W. Giambelluca, H. F. Diaz, and M. S. A. Luke, 2008 "Secular Temperature Changes in Hawai'i" Geophysical Research Letters, 35:L12702.)

Another concern is changes in the process of evapotranspiration. While rainfall and cloud water are the sources of water to the ecosystem, evaporation and transpiration (the emission of water vapor through the leaves of plants) return water to the atmosphere, thus reducing the amount going into streams and groundwater. Effects of warming on evapotranspiration are as yet unknown, but changes could further impact water resources already being affected by reduced rainfall.

Because surface air temperature, cloudiness, and rainfall depend on the trade winds, forecasting Hawai'i climate is dependent on accurately modeling trade wind changes. Intergovernmental Panel on Climate Change models do not agree on these aspects of climate for the region around Hawai'i. Other modeling has shown that wind and rainfall responses to warming around the Pacific are not uniform, and depend strongly on the climate model being used. Skillful projections of island climate must take into account the interaction of trade winds with island topography and will rely on continued and enhanced monitoring of key climate variables.

Shifts in the Hadley circulation that governs winds and rains in this part of the Pacific need further understanding, and managers lack knowledge of how climate change will alter the El Niño and Pacific Decadal Oscillation systems. These impacts need refinement, and they need to be assigned probabilities so that managers have solid targets on which to make plans. Our economy, environment, and way of life are going to change with global warming. It is important that a concerted effort be made to forecast these possibilities as soon as possible.

BACK TO THE FUTURE: MINING THE HAWAI'I OF YESTERDAY FOR A PROSPEROUS TOMORROW

Perhaps the tide of resource depletion and environmental decay can be turned by recalling concepts emerging from cultural practices. Sustainability is a resurgent notion in Hawai'i, but has it truly permeated economic practices? Despite a decade of public discussion in support of "diversified agriculture" significant island food production remains a distant hope.[14] It is important to do more than give lip service.

Our stewardship system can be most effective if it is grounded in *pili 'āina*. But sustainable practices need scientific observations to define their parameters; otherwise, as has largely been the case already, the community discussion becomes a test of wills and opinions. Sustainability parameters change from one place to another, and observations, and governance, can be scaled to this reality. As scientists we are vulnerable to viewing solutions through the lens that science can solve our woes, an attitude that oversimplifies complicated issues. Nonetheless, an advantage is that obtaining comprehensive scientific observations of a watershed and its reef can be logistically feasible, and by starting small a sufficient database can be collected within a short period, especially with the leadership of a local community possessing a memory of the old ways. There is abundant scientific talent in Hawai'i to accomplish this goal. A disadvantage is that money is short, and the system of laws and rules is not easily changed to accommodate new information.

Let us remember the lessons of the *ahupua'a* and the basic tenets that underlie this management system: stewardship, cooperation, and respect—not just for one another but between humans and the land, the water, the winds, and the ocean. All of these concepts must be incorporated within our land-management system if Hawai'i is to choose a sustainable future. *Ahupua'a* has now taken hold as the guiding principle of the Hawai'i Ocean Resource Management Plan, the official state effort to unify and coordinate coastal and ocean-related activities and missions among government agencies. The Honolulu Board of Water Supply and several other agencies have also begun to redefine their activities and planning in the stewardship model. Approached with commitment, an updated and refined *ahupua'a* system within a twenty-first century setting can provide a template for sustainable living on the shores of Hawai'i.

Notes

Chapter 1: Introduction

1. For an overview of state information and specific Hawaiian environmental issues, see Federal Emergency Management Agency, State of Hawai'i: http://www.fema.gov/plan/ehp/regionix/hawaii.shtm (last accessed October 17, 2009).

2. There are many sources of information on how to improve the way we build and design communities in the face of natural hazards. For example, see the following reports available from the University of Hawai'i Sea Grant College: *Homeowner's Handbook to Prepare for Natural Hazards:* http://www.soest.hawaii.edu/SEA GRANT/communication/NaturalHazardsHandbook/naturalhazardprepbook.htm (last accessed October 17, 2009); *Hawai'i Coastal Hazard Mitigation Guidebook:* http://www.soest.hawaii.edu/SEAGRANT/communication/HCHMG/hchmg.htm (last accessed October 17, 2009); *Purchasing Coastal Real Estate in Hawai'i:* http://www.soest.hawaii.edu/SEAGRANT/communication/pdf/Purchasing%20 Coastal%20Real%20Estate.pdf (last accessed October 17, 2009).

3. For a review of early human impacts to island ecosystems see *"'Āina* and *Ahupua'a,"* pp. 313–324, in J. L. Culliney, 2006, *Islands in a Far Sea: The Fate of Nature in Hawai'i,* revised ed., University of Hawai'i Press, Honolulu.

4. In 1997 a paper was published demonstrating that about 25% of the total length of beaches around O'ahu has been lost due to chronic erosion in front of seawalls: C. H. Fletcher, R. A Mullane, and B. M. Richmond, 1997, "Beach Loss along Armored Shorelines of Oahu, Hawaiian Islands," *Journal of Coastal Research* 13:209–215.

5. Andrew Gomes, 2007, "Developing Trend Offers Ownership of Homes for 60 Days Each Year," *The Honolulu Advertiser,* June 24.

6. Karen Evans, David Woodside, and Marie Bruegman, 1994, "A survey of endangered waterbirds on Maui and O'ahu and assessment of potential impacts to waterbirds from the proposed Hawai'i Geothermal Project Transmission Corridor," U.S. Fish and Wildlife Service Pacific Islands Office Ecological Services, Final Report, p. 4.

Chapter 2: History of the Land

1. See discussion in Wikipedia entry: http://en.wikipedia.org/wiki/Ancient_Hawaii (last accessed October 27, 2009).

2. For more information, see Kamehameha Schools Hawaiian Studies Institute Staff, 1994, *Life in Early Hawai'i: The Ahupua'a*, 3rd ed., Kamehameha Schools Press, Honolulu.

3. For more information, see Herb Kawainui Kāne, "The 'Aumakua: Hawaiian Ancestral Spirits," Department of Land and Natural Resources, Department of Aquatic Resources: http://hawaii.gov/dlnr/dar/sharks/aumakua.html (last accessed October 27, 2009).

4. U.S. Geological Survey, Hawaiian Volcano Observatory, "Kīlauea: Perhaps the World's Most Active Volcano": http://hvo.wr.usgs.gov/kilauea/ (last accessed October 27, 2009).

5. Hawai'i 2050 Sustainability Plan, p. 19: http://hawaii2050.org/ (last accessed October 27, 2009).

6. See Encyclopedia Britannica entry: http://www.britannica.com/EBchecked/topic/634095/Mount-Waialeale (last accessed October 27, 2009).

7. National Oceanic and Atmospheric Administration (NOAA) National Weather Service, "Daily Climatic Normals for Hawai'i from 1961 to 1990": http://www.prh.noaa.gov/hnl/pages/climnormals.php (last accessed October 27, 2009).

8. Marie M. Bruegmann, Vickie Caraway, and Mike Maunder, 2002, "A Safety Net for Hawai'i's Rarest Plants," *Endangered Species Bulletin* (July–Aug.) 27 (3): 8–11.

9. Ann Fielding and Ed Robinson, 1989, *An Underwater Guide to Hawai'i*, University of Hawai'i Press, Honolulu.

10. Lucius G. Eldridge and Neal L. Evenhuis, 2003, "Hawai'i's Biodiversity: A Detailed Assessment of the Numbers of Species in the Hawaiian Islands," *Bishop Museum Occasional Papers*, No. 76, Bishop Museum Press, Honolulu.

11. Department of Land and Natural Resources, Division of Forestry, "The Hawai'i Endangered Bird Conservation Program": http://www.state.hi.us/dlnr/dofaw/captiveprop/consprog.htm (last accessed October 27, 2009).

12. U.S. Census Bureau, "United States and Puerto Rico—Metropolitan Area: Population, Housing Units, Area, and Density: 2000": http://factfinder.census.gov/servlet/GCTTable?_bm=y&-ds_name=DEC_2000_SF1_U&-CONTEXT=gct&-mt_name=DEC_2000_SF1_U_GCTPH1_US25&-redoLog=false&-_caller=geoselect&-geo_id=&-format=US-10|US-10S&-_lang=en (last accessed October 27, 2009).

13. State of Hawai'i, Department of Business, Economic Development, and Tourism, "Historical Visitor Statistics Spreadsheet": http://hawaii.gov/dbedt/info/visitor-stats/ (last accessed October 27, 2009).

14. Gavan Daws, 1968, *Shoal of Time: A History of the Hawaiian Islands*, University of Hawai'i Press, Honolulu, pp. 32–44.

15. B. E. Hope and J. H. Hope, 2003, "Native Hawaiian Health in Hawaii: Historical Highlights," *Californian Journal of Health Promotion*, vol. 1, Special Issue: Hawaii, pp. 1–9.

16. R. C. Schmitt, 1968, *Demographic Statistics of Hawaii: 1778–1965*, University of Hawai'i Press, Honolulu.

17. Noel J. Kent, 1993, *Hawaii: Islands under the Influence*, University of Hawai'i Press, Honolulu, p. 22.

18. Thomas Kemper Hitch, 1993, *Islands in Transition: The Past, Present, & Future of Hawaii's Economy*, University of Hawai'i Press, Honolulu, p. 39.

19. Ibid., p. 37.

20. Culliney, 2006, p. 316 (see chap. 1, n3).

21. Andrew Gomes, 2009, "New Growth for Pineapple Farming," *Honolulu Advertiser,* Sunday, February 22.

22. U.S. Public Law 103-150, To Acknowledge the 100th Anniversary of the January 17, 1893 Overthrow of the Kingdom of Hawaii, And to Offer an Apology to Native Hawaiians on Behalf of the United States for the Overthrow of the Kingdom of Hawaii, 103d Congress Joint Resolution 19, November 23, 1993.

23. See, for example, Daws, 1968, pp. 294–295.

24. Haunani-Kay Trask, 2002, "Stealing Hawai'i: The War Machine at Work," *Honolulu Weekly,* July 17.

25. Fletcher et al., 1997, pp. 209–215 (see chap. 1, n4).

26. Evans et al., 1994 (see chap. 1, n6).

27. U.S. Census Bureau, Census 2000 Redistricting Data (Public Law 94-171), Summary File; figures compiled and calculated by the Hawai'i State Department of Business, Economic Development, and Tourism, Hawai'i State Data Center: http://hawaii.gov/dbedt/info/census/Folder.2005-10-13.2927/pl94-171/pltable2.pdf (last accessed September 18, 2009).

28. Schmitt, 1968; out of a population estimate of 1,108,229 for the state of Hawai'i in 1990, 138,742 (12.52%) were Native Hawaiian. U.S. Census Bureau, "1990 Census of Population. General Population Characteristics, Hawai'i," 1990 CP-1-13 (Washington 1992). Note: Hawaiian as defined by the U.S. Census Bureau: "The concept of race as used by the Census Bureau reflects self-identification; it does not denote any clear-cut scientific definition of biological stock."

29. NOAA, "The Economics and Social Benefits of NOAA Data and Products": http://www.economics.noaa.gov/?goal=ecosystems&file=users/business/tourism (last accessed October 27, 2009).

30. Alan Friedlander, "The State of Coral Reef Ecosystems of the Main Hawaiian Islands": http://ccma.nos.noaa.gov/ecosystems/coralreef/coral2008/pdf/Hawaii.pdf (last accessed October 27, 2009); see also discussion in "Hawaii Reefs Showing the Strain," *The Honolulu Advertiser,* April 26, 2009: http://www.honoluluadvertiser.com/section/manlandsea (last accessed October 27, 2009).

31. P. V. Kirch, 1982, "Transported Landscapes," *Natural History* 91 (12): 32–35.

32. For more information see S. M. Gon III, Hawaiian High Islands Ecoregion, Hawai'i Culture and Conservation, "The Nature Conservancy Ecoregion Project": http://www.hawaiiecoregionplan.info/culture.html (last accessed October 27, 2009).

33. Ibid.

34. For more information, see Kamehameha Schools Hawaiian Studies Institute Staff, 1994.

35. For more information, see Jon J. Chinen, 1958, *The Great Mahele: Hawaii's Land Division of 1848,* University of Hawai'i Press, Honolulu.

36. For more information, see Office of Hawaiian Affairs, 2007, *Ka Wai Ola* 24 (12).

Chapter 3: Volcanism among the Islands

1. C. J. Wolfe, S. C. Solomon, E. H. Hauri, G. Laske, J. A. Orcutt, J. A. Collins, R. S. Detrick, and D. Bercovici, 2009, "Mantle Shear-wave Velocity Structure beneath the

Hawaiian Hotspot," Science 326:1388–1390; see also the news account of this research at ScienceDaily: http://www.sciencedaily.com/releases/2009/12/091203141905. htm (last accessed December 28, 2009).

2. U.S. Geological Survey, Hawaiian Volcano Observatory, "Mauna Loa: Earth's Largest Volcano": http://hvo.wr.usgs.gov/maunaloa/ (last accessed October 28, 2009); and University of Hawai'i, Institute for Astronomy, "About Mauna Kea Observatories": http://www.ifa.hawaii.edu/mko/about_maunakea.htm (last accessed October 28, 2009).

3. For more information on the Hawaiian hot spot, see U.S. Geological Survey, Hawaiian Volcano Observatory, "Hotspots": http://hvo.wr.usgs.gov/volcano-watch/1995/95_04_14.html (last accessed October 28, 2009).

4. See the first still image of a deep-sea volcano actively erupting molten lava on the seafloor at the ScienceDaily Web site: http://www.sciencedaily.com/releases/2009/12/091217183101.htm (last accessed December 28, 2009).

5. For more information on the demigod Māui, see W. D. Westervelt, 2004, *Legends of Maui*, Kessinger Publishing Co., Whitefish, Montana.

6. For a map showing the tectonic motion in the Pacific Basin, see NASA Crustal Dynamics Data Information System, "Tectonic Motion in the Pacific Basin": http://cddis.nasa.gov/926/swpactect.html (last accessed October 29, 2009).

7. There is a rich body of research on the formation of atolls. Here are a few examples: K. O. Emery, J. I. Tracy, and H. S. Ladd, 1954, "Geology of Bikini and Nearby Atolls," U.S. Geological Survey Professional Paper 260-A; R. F. McLean and C. D. Woodroffe, 1994, "Coral Atolls," pp. 267–302 in R. W. G. Carter and C. D. Woodroffe, eds., *Coastal Evolution: Late Quaternary Shoreline Morphodynamics*, Cambridge University Press, Cambridge; E. G. Purdy and E. L. Winterer, 2001, "Origin of Atoll Lagoons," *Geological Society of America Bulletin* 113:837–854; W. R. Dickinson, 2004, "Impacts of Eustasy and Hydro-isostasy on the Evolution and Landforms of Pacific Atolls," *Palaeogeography, Palaeoclimatology, Palaeoecology* 213:251–269; E. G. Purdy and E. L. Winterer, 2006, "Contradicting Barrier Reef Relationships for Darwin's Evolution of Reef Types," *International Journal of Earth Sciences* 95:143–167; C. D. Woodroffe, 2008, "Reef-island Topography and the Vulnerability of Atolls to Sea-level Rise," *Global and Planetary Change* 62:77–96, doi:10.1016/j.gloplacha.2007.11.001; W. R. Dickinson, 2009, "Pacific Atoll Living: How Long Already and until When?," *GSA Today* (Geological Society of America) 19 (3), doi:10.1130/GSATG35A.1.

8. R. W. Grigg, 1982, "Darwin Point: A Threshold for Atoll Formation," *Coral Reefs* 1:29–34.

9. Ibid., 1997, "Paleoceanography of Coral Reefs in the Hawaiian-Emperor Chain—Revisited," *Coral Reefs* 16:S33–S38.

10. University of Hawai'i, School of Ocean and Earth Science and Technology, Hawai'i Center for Volcanology, "Volcanic History of Kīlauea": http://www.soest. hawaii.edu/GG/HCV/kil-general.html (last accessed October 29, 2009).

11. The U.S. Geological Survey has an information sheet on these eruptions called "Explosive Eruptions at Kīlauea Volcano, Hawai'i?": http://pubs.usgs.gov/fs/fs132-98/ (last accessed October 29, 2009).

12. U.S. Geological Survey, Hawaiian Volcano Observatory, "The 1924 Ex-

plosions of Kīlauea": http://hvo.wr.usgs.gov/kilauea/history/1924May18/ (last accessed October 29, 2009).

13. G. A. Macdonald, A. T. Abbott, and F. L. Peterson, 1983, *Volcanoes in the Sea: The Geology of Hawaii,* 2nd ed., University of Hawai'i Press, Honolulu, p. 164. See also the U.S. Geological Survey Fact Sheet 132-98: "Explosive Eruptions at Kīlauea Volcano, Hawai'i?": http://pubs.usgs.gov/fs/fs132-98/ (last accessed October 29, 2009).

14. See, for example, Ragnar Carlson, 2008, "Big Island Rocked by Cloud of Uncertainty," *Honolulu Weekly,* August 27, pp. 6–7.

15. The Hawai'i Interagency Task Force on Volcanic Emissions has a Web site: http://hawaii.gov/gov/vog (last accessed October 29, 2009).

16. Interview with Aisha Morris, geologist, in Honolulu, Hawai'i, November 2002.

17. For more information on the various issues at Kapoho, see Dennis H. Hwang, 2007, "Coastal subsistence in Kapoho, Puna, Island and State of Hawai'i," Report for the Hawai'i County Planning Department, County of Hawai'i, January.

18. U.S. Geological Survey, Hawaiian Volcano Observatory, "1984 Eruption: March 25–April 15": http://hvo.wr.usgs.gov/maunaloa/history/1984.html (last accessed October 29, 2009).

19. The U.S. Geological Survey volcano hazards map can be found at this site: http://pubs.usgs.gov/gip/hazards/maps.html (last accessed October 29, 2009).

Chapter 4: Earthquakes and Tsunamis

1. G. Fryer, personal communication.

2. U.S. Geological Survey, Hawaiian Volcano Observatory, "Tsunami Generated by Magnitude 7.2 Earthquake on November 29, 1975": http://hvo.wr.usgs.gov/earthquakes/destruct/1975Nov29/tsunami.html (last accessed October 30, 2009).

3. For more information on tsunami safety, see National Oceanic and Atmospheric Administration (NOAA) National Weather Service, Pacific Tsunami Warning Center, "Tsunami Safety and Preparedness in Hawai'i": http://www.prh.noaa.gov/ptwc/hawaii.php (last accessed October 30, 2009).

4. See the NOAA Pacific Tsunami Warning Center's Web site for tsunami warnings to Hawai'i: http://www.prh.noaa.gov/ptwc/?region=2 (last accessed October 30, 2009).

5. For more information on what to do before, during, and after a tsunami and for tsunami evacuation maps, see Hawai'i State Civil Defense: http://www.scd.state.hi.us/ (last accessed October 30, 2009).

6. Sandy Ward and Catherine Main, 1998, NOAA State of the Coast Report, "Population at Risk from Natural Hazards": http://oceanservice.noaa.gov/web sites/retiredsites/sotc_pdf/PAR.PDF (last accessed December 27, 2009).

7. For more information on the 1868 earthquake, see U.S. Geological Survey, Earthquake Hazards Program, "Historic Earthquakes": http://earthquake.usgs.gov/regional/states/events/1868_04_03.php (last accessed October 30, 2009). For information regarding the safety of Ka'ū-Puna, see 2008, "Rapid Growth in Puna, Ka'u Ignores Major Lava-Flow Risk," *The Honolulu Advertiser,* January 13.

8. For more information on the earthquake, see U.S. Geological Survey, Hawaiian Volcano Observatory, "Two Damaging Earthquakes with Revised Magnitudes 6.7 and 6.0 Occur on Northwest Side of Hawai'i Island": http://hvo.wr.usgs.gov/archive/2006_10_15.html (last accessed October 30, 2009); see also Gary Chock (ed.), 2006, *Compilation of Observations of the October 15, 2006, Kīloho Bay (Mw 6.7) and Māhukona (Mw 6.0) Earthquakes, Hawai'i,* Earthquake Engineering Research Institute, Hawaii State Earthquake Advisory Committee of State Civil Defense: http://www.seaoh.org/attach/2006-10-15_Kiholo_Bay_Hawaii.pdf (last accessed December 27, 2009).

9. U.S. Geological Survey, Earthquake Hazards Program, "Historic Earthquakes": http://earthquake.usgs.gov/regional/states/events/1983_11_16.php (last accessed November 2, 2009).

10. The County of Kaua'i has adopted the 2003 International Building Code (IBC) and 2003 International Residential Codes (IRC) with amendments. The 2003 IBC and 2003 IRC became effective approximately January 22, 2008. Copies of Kaua'i County's amendments to the 2003 IBC and 2003 IRC are available for purchase at the Office of the Kaua'i County Clerk, Kaua'i County Courthouse, Nāwiliwili, Hawai'i.

11. Hawai'i State Civil Defense, 2007, State of Hawai'i Multi-Hazard Mitigation Plan: http://www.scd.hawaii.gov/HazMitPlan/contents.pdf (last accessed December 27, 2009).

12. Last accessed December 27, 2009.

13. U.S. Geological Survey, 1999, Hawaiian Volcano Observatory, Volcano Watch, July 29, "Here Today, Gone to Maui": http://hvo.wr.usgs.gov/volcanowatch/1999/99_07_29.html (last accessed December 27, 2009).

14. For information on the national system of state geological surveys (Hawai'i is the only state without one), see the following: Association of American State Geologists: http://www.stategeologists.org/ (last accessed November 2, 2009); Oregon Department of Geology and Mineral Industries: http://www.oregongeology.org/sub/earthquakes/earthquakehome.htm (last accessed November 2, 2009); also see the Delaware Geological Survey home page for an example from another small state: http://www.dgs.udel.edu/ (last accessed November 2, 2009).

15. U.S. Geological Survey, Earthquake Hazards Program, "Magnitude 9.1 - Off the West Coast of Northern Sumatra": http://earthquake.usgs.gov/eqcenter/eqinthenews/2004/usslav/#details (last accessed November 2, 2009). See also Wikipedia entry: http://en.wikipedia.org/wiki/2004_Indian_Ocean_earthquake (last accessed November 2, 2009).

16. Pacific Disaster Center, "Hawai'i Tsunami Events": http://www.pdc.org/iweb/tsunami_history.jsp?subg=1 (last accessed November 2, 2009).

17. Ibid.

18. Ward and Main, 1998, NOAA State of the Coast Report, "Population at Risk from Natural Hazards."

19. For more information see the NOAA National Weather Service, Pacific Tsunami Warning Center home page: http://www.prh.noaa.gov/ptwc/ (last accessed November 3, 2009).

20. Mike Gordon, 2005, "Tsunami Warning Signs Not So Clear," *The Honolulu Advertiser,* September 9.

21. On May 7, 1986, a magnitude 8.0 earthquake struck the central Aleutian Islands in Alaska, triggering a tsunami warning along the coasts of Hawai'i. "The tsunami that ultimately struck the Hawai'i coastline, however, was less than a foot in height and caused no damage. The average cost of lost business and productivity because of the evacuation was estimated by the State of Hawai'i's Department of Business, Economic Development and Tourism to be $40 million." *NOAA Magazine Online,* "NOAA Tsunami Warning System Receives High Marks": http://www.magazine.noaa.gov/stories/mag153.htm (last accessed November 3, 2009).

22. "Plain Text: Technology Alone Can't Save Us," *Newsweek,* December 30, 2004: http://www.newsweek.com/id/55539?tid=relatedcl (last accessed November 3, 2009).

Chapter 5: Hurricanes

1. Joseph A. Angelo, 2003, *Space Technology,* Greenwood Publishing Group, Santa Barbara, California, p. 213.

2. Kerry A. Emanuel, 2005, *Divine Wind: The History and Science of Hurricanes,* Oxford University Press, Oxford, England, p. 18.

3. See Wikipedia entry for discussion of tropical cyclones: http://en.wikipedia.org/wiki/Tropical_cyclone (last accessed November 3, 2009).

4. For more information about the scale, see the National Hurricane Center: http://www.nhc.noaa.gov/aboutsshs.shtml (last accessed November 4, 2009).

5. The Central Pacific Hurricane Center is run by the National Oceanic and Atmospheric Administration (NOAA) National Weather Service. For more information see http://www.prh.noaa.gov/hnl/cphc/ (last accessed November 4, 2009).

6. See T. T. Fujita, 1985, "The Downburst: Microburst and Macroburst," Satellite and Mesometeorology Research Project Research Paper 210, Department of Geophysical Sciences, University of Chicago.

7. City and County of Honolulu, Hawai'i Hazard Mitigation Forum, O'ahu: http://www.mothernature-hawaii.com/county_honolulu/county_honolulu.htm (last accessed December 28, 2009).

8. C. H. Fletcher, E. E. Grossman, B. M. Richmond, and A. E. Gibbs, 2002, *Atlas of Natural Hazards in the Hawaiian Coastal Zone,* U.S. Geological Survey, Denver, Colorado, Geologic Investigations Series I-2761: http://geopubs.wr.usgs.gov/i-map/i2761/ (last accessed 5 November 2009).

9. C. H. Fletcher, B. M. Richmond, G. M. Barnes, and T. A Schroeder, 1995, "Marine Flooding on the Coast of Kauai during Hurricane Iniki: Hindcasting Inundation Components and Delineating Washover," *Journal of Coastal Research* 10:890–907. See also C. H. Fletcher, T. A. Schroeder, and B. M. Richmond, 1994, "Impact of Hurricane Iniki Overwash on the Coastal Zone of Kauai," pp. 746–755, in *Proceedings of the American Society of Civil Engineers Conference on Hurricanes of 1992,* Miami. Both available at: http://www.soest.hawaii.edu/coasts/publications/index.html (last accessed November 4, 2009).

10. See chapter 6.

11. Jack Williams, 2005, "Background: California's Tropical Storms," *USA Today,* May 17.

12. Ibid.

13. M. Chenoweth and C. Landsea, 2004, "The San Diego Hurricane of 2 October 1858," *Bulletin of the American Meteorological Society* 85:1689–1697.

14. NOAA, National Weather Service, Central Pacific Hurricane Center, "The 1994 Central Pacific Tropical Cyclone Season": http://www.prh.noaa.gov/cphc/summaries/1994.php (last accessed November 4, 2009).

15. To see the course of Tropical Storm Allison: http://www.hpc.ncep.noaa.gov/tropical/rain/allison1989.html (last accessed November 4, 2009).

16. Bryan Norcross, 2007, *Hurricane Almanac: The Essential Guide to Storms Past, Present, and Future*, Macmillan, New York, p. 74.

17. Fletcher et al., 1995.

18. P.-S. Chu and J. X. Wang, 1997, "Tropical Cyclone Occurrences in the Vicinity of Hawaii: Are the Differences between El Niño and Non-El Niño Years Significant?" *Journal of Climate* 10:2683–2689.

19. Kerry Emanuel, 2005, "Increasing Destructiveness of Tropical Cyclones over the Past 30 Years," *Nature* 436:686–688.

20. John Roach, 2005, "Is Global Warming Making Hurricanes Worse?" National Geographic News, August 4, p. 1: http://news.nationalgeographic.com/news/2005/08/0804_050804_hurricanewarming.html (last accessed December 28, 2009).

21. K. Emanuel, R. Sundararajan, and J. Williams, 2008, "Hurricanes and Global Warming: Results from Downscaling IPCC AR4 Simulations," Bulletin of the American Meteorological Society 89:347–367.

22. P. J. Webster, G. J. Holland, J. A. Curry, and H.-R. Chang, 2005, "Changes in Tropical Cyclone Number, Duration, and Intensity in a Warming Environment," *Science* 309:1844–1846.

23. Intergovernmental Panel on Climate Change, 2007, Fourth Assessment Report (IPCC-AR4): Summary for Policymakers, p. 15, in S. Solomon, D. Qin, M. Manning, Z. Chen, M. Marquis, K. B. Averyt, M. Tignor, and H. L. Miller, eds., *Climate Change 2007: The Physical Science Basis. Contribution of Working Group I to the Fourth Assessment Report (AR4) of the Intergovernmental Panel on Climate Change.* Cambridge University Press, Cambridge, United Kingdom.

24. Ibid., p. 9.

25. James B. Elsner, James P. Kossin, and Thomas H. Jagger, 2008, "The Increasing Intensity of the Strongest Tropical Cyclones," *Nature* 455:92–95.

26. Pew Center on Global Climate Change, Hurricane and Global Warming FAQs, Is the Frequency of Hurricanes Increasing?: http://www.pewclimate.org/hurricanes.cfm#freq (last accessed November 4, 2009).

27. See NOAA, National Weather Service, Central Pacific Hurricane Center, "Climatology of Tropical Cyclones in the Central Pacific Basin": http://www.prh.noaa.gov/cphc/pages/climatology.php (last accessed November 4, 2009).

28. See Wikipedia entry for Hurricane Iniki: http://en.wikipedia.org/wiki/Hurricane_Iniki (last accessed November 4, 2009). See also County of Kaua'i, Hawai'i Hazard Mitigation Forum, Hurricane: http://www.mothernature-hawaii.com/county_kauai/county_kauai.htm (last accessed December 28, 2009).

29. NOAA, National Weather Service, Central Pacific Hurricane Center, 1994.

30. Federal Emergency Management Agency, 1993, "Building Performance: Hurricane Iniki," FIA-23, Washington D.C.

31. NOAA, National Weather Service, Central Pacific Hurricane Center, 1994.

32. Anthony Sommer, 2002, "A Decade after the Disaster Iniki," *Honolulu Star-Bulletin*, September 8.

33. See http://www.mothernature-hawaii.com/index.html (last accessed November 4, 2009).

34. See chapters 6 and 9 for information on climate change.

35. See http://starbulletin.com/2002/05/19/news/story2.html (last accessed December 28, 2009).

36. See http://www.mothernature-hawaii.com/index.html (last accessed December 28, 2009).

Chapter 6: Climate and Water Resources

1. City and County of Honolulu, Board of Water Supply: http://www.hbws.org/cssweb/ (last accessed November 5, 2009).

2. D. S. Oki, 2004, "Trends in Streamflow Characteristics at Long-term Gaging Stations, Hawai'i," U.S. Geological Survey Scientific Investigations Report 2004-5080. See also "No Rain Could Spell Plenty Pain," *Honolulu Star-Bulletin*: http://archives.starbulletin.com/2003/05/30/news/story1.html (last accessed December 29, 2009).

3. See the PowerPoint by Ramsay Taum, University of Hawai'i: http://www.wrrc.hawaii.edu/seminars/Wai-6perpage.pdf (last accessed December 29, 2009).

4. Ibid.

5. Lawrence H. Miike, 2004, *Water and the Law in Hawai'i*, University of Hawai'i Press, Honolulu, p. 118.

6. Mary Kawena Pukui and Samuel H. Elbert, 1986, *Hawaiian Dictionary*, revised and enlarged ed., University of Hawai'i Press, Honolulu, p. 362.

7. Oki, 2004.

8. Pao-Shin Chu and Huaiqun Chen, 2005, "Interannual and Interdecadal Rainfall Variations in the Hawaiian Islands," *Journal of Climate* 18:4796–4813.

9. David E. Alexander and Rhodes Whitmore Fairbridge, 1999, *Encyclopedia of Environmental Science*, Springer, New York, p. 181.

10. S. Hastenrath, 1991, *Climate Dynamics of the Tropics*, Springer, New York, p. 157.

11. See Peter Leroy Grimm, 1987, *Climatic Features of the North Pacific High*, University of Wisconsin-Madison.

12. R. E. Munn, ed., 2001, *Encyclopedia of Global Environmental Change*, Wiley, Hoboken, New Jersey, p. 592.

13. David Greenland, Douglas G. Goodin, and Raymond C. Smith, eds., 2003, *Climate Variability and Ecosystem Response at Long-Term Ecological Research Sites*, Oxford University Press, Oxford, United Kingdom, p. 257.

14. Oki, 2004.

15. For an account of Hawaiian wayfinding using only natural signs of the sea, skies, and land, see the Polynesian Voyaging Society Web page: http://pvs.kcc.hawaii.edu/navigate/wayfind.html (last accessed November 5, 2009).

16. Derived from wind descriptions found on the Web sites of the Pacific Disaster Center: http://www.pdc.org/iweb/pdchome.html (last accessed November 5,

2009), and the Polynesian Voyaging Society: http://pvs.kcc.hawaii.edu/ (last accessed November 5, 2009).

17. Pacific Disaster Center, see note 16.

18. From the Polynesian Voyaging Society Web site on Traditional Tahitian Navigation: http://pvs.kcc.hawaii.edu/andia.html (last accessed November 5, 2009).

19. William J. Kotsch, 1983, *Weather for the Mariner,* 3rd ed., Naval Institute Press, Annapolis, Maryland.

20. Such as Mālama Maunalua (Bay), a community-based stewardship initiative: http://www.nature.org/wherewework/northamerica/states/hawaii/marine/art21062.html (last accessed November 5, 2009).

21. The Hawai'i Coastal Zone Management Program staffs the process of updating the Ocean Resources Management Plan and compiles annual reports to the legislature on its implementation. The report is available at: http://hawaii.gov/dbedt/czm/ormp/ormp.php (last accessed November 5, 2009).

22. Ruth M. Tabrah, 1985, *Hawaii: A History,* Norton, New York, p. 17.

23. Beatrice H. Krauss, 1993, *Plants in Hawaiian Culture,* University of Hawai'i Press, Honolulu.

24. Eileen K. Schofield, 1989, "Effects of Introduced Plants and Animals on Island Vegetation: Examples from the Galapagos Archipelago," Conservation Biology 3:227–238.

25. See Gordon Tribble, 2008, "Groundwater on Tropical Pacific Islands: Understanding a Vital Resource," U.S. Geological Survey Circular 1312: http://pubs.usgs.gov/circ/1312/ (last accessed November 5, 2009).

26. Oki, 2004.

27. Jocelyn Linnekin, 1983, "The *Hui* Lands of Keanae: Hawaiian Land Tenure and the Great Mahele," *Journal of the Polynesian Society* 92:169–188.

28. William Alanson Bryan, 1915, *Natural History of Hawaii: Being an Account of the Hawaiian People, the Geology and Geography of the Islands, and the Native and Introduced Plants and Animals of the Group,* The Hawaiian Gazette Co., Ltd., p. 126.

29. John Mink and Glenn Bauer, 1998, "Water," p. 90, in Sonia P. Juvik and James O. Juvik, eds., *Atlas of Hawai'i,* 3rd ed., University of Hawai'i Press, Honolulu.

30. Ibid.

31. Ibid.

32. Ibid., p. 91.

33. Ibid.

34. Ibid., p. 178.

35. City and County of Honolulu, Board of Water Supply, "O'ahu's Water History": http://www.hbws.org/cssweb/display.cfm?sid=1106 (last accessed November 5, 2009).

36. Ibid.

37. Ibid.

38. Ibid.

39. Miike, 2004, p. 115.

40. Office of the Auditor, 1996, State of Hawai'i, "Management Audit of the Commission on Water Resource Management," Report No. 96-3.

41. Senate Bill No. 503, Twenty-third Legislature, State of Hawai'i, 2005:

Amends the State Constitution to specify that the counties have the responsibility of setting overall water conservation, quality, and use policies.

42. U.S. Water News Online, 2005, "Top Administrator of Hawaii Water Commission Resigns": http://www.uswaternews.com/archives/arcpolicy/5topxadmi2.html (last accessed November 5, 2009).

43. For example, see Chris Hamilton, 2009, "Plan Could Rejuvenate Waters of Na Wai Eha." A *Maui News* report of court decisions regarding plantation diversion of water on West Maui: http://www.mauinews.com/page/content.detail/id/517136.html (last accessed November 5, 2009).

44. Timothy Hurley, 2004, "Water Problem 'Dire' on Maui," *The Honolulu Advertiser*, February 28.

45. Ibid.

46. Ibid.

47. Ibid.

48. A discussion of Maui water issues is presented on the Maui Tomorrow Web site: http://www.maui-tomorrow.org/wai/wainews.php (last accessed November 5, 2009). Details of Maui water science are presented by the U.S. Geological Survey, Pacific Islands Water Science Center: http://hi.water.usgs.gov/lahaina/lahaina_tab.htm (last accessed November 5, 2009) and other U.S. Geological Survey pages. The County of Maui Department of Water Supply Web site also provides information: http://co.maui.hi.us/index.aspx?NID=126 (last accessed November 5, 2009). East Maui Watershed Partnership: http://eastmauiwatershed.org/ (last accessed November 5, 2009) and West Maui Mountains Watershed Partnership: http://www.westmauiwatershed.org/ (last accessed November 5, 2009) provide additional information.

49. Diana Leone, 2003, "Federal Report Warns of Isle Water Shortage," *Honolulu Star-Bulletin*, July 11: http://archives.starbulletin.com/2003/07/11/news/story3.html (last accessed November 5, 2009).

50. Rocky Mountain Institute, 2008, "Hawaii's Energy Future: Policy Recommendations," Report for Hawai'i Department of Business, Economic Development, and Tourism, Boulder, Colorado: http://hawaii.gov/dbedt/info/energy/publications/HEPR%20Full%20Report%20080407.pdf (last accessed December 29, 2009).

51. State of Hawai'i Data Book: http://hawaii.gov/dbedt/info/economic/databook/ (last accessed February 3, 2010)

52. Ibid.

53. Tribble, 2008.

54. Oki, 2004.

55. Ker Than, 2006, "Global Warming Weakens Pacific Trade Winds," *msnbc Technology and Science:* http://www.msnbc.msn.com/id/12612965/ (last accessed November 6, 2009); based on G. A. Vecchi, B. J. Soden, A. T. Wittenberg, I. M. Held, A. Leetmaa, and M. J. Harrison, 2006, "Weakening of Tropical Pacific Atmospheric Circulation due to Anthropogenic Forcing," *Nature* 441:73–76.

56. Thomas Karl, Jerry Melillo, Thomas Peterson, and Susan Hassol, eds., 2009, *Global Climate Change Impacts in the United States,* U.S. Global Change Research Program, Cambridge University Press, New York: http://www.globalchange.gov/ (last accessed December 29, 2009); see also Oliver Timm and Henry Diaz, 2009, "Synoptic-Statistical Approach to Regional Downscaling of IPCC Twenty-first

Century Climate Projections: Seasonal Rainfall over the Hawaiian Islands," *Journal of Climate* 22:4261–4280.

57. See note 55.

58. See note 56.

59. See *ScienceDaily,* 2008, "Climate Change: When It Rains It Really Pours": http://www.sciencedaily.com/releases/2008/08/080807144240.htm (last accessed November 6, 2009); based on R. P. Allan and B. J. Soden, 2008, "Atmospheric Warming and the Amplification of Precipitation Extremes," *Science* 321:1481–1484; see also Karl et al., 2009, note 56.

60. Joan Conrow, 2008, "Groundwater Sheds Not Recharging Like They Used to," *Honolulu Weekly,* April 16–22, p. 19.

61. Hawai'i Revised Statutes § 174C-3.

62. Mink and Bauer, 1998, p. 91.

63. Honolulu Board of Water Supply pamphlet, "Frequently Asked Questions; Water Recycling Program": http://www.boardofwatersupply.com/cssweb/display.cfm?sid=2076 (last accessed November 6, 2009).

64. Ibid.

65. See http://www.epa.gov/region09/water/recycling/index.html (last accessed November 6, 2009).

66. Hawai'i State Department of Health, 2002, Wastewater Branch, Guidelines for the Treatment and Use of Recycled Water, May 15.

67. See note 55.

68. Timm and Diaz, 2009, see note 56.

69. S. Solomon, D. Qin, M. Manning, Z. Chen, M. Marquis, K. B. Avery, M. Tignor, and H. L. Miller, eds., Intergovernmental Panel on Climate Change, "The Physical Science Basis," pp. 847–940, in *Climate Change 2007,* Contribution of Working Group I to the Fourth Assessment Report (AR4) of the IPCC, Cambridge University Press, New York: http://www.ipcc.ch/ipccreports/ar4-wg1.htm (last accessed November 6, 2009).

70. P. Ya. Groisman, R. W. Knight, T. R. Karl, D. R. Easterling, B. Sun, J. H. Lawrimore, 2004 "Contemporary changes of the hydrological cycle over the contiguous United States, trends derived from *in situ* observations" *Journal of Hydrometeorology* 5(1): 64–85. Cited in: *Global Climate Change Impacts in the United States,* T. R. Karl, J. M. Melillo, T. C. Peterson, (eds.). Cambridge University Press, 2009.

71. See, for example, http://www.usatoday.com/weather/news/2004-11-01-hawaii-flood_x.htm (last accessed November 6, 2009).

72. See note 59.

73. See article by Harry Eagar, 2008, "Na Wai Eha Designated," *The Maui News,* March 14: http://www.mauinews.com/page/content.detail/id/501447.html (last accessed November 6, 2009).

74. The Honolulu Board of Water Supply offers several online public education documents: http://www.boardofwatersupply.com/cssweb/display.cfm?sid=2076 (last accessed November 6, 2009).

Chapter 7: Stream Flooding and Mass Wasting

1. Titus Coan, 1882, *Life in Hawaii: An Autobiographic Sketch of Mission Life and Labors, 1835–1881,* A. D. F. Randolf and Co., New York, p. 32.

2. Ed Darack, 2007, "Kauai's Mount Waialeale," *Weatherwise* 60 (6): 16–17.

3. National Weather Service, Hydrology in Hawai'i; see these Web sites for information on rainfall around the state: Routine Products, http://www.prh.noaa.gov/hnl/pages/hydrology.php; Hydrology Summary, http://www.prh.noaa.gov/hnl/hydro/pages/hawaii-hydrology.php; Rainfall Graphs, http://www.prh.noaa.gov/hnl/pages/rra_maps.php (last accessed January 1, 2010).

4. National Oceanic and Atmospheric Administration (NOAA) National Weather Service, "Flash Floods in Hawaii: Event Statistics": http://www.prh.noaa.gov/hnl/pages/weather_hazards_stats.pdf (last accessed November 7, 2009).

5. Ibid.

6. Much of the island-by-island discussion is from the Hawai'i Statewide Hazard Mitigation Forum Web site: http://www.mothernature-hawaii.com/aboutus.html (last accessed November 7, 2009).

7. For planning activities related to mass-wasting hazard mitigation, see the State and County Hazard Mitigation Plan, 2007: http://www.mothernature-hawaii.com/hazmit_planning_toc2007.htm (last accessed January 1, 2010).

8. Ibid.

9. Mary Vorsino and Crystal Kua, 2006, "Floods Plague Windward Oahu," *Honolulu Star-Bulletin*, March 3.

10. See Kolja Rotzoll, 2007, "Hydraulic Parameter Estimation Using Aquifer Tests, Specific Capacity, Ocean Tides, and Wave Setup for Hawaii Aquifers," Thesis, University of Hawai'i, Department of Geology and Geophysics, see "PhD, 2007" at http://www.soest.hawaii.edu/asp/GG/academics/gradtheses.asp (last accessed January 1, 2010).

11. See note 6.

12. See John Dracup, E. D. H. Cheng, J. M. Nigg, and T. A. Schroeder, 1991, *The New Year's Eve Flood on Oahu, Hawaii, December 31, 1987–January 1, 1988*, National Academies Press: http://www.nap.edu/openbook.php?record_id=1748&page=R1 (last accessed November 7, 2009).

13. See note 6.

14. U.S. Army Corp of Engineers, 2006, Hydrology and Hydraulics Study of Flood of October 30, 2004, Mānoa Stream, Honolulu, Hawai'i: http://www.scd.state.hi.us/HazMitPlan/chapter_3_appA.pdf (last accessed January 1, 2010).

15. See note 3.

16. Andy Nash, Nezette Rydell, and Kevin Kodama, 2006, "Unprecedented Extended Wet Period across Hawaii," National Weather Service, May 11: http://www.prh.noaa.gov/hnl/pages/events/weeksrain/weeksrainsummary.php (last accessed November 7, 2009).

17. See note 6.

18. Oahu Civil Defense Agency, 2003.

19. Ibid.

20. See note 6.

21. Tom Finnegan, 2006, "Kauai Dam Breach Leaves Doubts about Spillway," *Honolulu Star-Bulletin*, March 22.

22. Stories about the Kaloko Dam failure can be found at several news sites: http://www.hawaiireporter.com/story.aspx?d4b463ac-91d0-48ad-a264-2a7b05f33143 (last accessed November 7, 2009); http://archives.starbulletin.com/2006/03/15/

news/story01.html (last accessed November 7, 2009); report of the independent investigation: http://www.kalokodam.net/report/Report.pdf; http://en.wikipedia. org/wiki/Ka_Loko_Reservoir (last accessed November 7, 2009).

23. See note 3.

24. Charles Fletcher, Eric E. Grossman, Bruce M. Richmond, and Ann E. Gibbs, 2004, *Atlas of Natural Hazards in the Hawaiian Coastal Zone: Island of Maui*, U.S. Geological Survey, Geological Investigations Series I-2761; available on the Web at: http://pubs.usgs.gov/imap/i2761/ (last accessed November 7, 2009).

25. Ibid.

26. County of Hawai'i Civil Defense Agency, 2005, Hawai'i County Multi-Hazard Mitigation Plan, chap. 2: Hazard Analysis.

27. Ibid.

28. See note 7.

29. Ibid.

30. See note 6.

31. Public briefing by the National Weather Service, October, 2008. See PowerPoint at http://www.prh.noaa.gov/hnl/pages/examples/WP_FF_Prog_Briefing_2008-SWP.ppt (last accessed January 1, 2010).

32. For more information, see Flood Insurance Information at: http://www. floodcontrol.co.riverside.ca.us/content/floodinsurance.htm (last accessed November 7, 2009).

33. Vorsino and Kua, 2006.

34. Mark Dorfman and Nancy Stoner, 2007, *Testing the Waters: A Guide to Water Quality at Vacation Beaches*, 17th ed., National Resources Defense Council, New York: http://www.nrdc.org/water/oceans/ttw/ttw2007.pdf (last accessed January 1, 2010). See recent reports at http://www.nrdc.org/water/oceans/ttw/titinx.asp (last accessed January 1, 2010).

35. Crystal Kua, 2006, "Waikiki Beaches Awash in Filth as Sewage Changes Course," *Honolulu Star-Bulletin*, March 29.

36. See County of Sacramento Public Works Agency, 2001, "Local Floodplain Management Plan for the County of Sacramento": http://www.msa.saccounty.net/ waterresources/drainage/docs/LocalFPMgmtPlan.pdf (last accessed November 7, 2009).

37. See S. K. Deb and A. I. El-Kadi, 2009, "Susceptibility Assessment of Shallow Landslides on Oahu, Hawaii, under Extreme-Rainfall Events," Journal of Geomorphology 108:219–233: http://www.soest.hawaii.edu/GG/FACULTY/aly/slide. html (last accessed November 7, 2009).

38. News Desk, 2006, "Rain Falls throughout Spring Break," *Ka Leo O Hawai'i*, April 4.

39. Rod Ohira, 2000, "Rock Slide Shuts Road at Waimea," *Honolulu Star-Bulletin*, March 6.

40. See note 6.

41. For more information on mass wasting, see G. A. Macdonald, A. T. Abbott, and F. L. Peterson, 1983, *Volcanoes in the Sea: The Geology of Hawaii*, 2nd ed., University of Hawai'i Press, Honolulu, chap. 9.

42. Randall W. Jibson and Rex L. Baum, 1999, "Assessment of Landslide Hazards in Kaluanui and Maakua Gulches, Oahu, Hawaii, following the 9 May

1999 Sacred Falls Landslide," U.S. Geological Survey, Open File Report 99–364.

43. William De Witt Alexander, 1899, *A Brief History of the Hawaiian People,* American Book Co., New York, p. 292.

44. U.S. Geological Survey, 1997, "Volcanic and Seismic Hazards on the Island of Hawaii." Online edition: http://pubs.usgs.gov/gip/hazards/ (last accessed January 1, 2010).

45. Oʻahu Civil Defense Agency, 2003, Multi-Hazard Pre-Disaster Mitigation Plan for the City and County of Honolulu, Geologic Hazards: 7: Landslides, Debris Flows and Rock Falls: http://www.mothernature-hawaii.com/files/honolulu_planning-11.pdf (last accessed November 7, 2009).

46. J. G. Moore, D. A. Clague, R. T. Holcomb, P. W. Lipman, W. R. Normark, and M. E. Torresan, 1989, "Prodigious Submarine Landslides on the Hawaiian Ridge," *Journal of Geophysical Research* 94:17465–17484: http://www.agu.org/pubs/crossref/1989/JB094iB12p17465.shtml (last accessed November 7, 2009).

47. U.S. Geological Survey, Western Coastal and Marine Geology, "Giant Hawaiian Underwater Landslides": http://walrus.wr.usgs.gov/posters/underlandslides.html (last accessed November 7, 2009).

48. Moore et al., 1989.

49. Will Kyselka and Ray E. Lanterman, 1980, *Maui: How It Came to Be,* University of Hawaiʻi Press, Honolulu, p. 119.

50. Deb and El-Kadi, 2009.

51. Oʻahu Civil Defense Agency, 2003.

52. For information on creep in Pālolo Valley, see Macdonald et al., 1983, p. 194.

53. Oʻahu Civil Defense Agency, 2003.

Chapter 8: Sewage Treatment and Polluted Runoff

1. Water Resources Research Center, University of Hawaiʻi at Mānoa: http://www.wrrc.hawaii.edu/bulletins/2001_05/WQWorkshop.html (last accessed November 8, 2009).

2. J. Brannon, 2007a, "Sewage Plant Waiver Denied," *The Honolulu Advertiser,* December 11, A1.

3. The city's largest sewage treatment plant on Sand Island has a bioconversion facility that turns sewage sludge into fertilizer pellets. See J. Brannon, 2007b, "High-tech Sewage Units Ready to Run," *The Honolulu Advertiser,* October 30, B3.

4. The Sand Island treatment plant has a disinfection unit that blasts treated wastewater with ultraviolet light to kill pathogens before it is discharged offshore. See Brannon, 2007b, October 30.

5. Frank Spellman, 2003, *Handbook of Water and Wastewater Treatment Plant Operations,* CRC Press, Boca Raton, Florida, p. 609.

6. Federal Water Pollution Control Act Amendments, PL 92-500, October 18, 1972, 86 Stat. 816.

7. Clean Water Act §43 (g).

8. Environment News Service, 2007, "Honolulu Faces $1.2 billion Wastewater Treatment Upgrade": http://www.ens-newswire.com/ens/dec2007/2007-12-10-096.asp (last accessed November 8, 2009).

9. J. Brannon, 2007c, "City to Fight EPA's Sand Island Ruling," *The Honolulu Advertiser*, December 11, A1.

10. Ibid.

11. Ibid.

12. Environment News Service, 2007.

13. Brannon, 2007c, December 11.

14. Additional discussion of this issue can be found in the following: Brannon, 2007b, October 30; R. K. Abe, 2008, "Sewage Treatment Waiver Backed by Science," *The Honolulu Advertiser*, March 11; P. Boylan, 2008a, "City Presses Feds to Ease Demands on Waste Plants, *The Honolulu Advertiser*, March 13; P. Boylan, 2009, "EPA Won't Renew Sewage Exemption," *The Honolulu Advertiser*, January 7.

15. For more information about the program, see http://www.epa.gov/OWM/cwfinance/construction.htm (last accessed November 9, 2009).

16. Harry Eagar, 2008, "Algae Blooms Gone Missing –Why?" *The Maui News:* http://www.mauinews.com/page/content.detail/id/511895.html (last accessed November 8, 2009).

17. Brannon, 2007b, October 30.

18. James Honke, H. Krock, J. Kumagai, and V. Moreland, 2007, "White Paper Recommending Approval of the City and County of Honolulu's Honouliuli Wastewater Treatment Plant Application for a Modified National Pollution Discharge Elimination System Permit under Section 301(h) of the Clean Water Act," Hawai'i Water Environment Association, p. 2.

19. Brannon, 2007c, December 11.

20. City and County of Honolulu, Department of Environmental Services: http://www.co.honolulu.hi.us/env/wwplants.htm (last accessed November 9, 2009).

21. Roger S. Fujioka, J. E. T. Moncur, and A. Russo, 2002, "Community Structure of Fish and Macrobenthos at Selected Sites, Wai'anae, O'ahu, Hawai'i, in Relation to the Wai'anae Deep Ocean Sewage Outfall, 1990–1998," Water Resources Research Center, University of Hawai'i, Honolulu.

22. John R. Clark, 1996, *Coastal Zone Management Handbook*, CRC Press, Boca Raton, Florida, p. 404.

23. C. D. Hunt, 2007, "Groundwater Nutrient Flux to Coastal Waters and Numerical Simulation of Wastewater Injection at Kihei, Maui, Hawaii," U.S. Geological Survey Scientific Investigations Report 2006-5283.

24. See Maui County Frequently Asked Questions: http://www.co.maui.hi.us/FAQ.aspx?QID=462 (last accessed November 9, 2009).

25. J. R. Pegg, 2004, "Crumbling U.S. Sewage System Undermines Public Health," Environment News Service, February 20.

26. Mark Dorfman, 2004, "Swimming in Sewage," Natural Resources Defense Council, NewYork: http://www.nrdc.org/water/pollution/sewage/contents.asp (last accessed January 1, 2010).

27. Pegg, 2004.

28. Diana Leone, 2006, "48 Million Gallons Spill in 6-day Sewer Break," *Honolulu Star-Bulletin*, March 31.

29. P. Boylan, 2008b, "Judge Finds Honolulu Liable for 297 Violations in Sewage Spills," *The Honolulu Advertiser*, August 20.

30. The Council for LAB/LAS Environmental Research, "LAS Biodegradation and Removal in Sewage Treatment": http://www.cler.com/facts/sewage.html (last accessed November 9, 2009).

31. K. L. Knee, J. Street, A. Santoro, C. Berg, A. Boehm, and A. Paytan, 2006, "Estimating the Importance of Submarine Groundwater Discharge in Hanalei Bay, Kaua'i," *Eos, Transactions, American Geophysical Union* 87 (36), Ocean Science Meeting, Suppl., Abstract OS15B-20.

32. S. Dollar, personal communication, University of Hawai'i, October 15, 2008.

33. Hawai'i Department of Health, Clean Water Branch, administers the Polluted Runoff Control Program. See http://hawaii.gov/health/environmental/water/cleanwater/prc/index.html (last accessed November 9, 2009).

34. Katina Henderson and J. Harrigan, 2002, "Final 2002 List of Impaired Waters in Hawaii Prepared under the Clean Water Act §303(d)," Hawai'i State Department of Health.

35. Linda Koch, J. Harrigan-Lum, and K. Henderson, 2004, "Final 2004 List of Impaired Waters in Hawaii Prepared under the Clean Water Act §303(d)," Hawai'i State Department of Health.

36. See http://hawaii.gov/health/environmental/env-planning/wqm/wqm.html.

37. Watershed planning guidance has been completed by the Hawai'i Department of Health and is available at: http://hawaii.gov/health/environmental/water/cleanwater/prc/prc/index.html (last accessed January 15, 2010).

38. See note 33.

39. See Board of Water Supply, Water Resources, "Watershed Management Plans": http://www.hbws.org/cssweb/display.cfm?sid=1406 (last accessed November 9, 2009).

40. Implementation Plan, Hawai'i's Polluted Runoff Control Program, Department of Health, Clean Water Branch, Polluted Runoff Control (July 2002): http://hawaii.gov/health/environmental/water/cleanwater/prc/implan-index.html (last accessed November 9, 2009).

41. Section 6217 of the federal Coastal Zone Management Act Reauthorization Amendments of 1990.

42. Clean Water Act Reauthorization Bill of 2004.

43. For more information, see http://hawaii.gov/health/environmental/water/cleanwater/about/aboutcwb.html (last accessed November 9, 2009).

44. A. Shileikis, 2000, "Hawaii Non-Point Source Program: Review of the Implementation Plan," Environmental Protection Agency.

45. Ibid.

46. A. M. Brasher and S. S. Anthony, 2000, "Occurrence of Organochlorine Pesticides in Stream Bed Sediment and Fish from Selected Streams on the Island of Oahu, Hawaii," 1998, Fact sheet 140-00: http://pubs.er.usgs.gov/usgspubs/fs/fs14000 (last accessed November 9, 2009.)

47. The contamination maps are posted at: http://hawaii.gov/health/environmental/water/sdwb/conmaps/conmaps.html (last accessed November 9, 2009).

48. For state law, see Hawai'i Water Pollution Law, Hawai'i Revised Statutes §342D and Hawai'i Constitution, art. VII, §9.

49. For more information on the Safe Drinking Water Act, see http://www.epa. gov/OGWDW/sdwa/laws_statutes.html (last accessed November 9, 2009).

50. See the Hawai'i Department of Health Groundwater Protection Program Web site for more information: http://hawaii.gov/health/environmental/water/ sdwb/conmaps/conmaps.html (last accessed November 9, 2009).

51. For the water-quality reports of the City and County of Honolulu, see http:// www.hbws.org/cssweb/display.cfm?sid=1081 (last accessed November 9, 2009).

52. Ibid.

53. Fox News.com—Health, 2008, "Study Finds Traces of Drugs in Drinking Water in 24 Major U.S. Regions," March 10: http://www.foxnews.com/ story/0,2933,336286,00.html (last accessed November 9, 2009).

54. Ibid.

55. Beverly Creamer and Loren Moreno, 2006, "'Horrible, Horrible Death' by Infection," *The Honolulu Advertiser,* April 8.

56. Christie Wilson, 2008, "How Safe Is the Water?" *The Honolulu Advertiser,* May 7.

57. Ibid.

58. Natural Resources Defense Council, 2008, "Testing the Waters 2008: Hawai'i."

59. Hawai'i Division of Forestry and Wildlife, "FY 1998–2000 Accomplishments": Participated in the spill response and cleanup of the Fishing Vessel *Van Loi* grounding on Kaua'i and the spill of 16,000 gallons of diesel fuel in April, 1999: www.state.hi.us/dlnr/dofaw/pubs/accomplish00.doc (last accessed November 9, 2009).

60. Kenneth C. Schiff, M. James Allen, Eddy Y. Zeng, and Steven M. Bay, 2000, "Seas at the Millennium: An Environmental Evaluation: Southern California," *Marine Pollution Bulletin* 41 (1–6): 76–93.

61. Natural Resources Defense Council, 2007, "Testing the Waters 2007: Hawai'i."

62. Surfrider Foundation, State of the Beach Report, "Surf Zone Water Quality," California: http://www.surfrider.org/stateofthebeach/08-fc/body.asp?sub=Surf ZoneWaterQuality (last accessed November 9, 2009).

63. Brannon, 2007c, December 11.

64. R. R. Colwell, G. T. Orlob, and J. R. Schubel, 1995, "Mamala Bay Study, Water Quality Management in Mamala Bay, Executive Summary," Mamala Bay Study Commission, December: http://openlibrary.org/books/OL15425504M/Mamala_ Bay_study?V=2 (last accessed November 9, 2009).

Chapter 9: Climate Change and Sea-Level Rise

1. National Oceanic and Atmospheric Administration (NOAA), National Climatic Data Center: http://www.ncdc.noaa.gov/oa/climate/globalwarming.html (last accessed November 10, 2009).

2. For example, see National Aeronautics and Space Administration (NASA), Goddard Institute for Space Studies, 2009, "Global Temperature Trends: 2009 Annual Summation": http://www.nasa.gov/topics/earth/features/temp-analysis-2009.html (last accessed February 3, 2010); see also the NOAA Mauna Loa Observatory home page: http://www.esrl.noaa.gov/gmd/ccgg/trends/ (last accessed November 10, 2009).

3. For example, see J. Hansen, M. Sato, R. Ruedy, K. Lo, D. W. Lea, and M. Medina-Elizade, 2006, "Global Temperature Change," *Proceedings of the National Academy of Sciences of the U.S.A.* 103:14288–14293, doi:doi:10.1073/pnas.0606291103; see also http://www.nasa.gov/topics/earth/features/temp-analysis-2009.html (last accessed February 3, 2010); see also F. J. Wentz, L. Ricciardulli, K. Hilburn, and C. Mears, 2007, "How Much More Rain Will Global Warming Bring?" *Science* 317:233–235.

4. NASA, http://www.nasa.gov/topics/earth/features/temp-analysis-2009.html (last accessed November 10, 2009).

5. Ibid.

6. NOAA, National Climatic Data Center, "Global Warming Frequently Asked Questions": http://www.ncdc.noaa.gov/oa/climate/globalwarming.html#q3 (last accessed November 10, 2009).

7. The Intergovernmental Panel on Climate Change (IPCC) provides authoritative reviews of the state of climate science for national governments throughout the world. A key aspect of its work has been the production of major assessment reports in 1990, 1995, 2001 (TAR), and 2007 (AR4). For more information, see http://www.ipcc.ch/ (last accessed November 10, 2009).

8. National Research Council, 2006, "Surface Temperature Reconstructions for the Last 2,000 Years,": see http://www.nap.edu/catalog.php?record_id=11676 (last accessed November 7, 2009).

9. Each of these organizations has global warming Web sites. For the National Academy, see http://www.nap.edu/catalog.php?record_id=11676; for the National Climatic Data Center, see http://www.ncdc.noaa.gov/oa/climate/globalwarming.html; for NASA, see http://data.giss.nasa.gov/gistemp/; for the United Kingdom Meteorological Office, see http://www.cru.uea.ac.uk/cru/info/warming/; for the Union of Concerned Scientists, see http://www.ucsusa.org/global_warming/global_warming_101/ (last accessed January 1, 2010).

10. J. K. Angell, 2009, "Global, Hemispheric, and Zonal Temperature Deviations Derived from Radiosonde Records," in *Trends Online: A Compendium of Data on Global Change,* Carbon Dioxide Information Analysis Center, Oak Ridge National Laboratory, U.S. Department of Energy, Oak Ridge, Tennessee.

11. K. Vinnikov and N. Grody, 2003, "Global Warming Trend of Mean Tropospheric Temperature Observed by Satellites," *Science* 302:269–272.

12. Hansen et al., 2006.

13. S. Levitus, J. I. Antonov, T. P. Boyer, R. A. Locarnini, H. E. Garcia, and A. V. Mishonov, 2009, "Global Ocean Heat Content 1955–2008 in Light of Recently Revealed Instrumentation Problems," Geophysical Research Letters 36: L07608.

14. See the NASA Goddard Institute for Space Studies Web site for description of their methods of data analysis: http://data.giss.nasa.gov/gistemp/updates/ (last accessed January 1, 2010).

15. See U.S. Global Change Research Program, 2009, "Global Climate Change Impacts in the United States": http://www.globalchange.gov/ (last accessed November 10, 2009).

16. P. T. Doran and M. K. Zimmerman, 2009, "Examining the Scientific Consensus on Climate Change," *EOS* 90 (3): http://tigger.uic.edu/~pdoran/012009_Doran_final.pdf (last accessed January 1, 2010).

17. N. Oreskes, 2004, "Beyond the Ivory Tower: The Scientific Consensus on Climate Change," *Science* 306:1686.

18. Archive for the Press, 2006, Climate Science, "2005 Warmest Year in Over a Century": http://www.climatechange.com.au/category/press-releases/press-climate-science/ (last accessed November 10, 2009).

19. See note 2.

20. A. K. Tripati, C. D. Roberts, and R. A. Eagle, 2009, "Coupling of CO_2 and Ice Sheet Stability over Major Climate Transitions of the Last 20 Million Years," *Science* 326:1394–1397: http://www.sciencemag.org/cgi/content/abstract/1178296 (last accessed November 8, 2009).

21. Intergovernmental Panel on Climate Change, 2007, "Climate Change 2007, The Physical Science Basis," Contribution of Working Group I to the Fourth Assessment Report (AR4) of the IPCC, S. Solomon, D. Qin, M. Manning, Z. Chen, M. Marquis, K. B. Avery, M. Tignor, and H. L. Miller, eds., Cambridge University Press, Cambridge: http://www.ipcc.ch/ipccreports/ar4-wg1.htm (last accessed November 10, 2009).

22. See note 2.

23. See note 21.

24. Ibid.

25. University of Colorado at Boulder, "Sea Level Change": http://sealevel.colorado.edu/ (last accessed November 10, 2009); see also J. A. Church and N. J. White, 2006, "A 20th Century Acceleration in Global Sea-level Rise," *Geophysical Research Letters* 33: L01602.

26. See note 15.

27. Ibid.

28. Ibid.

29. Camille Parmesan and Hector Galbraith, 2004, *Observed Impacts of Climate Change in the United States*, Pew Center on Global Climate Change, Arlington, Virginia: http://www.pewclimate.org/global-warming-in-depth/all_reports/observed-impacts (last accessed January 2, 2010).

30. S. R. Loarie, P. B. Duffy, H. Hamilton, G. P. Asner, C. B. Field, and D. D. Ackerly, 2009, "The Velocity of Climate Change," *Nature* 462:1052–1055.

31. W. F. Ruddiman, 2005a, *Plows, Plagues & Petroleum: How Humans Took Control of Climate*, Princeton University Press, Princeton, New Jersey.

32. Ibid., 2005b, "How Did Humans First Alter Global Climate?" *Scientific American* 292:46–53.

33. A. E. Carlson, A. N. LeGrande, D. W. Oppo, R. E. Came, G. A. Schmidt, F. S. Anslow, J. M. Licciardi, and Elizabeth A. Obbink, 2008, "Rapid Early Holocene Deglaciation of the Laurentide Ice Sheet," *Nature Geoscience* 1:620–624.

34. W. F. Ruddiman, 2006, "The Early Anthropogenic Hypothesis: Challenges and Responses," *Reviews of Geophysics* 45: RG4001, doi:10.1029/2006RG000207. See also http://www.newscientist.com/article/dn4464-early-farmers-warmed-earths-climate.html for additional discussion (last accessed November 10, 2009).

35. D. R. Easterling and M. F. Wehner, 2009, "Is the Climate Warming or Cooling? *Geophysical Research Letters* 36: L08706.

36. S. Borenstein, 2009, "Statisticians Reject Global Cooling," Associated Press,

October 26. See, for example, http://www.foxnews.com/scitech/2009/10/26/statisticians-reject-global-cooling-earth-heating/ (last accessed January 3, 2010).

37. See note 15.

38. D. S. Oki, 2004, "Trends in Streamflow Characteristics at Long-term Gaging Stations, Hawai'i," U.S. Geological Survey Scientific Investigations Report 2004-5080. See also 2003, "No Rain Could Spell Plenty Pain," *Honolulu Star-Bulletin*: http://archives.starbulletin.com/2003/05/30/news/story1.html (last accessed December 29, 2009).

39. T. W. Giambelluca, H. F. Diaz, and M. S. A. Luke, 2008, Secular Temperature Changes in Hawai'i," *Geophysical Research Letters* 35: L12702, doi: 10.1029/2008GL034377.

40. See note 15.

41. See R. Chowdhury, P. S. Chu, T. A. Schroeder, and X. Zhao, 2008, "Variability and Predictability of Sea-level Extremes in the Hawaiian and U.S.-Trust Islands—A Knowledge Base for Coastal Hazards Management," *Journal of Coastal Conservation* 12:93–104.

42. See note 15. See also James B. Elsner, James P. Kossin, and Thomas H. Jagger, 2008, "The Increasing Intensity of the Strongest Tropical Cyclones," *Nature* 455:92–95.

43. Ker Than, 2006, "Global Warming Weakens Pacific Trade Winds," *MSNBC Technology and Science*: http://www.msnbc.msn.com/id/12612965/ (last accessed November 10, 2009), based on G. A. Vecchi, B. J. Soden, A. T. Wittenberg, I. M. Held, A. Letmaa, and M. J. Harison, 2006, "Weakening of Tropical Pacific Atmospheric Circulation due to Anthropogenic Forcing," *Nature* 441:73–76, doi:10.1038/nature04744.

44. M. A. Merrifield, Y. L. Firing, and J. J. Marra, 2004, "Annual Climatology of Extreme Water Levels," pp. 27–32, in P. Müller, C. Garrett, and D. Henderson, eds., *Extreme Events,* the 15th 'Aha Huliko'a Hawaiian Winter Workshop, University of Hawai'i at Mānoa, January 23–26, 2007, SOEST Special Publication, University of Hawai'i, Honolulu: http://www.soest.hawaii.edu/PubServices/Aha_2007_final.pdf (last accessed January 2, 2010); see also Y. Firing and M. A. Merrifield, 2004, "Extreme Sea Level Events at Hawaii: Influence of Mesoscale Eddies," *Geophysical Research Letters* 31: L24306, doi: 10.1029/2004GL021539.

45. C. H. Fletcher and B. Richmond, 2009, "Climate Change in the Federated States of Micronesia," *Ka Pili Kai*, Hawai'i Sea Grant Magazine, Fall, pp. 3–5: http://www.soest.hawaii.edu/SEAGRANT/communication/kapilikai/kapilikai.html (last accessed January 3, 2010).

46. AR4 (see note 21); U.S. Global Change (see note 15).

47. Elsner et al., 2008 (see note 42).

48. Vecchi et al., 2006 (see note 43).

49. B. F. Chao, Y. H. Wu, and Y. S. Li, 2008, "Impact of Artificial Reservoir Water Impoundment on Global Sea Level," *Science* 320:212–214.

50. S. Levitus, T. Boyer, J. Antonov, H. Garcia, and R. Locarnini, 2005, "Ocean Warming 1955–2003," poster presented at the U.S. Climate Change Science Program Workshop, 14–16 November 2005, Arlington, Virginia, Climate Science in Support of Decision Making: http://www.climatescience.gov/workshop2005/default.htm (last accessed November 11, 2009).

51. C. M. Domingues, J. A. Church, N. J. White, P. J. Gleckler, S. E. Wijffels, P. M. Barker, and J. R. Dunn, 2008, "Improved Estimates of Upper-ocean Warming and Multi-decadal Sea-level Rise," *Nature* 453:1090–1093, doi: 10.1038/nature07080.

52. G. C. Johnson and S. C. Doney, 2006, "Recent Western South Atlantic Bottom Water Warming," *Geophysical Research Letters* 33: L14614, doi:10.1029/2006GL026769; C. Johnson, S. Mecking, B. M. Sloyan, and S. E. Wijffels, 2007, "Recent Bottom Water Warming in the Pacific Ocean," *Journal of Climate* 13:2987–3002.

53. AR4 (see note 21).

54. Goddard Space Flight Center, 2009, "Sea Ice Yearly Minimum 1979–2007": http://svs.gsfc.nasa.gov/vis/a000000/a003400/a003464/index.html (last accessed November 11, 2009).

55. M. Wang and J. E. Overland, 2009, "A Sea Ice Free Summer Arctic within 30 Years?" *Geophysical Research Letters* 36: L07502, doi:10.1029/2009GL037820.

56. E. Rignot, J. Bamber, M. van den Broeke, C. Davis, Y. Li, W. van de Berg, and E. van Meijgaard, 2008a, "Recent Mass Loss of the Antarctic Ice Sheet from Dynamic Thinning," *Nature Geoscience*, doi: 10.1038/ngeo102.

57. E. J. Steig, D. P. Schneider, D. R. Scott, M. E. Mann, C. C. Josefino, and D. T. Shindell, 2009, "Warming of the Antarctic Ice-sheet Surface since the 1957 International Geophysical Year," *Nature* 457:459–462.

58. E. Rignot, J. E. Box, E. Burgess, and E. Hanna, 2008b, "Mass Balance of the Greenland Ice Sheet from 1958 to 2007," *Geophysical Research Letters* 35: L20502.

59. NASA, Earth Observatory, "Melting Anomalies on Greenland in 2007": http://earthobservatory.nasa.gov/Newsroom/NewImages/images.php3?img_id=17846 (last accessed November 11, 2009).

60. E. Rignot and P. Kanagaratnam, 2006, "Changes in the Velocity Structure of the Greenland Ice Sheet," *Science* 311:986–989.

61. National Snow and Ice Data Center, 2009a, "Arctic Sea Ice News and Analysis": http://nsidc.org/arcticseaicenews/ (last accessed November 11, 2009); ibid., 2009b, "State of the Cryosphere": http://nsidc.org/sotc/glacier_balance.html (last accessed November 11, 2009).

62. M. F. Meier, Mark B. Dyurgerov, Ursula K. Rick, Shad O'Neel, W. Tad Pfeffer, Robert S. Anderson, Suzanne P. Anderson, and Andrey F. Glazovsky, 2007, "Glaciers Dominate Eustatic Sea-level Rise in the 21st Century," *Science* 317:1064–1067. See also World Wildlife Fund Nepal Program, "An Overview of Glaciers, Glacier Retreat, and Subsequent Impacts in Nepal, India and China," March 2005.

63. M. B. Dyurgerov and M. F. Meier, 2005, "Glaciers and the Changing Earth System: A 2004 Snapshot," Occasional Paper 58, Institute of Arctic and Alpine Research, University of Colorado: http://instaar.colorado.edu/other/occ_papers.html (last accessed November 11, 2009).

64. E. W. Leuliette, R. S. Nerem, and G. T. Mitchum, 2004, "Calibration of TOPEX/Poseidon and Jason Altimeter Data to Construct a Continuous Record of Mean Sea Level Change," *Marine Geodesy* 27:79–94. See also B. D. Beckley, F. G. Lemoine, S. B. Luthcke, R. D. Ray, and N. P. Zelensky, 2007, "A Reassessment of Global and Regional Mean Sea Level Trends from TOPEX and Jason-1 Altimetry Based on Revised Reference Frame and Orbits," *Geophysical Research Letters* 34: L14608.

65. NASA, Jet Propulsion Laboratory, 2009, "Ocean Surface Topography from Space": http://sealevel.jpl.nasa.gov/ (last accessed November 11, 2009); ibid., Global Climate Change, "Rising Waters: New Map Pinpoints Areas of Sea-level Increase": http://climate.jpl.nasa.gov/news/index.cfm?FuseAction=ShowNews&NewsID=16 (last accessed November 11, 2009).

66. The Pacific Decadal Oscillation is discussed in chapter 6. See also N. J. Mantua, S. R. Hare, Y. Zhang, J. M. Wallace, and R. C. Francis, 1997, "A Pacific Interdecadal Climate Oscillation with Impacts on Salmon Production," *Bulletin of the American Meteorological Society* 78:1069–1079; Y. Zhang, J. M. Wallace, and D. S. Battisti, 1997, "ENSO-like Interdecadal Variability: 1900–1993," *Journal of Climate* 10:1004–1020; F. Biondi, A. Gershunov, and D. R. Cayan, 2001, "North Pacific Decadal Climate Variability since 1661," Journal of Climate 14:5–10.

67. Church and White, 2006 (see note 25).

68. AR4 (see note 21).

69. Church and White, 2006 (see note 25).

70. S. Jevrejeva, J. C. Moore, A. Grinsted, and P. L. Woodworth, 2008, "Recent Global Sea Level Acceleration Started over 200 Years Ago?" *Geophysical Research Letters* 35:L08715, doi:10.1029/2008GL033611.

71. AR4 (see note 21).

72. M. A. Merrifield, S. T. Merrifield, and G. T. Mitchum, 2009, "An Anomalous Recent Acceleration of Global Sea Level Rise," *Journal of Climate* 22:5772–5781.

73. C. M. Domingues, J. A. Church, N. J. White, P. J. Gleckler, S. E. Wijffels, P. M. Barker, and J. R. Dunn, 2008, "Improved Estimates of Upper-ocean Warming and Multi-decadal Sea-level Rise," *Nature* 453:1090–1093, doi:10.1038/nature07080.

74. See University of Colorado at Boulder, Sea Level Change: http://sealevel.colorado.edu/ (last accessed November 11, 2009), which periodically releases interpretations of the mean trend of satellite altimetry data recording global sea level. See also NASA, Jet Propulsion Laboratory, Global Climate Change": http://global climatechange.jpl.nasa.gov/news/index.cfm?FuseAction=ShowNews&NewsID=16 (last accessed November 11, 2009).

75. NASA, 2005, "Scientists Get a Real 'Rise' Out of Breakthroughs in How We Understand Changes in Sea Level," July 7: http://www.nasa.gov/vision/earth/environment/sealevel_feature.html (last accessed November 11, 2009).

76. See description in Wikipedia entry: http://en.wikipedia.org/wiki/Milanko vitch_cycles (last accessed November 11, 2009).

77. Carlson et al., 2008.

78. Ruddiman, 2005a, b.

79. The last interglacial period is known in Europe as the "Eemian," a term that is taking on global usage. See Wikipedia entry: http://en.wikipedia.org/wiki/Eemian (last accessed November 11, 2009).

80. D. R. Muhs, K. R. Simmons, and B. Steinke, 2002, "Timing and Warmth of the Last Interglacial Period: New U-series Evidence from Hawaii and Bermuda and a New Fossil Compilation for North America," *Quaternary Science Reviews* 21:1355–1383; see also Daniel Muhs, 2006, "Last Interglacial: Timing and Environment," U.S. Geological Survey, Earth Surface Processes: http://esp.cr.usgs.gov/info/lite/ (last accessed November 11, 2009).

81. K. M. Cuffey and S. J. Marshall, 2000, "Substantial Contribution to Sea

Level Rise during the Last Interglacial from the Greenland Ice Sheet," *Nature* 404:591–594. See also R. E. Kopp, F. J. Simons, J. X. Mitrovica, A. C. Maloof, and M. Oppenheimer, 2009, "Probabilistic Assessment of Sea Level during the Last Interglacial Stage," *Nature* 462:863–867.

82. Cuffey and Marshall, 2000.

83. D. R. Muhs and B. J. Szabo, 1994, "New Uranium-series Ages of the Waimanalo Limestone, Oʻahu, Hawaii: Implications for Sea Level during the Last Interglacial Period," *Marine Geology* 118:315–326; Muhs et al., 2002; D. R. Muhs, J. F. Wehmiller, K. R. Simmons, and L. L. York, 2003, "Quaternary Sea-level History of the United States," pp. 147–153, in A. R. Gillespie, B. F. Atwater, and S. C. Porter, eds., *Quaternary Period in the United States: Developments in Quaternary Science*, Elsevier Science, Amsterdam.

84. C. H. Fletcher, C. Bochicchio, C. L. Conger, M. Engels, E. J. Feirstein, E. E. Grossman, R. Grigg, J. N. Harney, J. Rooney, C. E. Sherman, S. Vitousek, K. Rubin, and C. V. Murray-Wallace, 2008, "Geology of Hawaii Reefs," chapter 11 in B. Riegl and R. E. Dodge, eds., *Coral Reefs of the USA*, Springer, New York.

85. Cuffey and Marshall, 2000.

86. B. L. Otto-Bliesner, J. Marshall, J. T. Overpeck, G. H. Miller, A. Hu, and CAPE Last Interglacial Project members, 2006, "Simulating Arctic Climate Warmth and Ice Field Retreat in the Last Interglaciation," *Science* 311:1751–1753; J. T. Overpeck, B. L. Otto-Bliesner, G. H. Miller, D. R. Muhs, R. B. Alley, and J. T. Kiehl, 2006, "Paleoclimatic Evidence for Future Ice-sheet Instability and Rapid Sea-level Rise," *Science* 311:1747–1750.

87. Rignot et al., 2008b.

88. C. H. Fletcher and A. T. Jones, 1996, "Sea-level High Stand Recorded in Holocene Shoreline Deposits on Oʻahu, Hawaii," *Journal of Sedimentary Research* 66 (3): 632–641; E. E. Grossman and C. H. Fletcher, 1998, "Sea Level 3500 Years Ago on the Northern Main Hawaiian Islands," *Geology* 26 (4): 363–366; E. Grossman, C. Fletcher, and B. Richmond, 1998, "The Holocene Sea-level High Stand in the Equatorial Pacific: Analysis of the Insular Paleosea-level Database," *Coral Reefs* (Special Issue on Holocene and Pleistocene Coral Reef Geology) 17:309–327.

89. Ian Lilley, 2006, *Archaeology of Oceania*, J. Wiley Publishing, Hoboken, New Jersey, p. 9.

90. Jevrejeva et al., 2008.

91. AR4 (see note 21).

92. K. Strohecker, 2008, "World Sea Levels to Rise 1.5 m by 2100," Scientists, Reuters, April 15.

93. Stefan Rahmstorf, Anny Cazenave, John A. Church, James E. Hansen, Ralph F. Keeling, David E. Parker, and Richard C. J. Somerville, 2007, "Recent Climate Observations Compared to Projections," *Science* 316:709.

94. Ibid.

95. M. Vermeer and S. Rahmstorf, 2009, "Global Sea Level Linked to Global Temperature," *Proceedings of the National Academy of Sciences of the USA*, Early Edition: www.pnas.org_cgi_doi_10.1073_pnas.0907765106.

96. A. Grinsted, J. C. Moore, and S. Jevrejeva, 2009, "Reconstructing Sea Level from Paleo and Projected Temperatures 200 to 2100 AD," *Climate Dynamics*, doi:10.1007/s00382-008-0507-2.

97. W. T. Pfeffer, J. T. Harper, and S. O'Neel, 2008, "Kinematic Constraints on Glacier Contributions to 21st Century Sea Level Rise," *Science* 321:1340–1343.

98. I. Allison, N. L. Bindoff, R. A. Bindschadler, P. M. Cox, N. de Noblet, M. H. England, J. E. Francis, N. Gruber, A. M. Haywood, D. J. Karoly, G. Kaser, C. Le Quéré, T. M. Lenton, M. E. Mann, B. I. McNeil, A. J. Pitman, S. Rahmstorf, E. Rignot, H. J. Schellnhuber, S. H. Schneider, S. C. Sherwood, R. C. J. Somerville, K. Steffen, E. J. Steig, M. Visbeck, and A. J. Weaver, 2009, *The Copenhagen Diagnosis, 2009: Updating the World on the Latest Climate Science,* The University of New South Wales Climate Change Research Centre (CCRC), Sydney, Australia.

99. S. Solomon, G.-K. Plattner, R. Knutti, and P. Friedlingstein, 2009, "Irreversible Climate Change Due to Carbon Dioxide Emissions," *Proceedings of the National Academy of Sciences of the U.S.A.* 106:1704–1709; T. M. L. Wigley, 2005, "The Climate Change Commitment," *Science* 307:1766–1769; G. A. Meehl, W. M. Washington, W. D. Collins, J. M. Arblaster, A. Hu, L. E. Buja, W. G. Strand, and H. Teng, 2005, "How Much More Global Warming and Sea Level Rise?" *Science* 307:1769–1772.

100. R. A. Pielke, 2008, "Climate Predictions and Observations," *Nature Geoscience* 1:206.

101. Overpeck et al., 2006 (see note 86).

102. Ibid.

103. California Executive Order S-13-08, 2008, Ordering the California Resources Agency to Complete the First California Sea Level Rise Assessment Report, November 14: http://gov.ca.gov/executive-order/11036/ (last accessed November 12, 2009).

104. See "The Impacts of Sea Level Rise on the California Coast": http://www.pacinst.org/reports/sea_level_rise/ (last accessed November 12, 2009).

105. California Executive Order S-13-08, 2008.

106. See James Hansen, 2007 "Scientific Reticence and Sea Level Rise," Environmental Research Letters 2: 024002, doi:10.1088/1748-9326/2/2/024002: http://www.iop.org/EJ/article/1748-9326/2/2/024002/erl7_2_024002.html (last accessed November 12, 2009).

107. Ibid.

108. June Watanabe, 2008, "Flooding Still a Problem in Mapunapuna," *Honolulu Star-Bulletin,* August 10.

109. See the Hawai'i Coastal Erosion Web site: http://www.soest.hawaii.edu/asp/coasts/ (last accessed November 12, 2009).

110. A discussion of "adjustable adaptation" to rising seas can be found at *Nature Geoscience:* http://www.nature.com/ngeo/journal/v2/n7/full/ngeo576.html (last accessed November 12, 2009).

111. There are numerous publications that discuss the consequences of various greenhouse-gas production scenarios. Some include Solomon et al., 2009; Wigley, 2005; Meehl et al., 2005 (see note 99).

112. See *ScienceDaily,* 2008, "Global Carbon Emissions Speed Up, beyond IPCC Projections," September 28: http://www.sciencedaily.com/releases/2008/09/080925072440.htm (last accessed November 12, 2009).

113. W. M. Washington, R. Knutti, G. A. Meehl, H. Teng, C. Tebaldi, D. Lawrence, L. Buja, and W. G. Strand, 2009, "How Much Climate Change

Can Be Avoided by Mitigation?" *Geophysical Research Letters* 36: L08703, doi:10.1029/2008GL037074.

114. D. Stammer, 2008, "Response of the Global Ocean to Greenland and Antarctic Ice Melting," *Journal of Geophysical Research* 113: C06022, doi:10.1029/2006JC004079.

115. Glenn A. Milne, W. Roland Gehrels, Chris W. Hughes, and Mark E. Tamisiea, 2009, "Identifying the Causes of Sea-level Change," *Nature Geoscience* advance online publication, 14 June, doi:10.1038/ngeo544.

116. See the Web site: http://shoals.sam.usace.army.mil/ (last accessed November 12, 2009).

117. C. H. Fletcher, R. A. Mullane, and B. M. Richmond, 1997, "Beach Loss along Armored Shorelines of Oʻahu, Hawaiian Islands," *Journal of Coastal Research* 13:209–215.

118. 2003, "Unusual South Swells Erode Beach at Kaanapali Resort," *Honolulu Star-Bulletin,* August 19.

119. Firing and Merrifield, 2004 (see note 44).

120. Ibid.

Chapter 10: Beach Erosion and Loss

1. C. H. Fletcher and A.T. Jones, 1996, "Sea-level High Stand Recorded in Holocene Shoreline Deposits on Oahu, Hawaii," *Journal of Sedimentary Research* 66:632–641; E. E. Grossman and C. H. Fletcher, 1998, "Sea Level 3500 years ago on the Northern Main Hawaiian Islands," *Geology* 26:363–366; E. Grossman, C. Fletcher, and B. Richmond, 1998, "The Holocene Sea-level High Stand in the Equatorial Pacific: Analysis of the Insular Paleosea-level Database," *Coral Reefs* (Special Issue on Holocene and Pleistocene Coral Reef Geology) 17:309–327.

2. The University of Hawaii Coastal Geology Group maintains a Web site that provides rates of coastal erosion for the beaches of Kauaʻi, Oʻahu, and Maui: http://www.soest.hawaii.edu/asp/coasts/index.asp (last accessed November 13, 2009). Erosion data can also be found on the County of Maui "Maui Shoreline Atlas" Web site: http://www.co.maui.hi.us/index.asp?nid=865 (last accessed November 13, 2009).

3. A. D. Short and C. L. Hogan, 1994, "Rip Currents and Beach Hazards: Their Impact on Public Safety and Implications for Coastal Management," pp. 197–209, in C. W. Finkl, ed., *Coastal Hazards, Journal of Coastal Research,* Special Issue 12; A. D. Short, 1999, *Handbook of Beach and Shoreface Morphodynamics,* John Wiley & Sons, New York; ibid., 2006, "Beach Hazards and Risk Assessment of Beaches," pp. 152–157, in J. J. L. M. Bierens, ed., *Handbook on Drowning,* Springer Verlag, Berlin.

4. For various discussions of shoreline responses to sea-level change, see the following: http://www.encora.eu/coastalwiki/Effect_of_climate_change_on_coast line_evolution (last accessed November 13, 2009); http://www.climatescience.gov/Library/sap/sap4-1/final-report/sap4-1-final-report-Appendix2.pdf (last accessed November 13, 2009).

5. There are several sources of sea-level data for Hawaiʻi. See University of Hawaiʻi Sea Level Center: http://uhslc.soest.hawaii.edu/ (last accessed November 13, 2009). See also the National Oceanic and Atmospheric Administration (NOAA)

tide gauge Web site: http://tidesandcurrents.noaa.gov/sltrends/sltrends_station. shtml?stnid=1612340, which reports long-term rates of water-level change for gauges both in Hawai'i and across the country and the world. This page provides the long-term rate for the Honolulu tide gauge: approximately 0.06 inch (1.50 ± 0.25 mm) per year (last accessed November 13, 2009).

6. C. Fletcher, J. Rooney, M., Barbee, S.-C. Lim, and B. Richmond, 2003, "Mapping Shoreline Change Using Digital Orthophotogrammetry on Maui, Hawaii," *Journal of Coastal Research*, Special Issue No. 38, pp. 106–124; O'ahu Civil Defense Agency, 2003, Multi-hazard Pre-disaster Mitigation Plan for the City and County of Honolulu, Hydrologic Hazards: Coastal Erosion, September; also calculated from the latest coastal erosion rate data on Kaua'i and O'ahu per the Web site maintained by the University of Hawai'i Coastal Geology Group (see note 2).

7. State of Hawai'i, Department of Land and Natural Resources, Office of Conservation and Coastal Lands, Kailua Beach Management Plan, p. 2; see also Z. Norcross, C. H. Fletcher, and M. Merrifield, 2002, "Annual and Interannual Changes on a Reef-fringed Pocket Beach: Kailua Bay, Hawaii," *Marine Geology* 190:553–580.

8. See, for example, Ocean Studies Board, 2007, *Mitigating Shore Erosion along Sheltered Coasts*, National Research Council, National Academy Press, Washington, D.C., stating, "Coastlines are perpetually changing—some from natural processes—some from human activities—many from both," p. 11.

9. See chapter 9. Several studies predict that sea-level rise has accelerated in the twenty-first century as compared with the twentieth century. See http://sealevel. colorado.edu/ (last accessed November 13, 2009); J. A. Church and N. J. White, 2006, "A 20th Century Acceleration in Global Sea-level Rise," *Geophysical Research Letters* 33: L01602. See also studies that predict future sea-level position that can be achieved only by acceleration in the rate of global sea-level rise compared with current rates. For example, the study by W. T. Pfeffer, J. T. Harper, and S. O'Neel, 2008, "Kinematic Constraints on Glacier Contributions to 21st-century Sea-level Rise," *Science* 321:1340–1343, predicts sea level approaching or exceeding 3.3 feet (1 m) above current level by the end of the twenty-first century, which would require global rates reaching a rate of ~ 0.4 inch (10 mm) per year.

10. Kimberly Miller, 2009, "State Farm Agrees to Stay," *The Palm Beach Post:* http://www.palmbeachpost.com/money/state-farm-agrees-to-stay-124557.html (last accessed January 4, 2010). See also: http://www.tampabay.com/news/business/banking/article970945.ece (last accessed November 14, 2009).

11. See Citizens Property Insurance Corporation: https://www.citizensfla.com/ (last accessed November 14, 2009).

12. Chapter 23, Section 1.2 (Purpose), Shoreline Setbacks, Revised Ordinances of Honolulu. For more information, see http://www.co.honolulu.hi.us/refs/roh/23.htm (last accessed November 14, 2009).

13. C. H. Fletcher, R. A. Mullane, and B. M. Richmond, 1997, "Beach Loss along Armored Shorelines of Oahu, Hawaiian Islands," *Journal of Coastal Research* 13:209–215.

14. Revised Ordinances of Honolulu (see note 12).

15. *County of Hawai'i v. Sotomura*, 55 Haw. 176, 517 P.2d 57 (1973).

16. Federal Emergency Management Agency (FEMA), 2000a, *Evaluation of*

Erosion Hazards, April: http://www.heinzctr.org/publications/PDF/erosnrpt.pdf (last accessed November 14, 2009).

17. FEMA, 2000b, "Significant Losses from Coastal Erosion Anticipated along U.S. Coastlines," June 27: http://www.fema.gov/news/newsrelease.fema?id=7708 (last accessed November 14, 2009).

18. FEMA, 2000a.

19. See FEMA, 2009, "Erosion": http://www.fema.gov/plan/prevent/flood plain/nfipkeywords/erosion.shtm (last accessed November 14, 2009).

20. Erosion resources are abundant; see the following: Hawai'i Department of Land and Natural Resources, Office of Conservation and Coastal Lands: http://hawaii.gov/dlnr/occl/ (last accessed November 14, 2009); County of Maui, Maui Shoreline Atlas: http://www.co.maui.hi.us/index.aspx?nid=865 (last accessed November 14, 2009); Coastal Geology Group, Hawai'i Coastal Erosion Web site: http://www.soest.hawaii.edu/asp/coasts/ (last accessed November 14, 2009).

21. R. S. Calhoun and C. H. Fletcher, 1996, "Late Holocene Coastal-plain Stratigraphy and Sea Level History at Hanalei, Kauai, Hawaiian Islands," *Quaternary Research* 45:47–58.

22. J. N. Harney and C. H. Fletcher, 2003, "A Budget of Carbonate Framework and Sediment Production, Kailua Bay, Oahu, Hawaii," *Journal of Sedimentary Research* 73:856–868; J. N. Harney, E. E. Grossman, B. M. Richmond, and C. H. Fletcher, 2000, "Age and Composition of Carbonate Shoreface Sediments, Kailua Bay, Oahu, Hawaii," *Coral Reefs* 19:141–154.

23. Beach loss due to coastal armoring is documented in several publications. See, for example, Fletcher et al., 1997; Fletcher et al., 2003.

24. S. Vitousek and C. H. Fletcher, 2008, "Maximum Annually Recurring Wave Heights in Hawai'i," *Pacific Science* 62:541–553. In that study the authors conducted a statistical analysis of the wave records from buoys in Hawaiian waters. They identified the annually recurring significant wave height (the average of the highest one-third of waves) to be 25 ± 0.9 feet (7.7 ± 0.28 m) and the top 10% and 1% wave heights during the annual swell to be 32 ± 1.1 feet and 42.3 ± 1.5 feet (9.8 ± 0.35 m and 12.9 ± 0.47 m), respectively, for open North and Northwest Pacific swells. Data are also provided (see Table 2 in the paper) for $H_{1/3}$, $H_{1/10}$, and $H_{1/100}$ for each of 12 30-degree compass windows around the Islands.

25. See Ocean Studies Board, 2007, *Mitigating Shore Erosion along Sheltered Coasts,* chap. 2, "Understanding Erosion on Sheltered Shores," National Research Council, National Academy Press, Washington, D.C.

26. Gary Kubota, 2002, "Weather, Seawalls Cited in Maui Beach Erosion," *Honolulu Star-Bulletin,* June 10.

27. J. Rooney and C. H. Fletcher, 2005, "Shoreline Change and Pacific Climate Oscillations in Kihei, Maui, Hawaii," *Journal of Coastal Research* 21:535–547.

28. J. F. Campbell and D. J. Hwang, 1982, "Beach Erosion at Waimea Bay, Oahu, Hawaii," *Pacific Science* 36:35–43.

29. These statements are made on the basis of data served on the Web site of the University of Hawai'i Coastal Geology Group (see note 2).

30. Fletcher et al., 1997.

31. Fletcher et al., 2003.

32. Kaua'i shoreline change study data available at Kaua'i Planning Depart-

ment. See also the following Web publications for discussion of Kaua'i coastal issues: http://raisingislands.blogspot.com/2008/02/new-kauai-shoreline-erosion-bill-among.html (last accessed November 15, 2009); http://kauaiworld.com/articles/2009/07/07/news/kauai_news/doc4a52f1edbee10708086642.txt (last accessed January 4, 2010).

33. Region 9, Environmental Protection Agency, news release, 2001, "EPA Issues Revised List of Polluted Waters in Hawaii," November 19: http://yosemite.epa.gov/opa/admpress.nsf/d2a3eb622562e96b85257359003d4809/bac39fa6181 9d766852570d8005e1462!OpenDocument (last accessed November 15, 2009).

34. See State of Hawai'i, 2000, Department of Land and Natural Resources, news release, "South Lanikai Beach Gets New Sand Through Partnership Project," March 23.

35. See http://www.fema.gov/rebuild/mat/fema55.shtm for FEMA's Coastal Construction Manual (last accessed November 15, 2009).

Chapter 11: Reefs and Overfishing

1. See C. H. Fletcher, C. Bochicchio, C. L. Conger, M. Engels, E. J. Feirstein, E. E. Grossman, J. N. Harney, J. Rooney, C. E. Sherman, S. Vitousek, K. Rubin, and C. V. Murray-Wallace, 2008, "Geology of Hawai'i Reefs," chapter 11 in B. Riegl and R. E. Dodge, eds. *Coral Reefs of the USA*, Springer, New York.

2. D. E. Alexander and R. W. Fairbridge, 1999, *Encyclopedia of Environmental Science*, Springer, New York, p. 99.

3. R. E. Munn, 2002, *Encyclopedia of Global Environmental Change*, Wiley, New York, p. 77.

4. Athline Clark and Dave Gulko, 1998, Hawai'i's *State of the Reefs,* State of Hawai'i, Department of Land and Natural Resources, Division of Aquatic Resources.

5. Ocean economic data compiled as part of the National Ocean Economics Program: www.oceaneconomics.org (last accessed November 16, 2009).

6. Herman S. J. Cesar and Pieter J. H. van Beukering, 2004, "Economic Valuation of the Coral Reefs of Hawai'i," *Pacific Science* 58:231–242; see also http://scholarspace.manoa.hawaii.edu/handle/10125/2723 (last accessed November 16, 2009).

7. *Protecting the Irreplaceable: Hawai'i Marine Conservation Strategy and Fund,* Fact Sheet distributed by Conservation International, SeaScape Project, Hawai'i, Melissa Bos (personal communication): www.conservation.org (last accessed November 16, 2009).

8. T. P. Hughes, A. H. Baird, D. R. Bellwood, M. Card, S. R. Connolly, C. Folke, R. Grosberg, O. Hoegh-Guldberg, J. B. C. Jackson, J. Kleypas, J. M. Lough, P. Marshall, M. Nyström, S. R. Palumbi, J. M. Pandolfi, B. Rosen, and J. Roughgarden, 2003, "Climate Change, Human Impacts, and the Resilience of Coral Reefs," *Science* 301:929–933.

9. See P. Jokiel and E. Brown, 2004, "Global Warming, Regional Trends and Inshore Environmental Conditions Influence Coral Bleaching in Hawai'i," *Global Change Biology* 10:1627–1641; see also P. Jokiel and S. Coles, 1990; "Response of Hawaiian and Other Indo-Pacific Reef Corals to Elevated Temperatures Associated with Global Warming," *Coral Reefs* 8:155–162.

10. A. Friedlander, G. Aeby, R. Brainard, E. Brown, K. Chaston, A. Clark, P. McGowan, T. Montgomery, W. Walsh, I. Williams, and W. Wiltse, with contributions from J. Asher, S. Balwani, E. Co, E. DeCarlo, P. Jokiel, J. Kenyon, J. Helyer, C. Hunter, J. Miller, C. Morshige, J. Rooney, H. Slay, R. Schroeder, H. Spalding, L. Wedding, and T. Work, 2008, *The State of Coral Reef Ecosystems of the Main Hawaiian Islands,* published online by the National Oceanographic and Atmospheric Administration (NOAA) at http://ccma.nos.noaa.gov/ecosystems/coralreef/coral2008/pdf/Hawaii.pdf (last accessed November 16, 2009).

11. See note 10. See also K. R. Weiss, 2008, "Hawaiian Islands' Reef Fish Declining, Study Finds": http://www.flmnh.ufl.edu/fish/innews/hawaii2008.html (last accessed November 16, 2009); A. M. Friedlander, E. K. Brown, and M. E. Monaco, 2007, "Defining Reef Fish Habitat Utilization Patterns in Hawai'i: Comparisons between Marine Protected Areas and Areas Open to Fishing," *Marine Ecology Progress Series* 351:221–233.

12. See note 10.

13. See a series of articles in *The Honolulu Advertiser* by Rob Perez: 2009, "Hawai'i's Reef Showing the Strain," April 26; 2009, "Island Reefs Await Extra Protection," April 27; 2009, "State Considers Fishing Ban to Help Revive Reef off Maui," May 5; 2009, "Sediment Is the Death That Keeps on Giving," May 31; 2009, "Molokai's Reef Is Choking to Death," June 1. Additional coverage of reef management challenges is found in *The Honolulu Advertiser:* Lynda Arakawa, 2008, "It's the Year of the Reef," January 24; Leanne Ta, 2008, "Hawai'i's Reef Fish in Peril," July 17.

14. See the Papahānaumokuākea Marine National Monument Web site: http://www.hawaiireef.noaa.gov/ (last accessed November 17, 2009).

15. See B. M. Riegl and R. E. Dodge, 2008, *Coral Reefs of the USA,* Springer, New York.

16. R. S. Scheltema, 1968, "Dispersal of Larvae by Equatorial Ocean Currents and Its Importance to the Zoogeography of Shoal Water Tropical Species," *Nature* 217:1159–1162; ibid., 1986, "Long-distance Dispersal by Planktonic Larvae of Shoal-water Benthic Invertebrates among Central Pacific Islands," *Bulletin of Marine Science* 39:241–256.

17. See note 10.

18. Scheltema, 1968.

19. R. W. Grigg, J. Polovina, A. M. Friedlander, and S. O. Rohmann, 2008, "Biology of Coral Reefs in the Northwestern Hawaiian Islands," chapter 14 in B. M. Riegl and R. E. Dodge, *Coral Reefs of the USA,* Springer, New York.

20. P. L. Jokiel, 2008, "Biology and Ecological Functioning of Coral Reefs in the Main Hawaiian Islands," chapter 12 in B. M. Riegl and R. E. Dodge, *Coral Reefs of the USA,* Springer, New York.

21. See Wikipedia entry "Coral," specifically the section "Evolutionary History": http://en.wikipedia.org/wiki/Coral (last accessed January 4, 2010).

22. J. Rooney, P. Wessel, R. Hoeke, J. Weiss, J. Baker, F. Parrish, C. H. Fletcher, J. Chojnacki, M. Garcia, R. Brainard, and P. Vroom, 2008, "Geology and Geomorphology of Coral Reefs in the Northwestern Hawaiian Islands," chapter 13 in B. M. Riegl and R. E. Dodge, *Coral Reefs of the USA,* Springer, New York.

23. For example, see R. W. Grigg, 1998, "Holocene Coral Reef Accretion

in Hawai'i: A Function of Wave Exposure and Sea Level History," *Coral Reefs* 17:263–272.

24. For a description of Hawai'i coral types, see D. Fenner, 2005, *Corals of Hawai'i*, Mutual Publishing, Honolulu. Also refer to http://www.marinelifephotography.com/corals/corals.htm (last accessed November 17, 2009).

25. See note 1, and Coastal Geology Group, Hawai'i's Coastline, O'ahu: http://www.soest.hawaii.edu/coasts/publications/hawaiiCoastline/oahu.html (last accessed November 17, 2009).

26. J. Rooney, C. H. Fletcher, E. E. Grossman, M. Engels, and M. E. Field, 2004, "El Niño Influence on Holocene Reef Accretion in Hawai'i," *Pacific Science* 58:305–324.

27. For more information see Wikipedia entry: http://en.wikipedia.org/wiki/Sea_level_rise (last accessed November 17, 2009).

28. For more information see Wikipedia entry: http://en.wikipedia.org/wiki/Milankovitch_cycles (last accessed November 17, 2009).

29. See R. G. Fairbanks, 1989, "A 17,000 Year Glacio-eustatic Sea Level Record: Influence of Glacial Melting Rates on the Younger Dryas Event and Deep Ocean Circulation," *Nature* 342:637–647; E. Bard, B. Hamelin, and R. G. Fairbanks, 1990, "U/Th Ages Obtained by Mass Spectrometry in Corals from Barbados: Sea Level during the Past 130,000 Years," *Nature* 346:456–458; Göran Burenhult, 1993, *The First Humans*, American Museum of Natural History, Harper, San Francisco, p. 82. A good discussion of the last ice age is found under Wikipedia entry "Ice age": http://en.wikipedia.org/wiki/Ice_age (last accessed November 17, 2009).

30. See Charles Sheppard, 2002, *Coral Reefs: Ecology, Threats, and Conservation*, Voyageur Press, Osceola, Wisconsin.

31. Ta, 2008 (see note 13); ibid., "Steep Decline of Reef Fish Points to Illegal Lay Nets," *The Honolulu Advertiser*, August 3; *The Honolulu Advertiser* Staff, 2008, "New Rules Would Protect Maui Fish That Graze on Invasive Limu Species," *The Honolulu Advertiser*, August 20.

32. C. H. Fletcher, C. Murray-Wallace, C. Glenn, B. Popp, and C. Sherman, 2005, "Age and Origin of Late Quaternary Eolianite, Kaiehu Point (Moomomi), Molokai, Hawai'i," *Journal of Coastal Research*, Special Issue 42:97–112.

33. Rooney, 2004.

34. *The Honolulu Advertiser* Staff, 2008, "Coral Thriving Best off Big Island," *The Honolulu Advertiser*, January 25.

35. See note 10.

36. Diana Leone, 2004a, "The Coral Connection: The City and Counties Prepare a Campaign to Protect Island Reefs," *Honolulu Star-Bulletin*, June 15.

37. See chapter 8 for a discussion of other land-based pollutants.

38. See chapter 8.

39. C. D. Hunt, 2007, *Groundwater Nutrient Flux to Coastal Waters and Numerical Simulation of Wastewater Injection at Kihei, Maui, Hawai'i*, U.S. Geological Survey Scientific Investigations Report 2006-5283.

40. See chapter 8, and note 10 (this chapter).

41. Leone, 2004a.

42. A. J. Soicher and F. L. Peterson, 1997, "Terrestrial Nutrient and Sediment Fluxes to the Coastal Waters of West Maui, Hawai'i," *Pacific Science* 51:221–232.

43. See note 10, p. 256.

44. Quotation from Eric Conklin, The Nature Conservancy, in R. Perez, 2009, "Hawaii Scientists Waging Uphill Battle," *The Honolulu Advertiser*, July 26.

45. For more information on the damage caused to coral reefs by certain land-use practices, see Hawai'i International Year of the Reef: http://www.iyor-hawaii.org/reef-facts/ (last accessed November 18, 2009).

46. O. Hoegh-Guldberg, P. J. Mumby, A. J. Hooten, R. S. Steneck, P. Greenfield, E. Gomez, C. D. Harvell, P. F. Sale, A. J. Edwards, K. Caldeira, N. Knowlton, C. M. Eakin, R. Iglesias-Prieto, N. Muthiga, R. H. Bradbury, A. Dubi, and M. E. Hatziolos, 2007, "Coral Reefs under Rapid Climate Change and Ocean Acidification," *Science* 318:1737–1742.

47. See note 10.

48. Discussion of ocean acidity is found in the following: Wikipedia entry: http://en.wikipedia.org/wiki/Ocean_acidification; J. A. Kleypas, R. W. Buddemeier, D. Archer, J. P. Gattuso, C. Langdon, and B. N. Opdyke, 1999a, "Geochemical Consequences of Increased Atmospheric Carbon Dioxide on Coral Reefs," *Science* 284:118–120; J. A. Kleypas, J. W. McManus, and L. A. B. Menez, 1999b, "Environmental Limits to Coral Reef Development: Where Do We Draw the Line? *American Zoologist* 39:146–159; D. F. Boesch, J. C. Field, and D. Scavia, eds., 2000, "The Potential Consequences of Climate Variability and Change on Coastal Areas and Marine Resources," Report of the Coastal Areas and Marine Resources Sector Team, U.S. National Assessment of the Potential Consequences of Climate Variability and Change, U.S. Global Change Research Program, NOAA Coastal Ocean Program Decision Analysis Series No. 21, NOAA Coastal Ocean Program, Silver Spring, Maryland.

49. Helen Altonn, 2005, "Reef Studies Gauge Global Warming Threats," *Honolulu Star-Bulletin*, October 18.

50. The NOAA Earth System Research Laboratory, Global Monitoring Division provides annual mean values of atmospheric carbon dioxide: ftp://ftp.cmdl.noaa.gov/ccg/co2/trends/co2_annmean_mlo.txt (last accessed November 18, 2009).

51. See chapter 9 for discussion of global warming.

52. "Acidic and basic are two extremes that describe chemicals, just like hot and cold are two extremes that describe temperature. Mixing acids and bases can cancel out their extreme effects; much like mixing hot and cold water can even out the water temperature. A substance that is neither acidic nor basic is neutral. The pH scale measures how acidic or basic a substance is. It ranges from 0 to 14. A pH of 7 is neutral. A pH less than 7 is acidic, and a pH greater than 7 is basic. Each whole pH value below 7 is 10 times more acidic than the next higher value. For example, a pH of 4 is 10 times more acidic than a pH of 5 and 100 times (10 times 10) more acidic than a pH of 6. The same holds true for pH values above 7, each of which is 10 times more alkaline—another way to say basic—than the next lower whole value. For example, a pH of 10 is 10 times more alkaline than a pH of 9. Pure water is neutral, with a pH of 7.0. When chemicals are mixed with water, the mixture can become either acidic or basic. Vinegar and lemon juice are acidic substances, while laundry detergents and ammonia are basic. Chemicals that are very basic or very acidic are called 'reactive.' These chemicals can cause severe burns. Automobile battery acid is an acidic chemical that is reactive. Automobile batteries contain a stronger form of

some of the same acid that is in acid rain. Household drain cleaners often contain lye, a very alkaline chemical that is reactive." (U.S. Environmental Protection Agency, 2007, Acid Rain, "What Is pH?": http://www.epa.gov/acidrain/measure/ph.html (last accessed November 18, 2009).

53. Tara Hicks, 2005, "Coral Reef Survival: New Research Predicts the Damage from Increased Carbon Dioxide in the Oceans," press release, University of Hawai'i, School of Ocean and Earth Science and Technology, Honolulu: http://www.soest.hawaii.edu/SOEST_News/PressReleases/Atkinson/index.htm (last accessed January 4, 2010).

54. See "Outcomes Overview and Call To Action," 11th International Coral Reef Symposium, 2008: http://www.nova.edu/ncri/11icrs/outcomes.html (last accessed November 18, 2009). See also "Honolulu Declaration on Ocean Acidification and Reef Management," 2008, U.S. Coral Reef Task Force meeting, Kona, Hawai'i: http://www.nature.org/pressroom/press/press3662.html (last accessed November 18, 2009).

55. Peter Glynn, 2006, "Coral Reef Bleaching: Facts, Hypotheses, and Implications," *Global Change Biology* 2 (6): 495–509.

56. See NASA World Book, "Global Warming": http://www.nasa.gov/world book/global_warming_worldbook.html (last accessed November 14, 2009).

57. Translated and edited by Martha Warren Beckwith, 1951, *The Kumulipo: A Hawaiian Creation Chant,* University of Hawai'i Press, Honolulu.

58. For more information on the International Coral Reef Initiative, see http://www.icriforum.org/secretariat/about_icri.html (last accessed November 18, 2009).

59. Clive Wilkinson, ed., 2004, *Status of Coral Reefs of the World: 2004,* Executive Summary, Global Coral Reef Monitoring Project: http://www.aims.gov.au/pages/research/coral-bleaching/scr2004/ (last accessed January 5, 2010).

60. Ibid., 2008, *Status of Coral Reefs of the World: 2008,* Executive Summary, Global Coral Reef Monitoring Project. See brochure: http://unesdoc.unesco.org/images/0017/001792/179244e.pdf (last accessed November 18, 2009).

61. For more information on the Hawai'i Coral Reef Initiative Research Program, see http://www.hawaii.edu/ssri/hcri/index.html (last accessed November 18, 2009).

62. Diana Leone, 2004b, "Program Hopes to Save Local Reefs," *Honolulu Star-Bulletin,* June 17.

63. See note 10.

64. Friedlander et al., 2007 (see note 11); C. Pala, 2007, "Fisheries Management: Conservationists and Fishers Face Off over Hawai'i's Marine Riches," *Science* 317:306–307: http://www.sciencemag.org/cgi/content/full/317/5836/306 (last accessed January 4, 2010); Helen Altonn, 2008, "Declining Fish Stocks Are Blamed on Anglers," *Honolulu Star-Bulletin,* October 30: http://www.starbulletin.com/news/20081030_Declining_fish_stocks_are_blamed_on_anglers.html#fullstory (last accessed November 19, 2009).

65. Pala, 2007 (see note 64), p. 1.

66. See the Western Pacific Regional Fishery Management Council discussion of the Hawaiian Moon calendar: http://www.wpcouncil.org/indigenous/Indig enous_Display-1.pdf (last accessed November 18, 2009).

67. See T. Dye and T. Graham, 2004, "Review of Archeological and Historical

Data Concerning Reef Fishing in Hawai'i and American Samoa," T. S. Dye and Colleagues, Archeologists, Inc., Honolulu, February 17.

68. Food and Agriculture Organization, 2005, "Review of the State of World Marine Fishery Resources," Fisheries Technical Paper 457, Rome: http://www.fao.org/docrep/009/y5852e/y5852e00.htm (last accessed November 18, 2009).

69. Ibid.

70. Pacific Fisheries Coalition, 2007, Linda Paul, Pacific Fisheries Coalition Annual Report, 2007: http://www.pacfish.org/ (last accessed November 18, 2009).

71. Ilima Loomis, 2006, "Full Tanks and Empty Reefs," *Honolulu Magazine*, September, pp. 43–50.

72. C. Dudley, 2005, "Top Fish Populations Being Replaced by Rays, Smaller Fish," Pew Institute for Ocean Science, April 4: http://www.oceanconservationscience.org/press/press-article_archive.php?ID=28 (last accessed November 18, 2009).

73. Ibid.

74. The following discussion draws primarily from two sources: "Managing Marine Fisheries of Hawai'i and the U.S. Pacific Islands – Past, Present and Future," 2003, Western Pacific Regional Fishery Management Council, K. M. Simonds Executive Director, Honolulu: http://www.wpcouncil.org/documents/WPRFM CDocument/WPRFCBrochure.pdf (last accessed January 5, 2010); A. Friedlander, B. Endreson, W. Aila, and L. Paul, 2002, "The Status of Hawai'i's Living Marine Resources at the Millennium," Pacific Fisheries Coalition white paper, Honolulu: http://www.pacfish.org/wpapers/ (last accessed January 5, 2010).

75. B. Endreson, W. Aila, and L. Paul, 2002, "Fishery Management Methods: Are Sustainable Fisheries Possible?" Pacific Fisheries Coalition white paper, Honolulu: http://www.pacfish.org/wpapers/ (last accessed January 5, 2010).

76. For more information, see State of Hawai'i, Department of Land and Natural Resources, Division of Aquatic Resources: http://hawaii.gov/dlnr/dar/index.html (last accessed November 18, 2009).

77. National Research Council, 1996, *Upstream: Salmon and Society in the Pacific Northwest*, National Academy Press, Washington, D.C., p. 304.

78. International Society of Reef Studies, 2006, "Marine Protected Areas in Management of Coral Reefs," Briefing Paper 1, Reef Encounter 30/31, July.

79. See MLCD page, Hawai'i Department of Land and Natural Resources, Division of Aquatic Resources: http://hawaii.gov/dlnr/dar/coral/mlcd.html (last accessed November 18, 2009).

80. Pala, 2007 (see note 64).

81. See note 79.

82. National Research Council, 2002, *Marine Protected Areas: Tools for Sustaining Ocean Ecosystems*, National Academy Press, Washington, D.C.

83. See study of global impacts of humans on the world ocean: Benjamin S. Halpern, Shaun Walbridge, Kimberly A. Selkoe, Carrie V. Kappel, Fiorenza Micheli, Caterina D'Agrosa, John F. Bruno, Kenneth S. Casey, Colin Ebert, Helen E. Fox, Rod Fujita, Dennis Heinemann, Hunter S. Lenihan, Elizabeth M. P. Madin, Matthew T. Perry, Elizabeth R. Selig, Mark Spalding, Robert Steneck, and Reg Watson, 2008, "A Global Map of Human Impact on Marine Ecosystems": http://www.sciencemag.org/cgi/content/abstract/319/5865/948?ijkey=.QBRU7cadgPCc&keytype=ref&siteid=sci (last accessed November 19, 2009).

84. To learn more about protecting Hawai'i's reefs and fisheries, see Hawai'i Department of Land and Natural Resources, Division of Aquatic Resources, "Local Action Strategies": http://hawaii.gov/dlnr/dar/coral/coral_las.html (last accessed January 5, 2010).

Chapter 12: A Responsibility to Nurture the Land

1. Taken from The Nature Conservancy, Hawaiian High Islands Ecoregion, "Hawaiian Culture and Conservation": http://www.hawaiiecoregionplan.info/culture.html (last accessed November 19, 2009).

2. T. W. Giambelluca, H. F. Diaz, and M. S. A. Luke, 2008, "Secular Temperature Changes in Hawai'i," *Geophysical Research Letters* 35: L12702, doi: 10.1029/2008GL034377.

3. Y. Firing and M. A. Merrifield, 2004, "Extreme Sea Level Events at Hawaii: Influence of Mesoscale Eddies," *Geophysical Research Letters* 31: L24306, doi:10.1029/2004GL021539.

4. C. H. Fletcher, R. A. Mullane, and B. M. Richmond, 1997, "Beach Loss along Armored Shorelines of Oahu, Hawaiian Islands," *Journal of Coastal Research* 13:209–215.

5. D. S. Oki, 2004, "Trends in Streamflow Characteristics at Long-term Gaging Stations, Hawai'i," U.S. Geological Survey Scientific Investigations Report 2004-5080. See also Pao-Shin Chu and Huaiqun Chen, 2005, "Interannual and Interdecadal Rainfall Variations in the Hawaiian Islands," *Journal of Climate* 18:4796–4813. See also P. Ya. Groisman, R. W. Knight, T. R. Karl, D. R. Easterling, B. Sun, J. H. Lawrimore, 2004 "Contemporary changes of the hydrological cycle over the contiguous United States, trends derived from *in situ* observations" *Journal of Hydrometeorology* 5(1): 64–85. Cited in: *Global Climate Change Impacts in the United States*, T. R. Karl, J. M. Melillo, T. C. Peterson, (eds.). Cambridge University Press, 2009.

6. At Station ALOHA (62 miles [100 km] north of Kahuku Point, Oahu), marine researchers at the University of Hawai'i and cooperating institutions have measured an increase of sea surface temperature of 0.22°F [0.12°C] per decade, a result that is consistent with other estimates: K. S. Casey, P. Cornillon, 2001 "Global and Regional Sea Surface Temperature Trends" *Journal of Climate* 14.18: 3801–3818. See also P. Jokiel and E. Brown, 2004, "Global Warming, Regional Trends and Inshore Environmental Conditions Influence Coral Bleaching in Hawai'i," *Global Change Biology* 10:1627–1641.

7. J. E. Dore, R. Lukas, D. W. Sadler, M. J. Church, and D. M. Karl, 2009, "Physical and Biogeochemical Modulation of Ocean Acidification in the Central North Pacific," *Proceedings of the National Academy of Sciences of the U.S.A.* 106:12235–12240. See also the press release: http://www.soest.hawaii.edu/soest_web/2009_news_PDFs/PNAS_Ocean_Acidification_release.pdf (last accessed January 6, 2010).

8. Ocean economic data compiled as part of the National Ocean Economics Program: www.oceaneconomics.org (last accessed November 19, 2009).

9. See Ko'olau Poko Watershed Management Plan, Honolulu Board of Water Supply: http://www.boardofwatersupply.com/cssweb/display.cfm?sid=2124 (last accessed November 19, 2009).

10. See, for example:
- U.S. Geological Survey, *Atlas of Natural Hazards in the Hawaiian*

Coastal Zone: http://pubs.usgs.gov/imap/i2761/ (last accessed November 19, 2009).

- *Purchasing Coastal Real Estate in Hawai'i:* http://www.soest. hawaii.edu/SEAGRANT/communication/pdf/Purchasing%20 Coastal%20Real%20Estate.pdf (last accessed November 19, 2009).
- *Hawai'i Coastal Hazard Mitigation Guidebook:* http://www.soest. hawaii.edu/SEAGRANT/communication/HCHMG/hchmg.htm (last accessed November 19, 2009).
- *Homeowner's Handbook to Prepare for Natural Hazards:* http:// www.soest.hawaii.edu/SEAGRANT/communication/Natural HazardsHandbook/naturalhazardprepbook.htm (last accessed November 19, 2009).

11. Surfrider Foundation, State of the Beach Report, "Surf Zone Water Quality, California": http://www.surfrider.org/stateofthebeach/08-fc/body.asp?sub= SurfZoneWaterQuality (last accessed November 19, 2009).

12. See U.S. Fish and Wildlife Service, 2009, "Climate Change in the Pacific Region": http://www.fws.gov/Pacific/Climatechange/changepi.html (last accessed January 6, 2010).

13. Oliver Timm and Henry F. Diaz, 2009, "Synoptic-Statistical Approach to Regional Downscaling of IPCC Twenty-first Century Climate Projections: Seasonal Rainfall over the Hawaiian Islands," *Journal of Climate* 22:4261–4280.

14. Hector Valenzuela, 2009, "Diversified Ag Gets Only Lip Service," *The Honolulu Advertiser,* June 29.

Index

About the Authors

Charles "Chip" Fletcher is a professor in the Department of Geology and Geophysics in the School of Ocean and Earth Science and Technology (SOEST) at the University of Hawai'i at Mānoa, Honolulu. He teaches and conducts research in the field of coastal geology. For his work in service to government agencies and public groups, Fletcher was the 2006 recipient of the Hung Wo and Elizabeth Lau Ching Foundation Award for Faculty Service to the Community given by the University of Hawai'i Board of Regents.

Robynne Boyd started writing about people's relationship to the Earth when living on the North Shore of Kaua'i, Hawai'i. Over a decade later, she continues this work as an editor and writer covering environment and energy. She lives in Atlanta, Georgia.

William "Bill" Neal is an emeritus professor and past chairman of the Geology Department, Grand Valley State University, Allendale, Michigan. He has conducted coastal studies for more than 30 years and has co-authored numerous books and papers on U. S. shorelines as well as other parts of the world (Colombia, Portugal, Puerto Rico). He is coeditor of Duke University Press's "Living with the Shore Series." Dr. Neal was co-recipient of the American Geological Institute's 1993 award for outstanding contribution to public understanding of geology.

Virginia Tice received her Juris Doctorate from the University of Hawai'i with Certificates in Environmental and International Law. She is now an attorney for Nature Iraq, working to improve environmental policies and land use management in Iraq and Iraqi Kurdistan.

PRODUCTION NOTES FOR
FLETCHER | LIVING ON THE SHORES OF HAWAI'I

Cover design by Julie Matsuo-Chun

Interior design and composition by Santos Barbasa Jr. in Galliard with display type in Cooper Std Black, Copperplate, and Seria Sans

Printing and binding by Everbest Printing Co., Ltd.

Printed on 105 gsm GoldEast Matt Art